Advances in Intelligent Systems and Computing

Volume 847

Series editor

Janusz Kacprzyk, Systems Research Institute, Polish Academy of Sciences,
Warsaw, Poland
e-mail: kacprzyk@ibspan.waw.pl

The series "Advances in Intelligent Systems and Computing" contains publications on theory, applications, and design methods of Intelligent Systems and Intelligent Computing. Virtually all disciplines such as engineering, natural sciences, computer and information science, ICT, economics, business, e-commerce, environment, healthcare, life science are covered. The list of topics spans all the areas of modern intelligent systems and computing such as: computational intelligence, soft computing including neural networks, fuzzy systems, evolutionary computing and the fusion of these paradigms, social intelligence, ambient intelligence, computational neuroscience, artificial life, virtual worlds and society, cognitive science and systems, Perception and Vision, DNA and immune based systems, self-organizing and adaptive systems, e-Learning and teaching, human-centered and human-centric computing, recommender systems, intelligent control, robotics and mechatronics including human-machine teaming, knowledge-based paradigms, learning paradigms, machine ethics, intelligent data analysis, knowledge management, intelligent agents, intelligent decision making and support, intelligent network security, trust management, interactive entertainment, Web intelligence and multimedia.

The publications within "Advances in Intelligent Systems and Computing" are primarily proceedings of important conferences, symposia and congresses. They cover significant recent developments in the field, both of a foundational and applicable character. An important characteristic feature of the series is the short publication time and world-wide distribution. This permits a rapid and broad dissemination of research results.

More information about this series at http://www.springer.com/series/11156

Lakhmi C. Jain · Valentina E. Balas
Prashant Johri
Editors

Data and Communication Networks

Proceedings of GUCON 2018

Springer

Editors
Lakhmi C. Jain
University of Technology Sydney
Sydney, Australia

and

University of Canberra
Canberra, Australia

and

Liverpool Hope University
Liverpool, UK

and

KES International
Shoreham-by-Sea, UK

Valentina E. Balas
Aurel Vlaicu University of Arad
Arad, Romania

Prashant Johri
School of Computing Science and Engineering
Galgotias University
Greater Noida, Uttar Pradesh, India

ISSN 2194-5357 ISSN 2194-5365 (electronic)
Advances in Intelligent Systems and Computing
ISBN 978-981-13-2253-2 ISBN 978-981-13-2254-9 (eBook)
https://doi.org/10.1007/978-981-13-2254-9

Library of Congress Control Number: 2018960207

This Springer imprint is published by the registered company Springer Nature Singapore Pte Ltd.
The registered company address is: 152 Beach Road, #21-01/04 Gateway East, Singapore 189721, Singapore

Preface

The book constitutes selected high-quality papers presented in International Conference on Computing, Power, and Communication Technologies 2018 (GUCON 2018) organized by Galgotias University, India, in September 2018. It discusses issues in electrical, computer, and electronics engineering and technologies. The selected papers are organized into three sections—cloud computing and computer networks; data mining and big data analysis; and machine learning and systems. In-depth discussions on various issues under topics provide an interesting compilation for researchers, engineers, and students.

We are thankful to all the authors that have submitted papers for keeping the quality of the GUCON 2018 at high levels. The editors of this book would like to acknowledge all the authors for their contributions and the reviewers. We have received an invaluable help from the members of the International Program Committee and the chairs responsible for different aspects of the workshop. We appreciate also the role of Special Sessions Organizers. Thanks to all of them, we had been able to collect many papers on interesting topics, and during the conference, we had very interesting presentations and stimulating discussions.

Our special thanks go to Janus Kacprzyk (Editor in Chief, Springer, Advances in Intelligent Systems and Computing Series) for the opportunity to organize this guest-edited volume.

We are grateful to Springer, especially to Dr. Thomas Ditzinger (Senior Editor, Applied Sciences and Engineering, Springer-Verlag), for the excellent collaboration, patience, and help during the evolvement of this volume.

We hope that the volume will provide useful information to professors, researchers, and graduated students in the area of soft computing techniques and applications, and all will find this collection of papers inspiring, informative, and useful. We also hope to see you at a future GUCON event.

Australia/UK	Lakhmi C. Jain
Romania	Valentina E. Balas
India	Prashant Johri
Sep 2018	

GUCON 2018 Organization

Chief Patrons

Mr. Suneel Galgotia, Chancellor, Galgotias University, India
Mr. Dhruv Galgotia, CEO, Galgotias University, India

Patron

Prof. Renu Luthra, Vice-Chancellor, Galgotias University, India

Co-Patrons

Prof. Harish C. Rai, EX Advisor, All India Council for Technical Education, India
Prof. Sibaram Khara, Pro Vice Chancellor, Galgotias University, India

General Chairs

Prof. S. N. Singh, Vice-Chancellor, MMMUT, Gorakhpur, India
Prof. J. Ramkumar, Immediate past Chairman IEEE UP Section, India
Prof. Priestly Shan, SECE, Galgotias University, India
Prof. Prabhakar Tiwari, MMMUT, Gorakhpur, India

Conference Chair and Convener

Prof. Rabindra Nath Shaw, India

Technical Program Committee Chairs

Prof. Lakhmi C. Jain, Australia
Prof. Prashant Johri, India

Publication Chairs

Prof. Valentina E. Balas, Romania
Prof. Jitendra Verma, India

Volume Editors

Prof. Lakhmi C. Jain, Australia
Prof. Valentina E. Balas, Romania
Prof. Prashant Johri, India

Advisory Committee

Prof. Bimal K. Bose, USA
Prof. Muhammad H. Rashid, USA
Prof. Muhammet Koksal, Turkey
Prof. Nasrudin Bin Abd Rahim, Malaysia
Prof. Hamid Ali Abed Al-asadi., Iraq
Prof. Tarek Bouktir, Algeria
Prof. Subramaniam Ganesan, USA
Prof. Claudio Moraga Roco, Spain
Prof. Suganthan P. N., Singapore
Prof. Sanjib Kumar P., Singapore
Prof. Ralph Kennel, Germany
Prof. Udaya Madawala, New Zealand
Prof. Dylan Lu, Australia, India
Prof. Yeap Peik Foong, India
Prof. Terence Karran, India
Prof. B. Chitti Babu, U.K
Prof. Jyh-Horng Chou, Taiwan
Prof. Saad Mekhilef, Malaysia
Prof. Yurii Boreisha, USA

Track Chairs

Cloud Computing, Computer Networks, Automata, Computer Architecture

Prof. Vincenzo Piuri, Italy
Prof. Anupam Chattopadhyay, Singapore

Data Mining and Big Data Analysis

Prof. George A. Tsihrintzis, Greece
Prof. Ajay K. Datta, USA

Machine Learning Algorithms

Prof. Farzin Deravi, UK
Prof. Maria Virvou, Greece

GUCON 2018 Editorial Board Members

Conference Chair (Covener)
Volume Editors
Technical Chairs
Publication Chairs
Track Chairs

Technical Program Committee and Reviewers

Dr. A. R. Abhyankar
Dr. Aditi Sharan
Dr. Ajay Mittal
Dr. Sudhir Kumar Sharma
Dr. Ajai Jain
Dr. Alok Kushwaha
Dr. Amit Agarwal
Dr. Amalendu Patnaik

Dr. Anil K. Ahlawat
Dr. Anil K. Singh
Dr. Anuradha
Dr. Arun Kumar Verma
Dr. Aseem Chandel
Dr. Asheesh K. Singh
Dr. Ashutosh Dixit
Dr. Asif Ekbal
Dr. B. Dushmanta Kumar Patro
Dr. Baij Nath Kaushik
Dr. Bhaskar Biswas
Dr. Bharat Singh Rajpurohit
Dr. C. Patvardhan
Dr. C. Rama Krishna
Dr. C. K. Nagpal
Dr. Chandra Sekaran
Dr. Chiranjeev Kumar
Dr. Chittaranjan Hota
Dr. D. Bhagwan Das
Dr. D. A. Mehta
Dr. D. S. Kushwaha
Dr. D. S. Yadav
Dr. Desh Deepak Sharma
Dr. Dhram Singh
Dr. Dimple J. Gupta
Dr. Diwakar Bhardwaj
Dr. Girish Patnaik
Dr. Jai Govind Singh
Dr. Joy Deep Mitra
Dr. K. V. Arya
Dr. Kiran Kumar Pattanaik
Dr. Kishor K. Bhoyar
Dr. Komal Kumar Bhatia
Dr. Lalit Kumar Awasthi
Dr. M. K. Dutta
Dr. M. P. Singh
Dr. Madhavi Sinha
Dr. Manisha Sharma
Dr. Mohd. Rihan
Dr. Mayank Pandey
Dr. Munesh C. Trivedi
Dr. N. Badal, Knit
Dr. Nanhay Singh
Dr. Narendra Kohli
Dr. Naresh Chauhan

Dr. Naveen Kumar
Dr. Neelam Duhan
Dr. Neeraj Tyagi
Dr. O. P. Verma
Dr. Pooja Jain
Dr. Pooja Pathak
Dr. Prabhat Ranjan
Dr. Prabhakar Tiwari
Dr. Prabin Panigrahi
Dr. Pragya Dwivedi
Dr. Pradeep Sharma
Dr. Pramod Kumar
Dr. Pramod Kumar Singh
Dr. Punam Bedi
Dr. R. K. Singh
Dr. R. S. Yadav
Dr. R. S. Rao
Dr. Rahul Rishi
Dr. Rajesh Prasad
Dr. Reena Dadhich
Dr. Ruchika Malhotra
Dr. S. P. Tripathi
Dr. Sapna Gambhir
Dr. Suneeta Agarwal
Dr. Sujoy Das
Dr. Sukomal Pal
Dr. Sunil Kumar Khatri
Dr. Tanveer Siddiqui
Dr. Tarun Shrimali
Dr. Vasudha Bhatnagar
Dr. Vishal Bhatnagar
Dr. Yashpal Singh
Prof. Herbert H. C. Lu
Dr. Senthilrajan Agni
Dr. Abhineet Anand
Dr. Anurag Baghel
Dr. Balamurugan Balusamy
Dr. Priti Bansal
Dr. Sonia Bansal
Dr. Annappa Basava
Dr. Rohit Beniwal
Dr. Vandana Bhasin
Dr. Rodrigo Bortoletto
Dr. John Moses Cyril
Dr. Pinaki Chakraborty

Dr. Sansar Chauhan
Dr. Rahul Chaurasiya
Dr. Surya Deo Choudhary
Dr. Anurag Dixit
Dr. Nripendra Narayan Das
Dr. Indrani Das
Dr. Aparna Datt
Dr. Parneeta Dhaliwal
Dr. Chandrakant Divate
Dr. Rajesh Dubey
Dr. Arman Faridi
Dr. Ankush Ghosh
Dr. Utkarsh Goel
Dr. Pallavi Goel
Dr. Amit Goel
Dr. Priyanka Goyal
Dr. Deepak Gupta
Dr. Suneet Gupta
Dr. Raza Haidri
Dr. Syed Shabih Hasan
Dr. Manas Hati
Dr. Brijesh Iyer
Dr. Manisha Jailia
Dr. Prashant Johri
Dr. Jegathesh Amalraj Joseph
Dr. Sandeep K. Singh
Dr. Vinay Kumar
Dr. Amita Kapoor
Dr. Sandhya Katiyar
Dr. Anvesha Katti
Dr. Ruqaiya Khanam
Dr. Aanchal Khatri
Dr. Shrawan Kumar
Dr. Devendra Kumar
Dr. Avneesh Kumar
Dr. Arun Kumar
Dr. Sanjeev Kumar
Dr. Vipin Kumar
Dr. Sanjay Kumar
Dr. Bhavnesh Kumar
Dr. Sandeep Kumar
Dr. Neetesh Kumar
Dr. Mohanraj M.
Dr. Ramakrishnan Malaichamy
Dr. Manas Kumar Mishra

Dr. Baibaswata Mohapatra
Dr. Thiyagarajan Muthunatesan
Dr. Rashid Mahmood
Dr. Yogendra Meena
Dr. Gitanjali Mehta
Dr. A. K. Mishra
Dr. Keshav Niranjan
Dr. Manoj Panda
Dr. Sanjeev Pippal
Dr. V. A. Sankar Ponnapalli
Dr. Shiv Prakash
Dr. Sheetla Prasad
Dr. Mohammed Abdul Qadeer
Dr. R. Gunasundari Ranganathan
Dr. Ranjeet Kumar Ranjan
Dr. Rohit Raja
Dr. Bharti Rana
Dr. Mukesh Rawat
Dr. Navaid Zafar Rizvi
Dr. Pravinth Raja S.
Dr. Anil Kumar Sagar
Dr. Rajeev Sharma
Dr. Birendra Kumar Sharma
Dr. Shrddha Sagar
Dr. Jyoti Sahni
Dr. Mohd. Saifuzzaman
Dr. Kavita Saini
Dr. Kamalesh Sethuramalingam
Dr. Priestly Shan
Dr. Gavaskar Shanmugam
Dr. Dilip Kumar Sharma
Dr. R. P. Sharma
Dr. Mayank Sharma
Dr. Sudhir Sharma
Dr. Lokesh Kumar Sharma
Dr. Vishnusharma
Dr. Jitendra Singh
Dr. Girish Singh
Dr. Karan Singh
Dr. Harikesh Singh
Dr. Prashant Singh
Dr. Neetay Singh
Dr. Ajay Shanker Singh
Dr. Arun Solanki
Dr. Subhranil Som

Dr. Ritesh Srivastava
Dr. Vijayalakshmi Subramanian
Dr. Hardeo Kumar Thakur
Dr. Pradeep Tomar
Dr. Shashi Kant Verma
Dr. Sohan Kumar Yadav
Dr. Vinod Yadav
Dr. Dileep Yadav
Dr. Chandra Yadav
Dr. Emre Yay
Dr. Aasim Zafar
Dr. Usha Chauhan
Dr. Chetna Dabas
Dr. Sanjoy Das
Dr. Sumithra Gavaskar
Dr. Vimal Kumar

Contents

About the Editors

Dr. Lakhmi C. Jain, Ph.D., M.E., B.E.(Hons) is Fellow (Engineers Australia) and is with the University of Canberra, Australia, University of Technology Sydney, Australia and Liverpool Hope University, Liverpool, UK. He founded the KES International for providing a professional community with the opportunities for publications, knowledge exchange, cooperation, and teaming. Involving around 5000 researchers drawn from universities and companies worldwide, KES facilitates international cooperation and generates synergy in teaching and research. KES regularly provides networking opportunities for professional community through one of the largest conferences of its kind in the area of KES. His interests focus on the artificial intelligence paradigms and their applications in complex systems, security, e-education, e-health care, unmanned air vehicles, and intelligent agents.

Valentina E. Balas is currently Full Professor in the Department of Automatics and Applied Software at the Faculty of Engineering, "Aurel Vlaicu" University of Arad, Romania. She holds a Ph.D. in Applied Electronics and Telecommunications from Polytechnic University of Timisoara. Dr. Balas is author of more than 280 research papers in refereed journals and international conferences. Her research interests are in Intelligent Systems, Fuzzy Control, Soft Computing, Smart Sensors, Information Fusion, Modeling and Simulation. She is the Editor-in Chief to International Journal of Advanced Intelligence Paradigms (IJAIP) and to International Journal of Computational Systems Engineering (IJCSysE), member in Editorial Board member of several national and international journals and is the director of Intelligent Systems Research Centre in Aurel Vlaicu University of Arad. She is a member of EUSFLAT, SIAM and a Senior Member IEEE, member in TC—Fuzzy Systems (IEEE CIS), member in TC—Emergent Technologies (IEEE CIS), member in TC—Soft Computing (IEEE SMCS).

Dr. Prashant Johri is working as Professor in School of Computing Science and Engineering, Galgotias University, Greater Noida, India. He completed his M.C.A. from Aligarh Muslim University and Ph.D. in Computer Science from Jiwaji University, Gwalior, India. He has also worked as a Professor and Director (M.C.A.),

Noida Institute of Engineering and Technology, (N.I.E.T.). He has served as Chair in many conferences and affiliated as member of program committee of many conferences. He has supervised 3 Ph.D. students and many M.Tech. students for their thesis. His area of research interests includes big data, data analytics, data retrieval and predictive analytics, information security, privacy protection, big data open platforms etc. He is actively publishing in these areas.

A Novel Hash-Based Mutual RFID Tag Authentication Protocol

Mansi Saxena, Rabindra Nath Shaw and Jitendra Kumar Verma

Abstract Radio frequency identification (RFID) is an integral part of our life. This term is coined for short-range radio communication technology. It is used to send and receive the digital information between stationary location and non-stationary object or between movable objects. It automates the wireless technology using radio waves to identify an object. This technology has widespread applications in the field of security, access control, transportation, etc. In this paper, we analyze an existing RFID-based protocol and demonstrate that it is insecure against impersonation attack, man-in-middle attack, server-masquerading attack, insider attack, and denial-of-service. We also propose a novel protocol, namely Encrypted Tag Identity and Secret Value Protocol, to overcome the shortcomings and loopholes existing in the surveyed protocols.

Keywords Impersonation attack · Server masquerading · Mutual authentication
Public and private keys · Secret value · Random numbers

1 Introduction

In modern world, radio frequency identification (RFID) has become an important and integral part of human life for promoting productivity and convenience [1, 2]. RFID is a wireless technology that uses radio frequency electromagnetic fields to transfer data from a Tag attached to an object and thus provides non-contact system for the purpose

M. Saxena
AbyM Technology, Noida 201301, Uttar Pradesh, India

R. N. Shaw
School of Electrical Electronics & Communication Engineering,
Galgotias University, Greater Noida 201310, Uttar Pradesh, India

J. K. Verma (✉)
School of Computing Science & Engineering,
Galgotias University, Greater Noida 201310, Uttar Pradesh, India
e-mail: jitendra.verma.in@ieee.org

L. C. Jain et al. (eds.), *Data and Communication Networks*, Advances in Intelligent Systems and Computing 847, https://doi.org/10.1007/978-981-13-2254-9_1

of automatic identification and tracking of moving objects. It has multidimensional applications like transportation, automated payment system, object traceability, and access control. RFID refers to a technology where digital data encoded in RFID Tags are captured by a RFID Reader via radio waves. An RFID system contains three key components which are as follows.

(i) RFID Tags: It is attached to the object which has to be identified by the help of carrying identification data.
(ii) RFID Readers: It reads or writes the identification information on Tags using Radio waves.
(iii) Back-end database: It collect the records related to tagged objects and associate them with the information that is going to be read by the Reader.

RFID has drawn a significant attention in recent years. Many researchers have contributed a lot of research and products to this technology. In recent years, many RFID-based protocols are proposed for secure implementation and deployment of this technology. Srivastava et al. [3] proposed a secure and robust hash-based mutual RFID Tag authentication protocol in telecare medicine information system. They claimed that their protocol provides mutual authentication and is secure against eavesdropping and replay attacks. But after analyzing we demonstrate that the given protocol is vulnerable to impersonation attack, insider attack, server-masquerading and man-in-middle attack. Therefore, we introduce a new protocol, namely, Encrypted Tag Identity and Secret Value Protocol (ETISVP) which fills the loopholes of the existing protocol and proved to be safe against the impersonation, server-masquerading, denial-of-service and man-in-middle attacks.

Rest of the paper is organized as follows. Section 2 presents related work, Sect. 3 presents state-of-the-arts of hash-based Tag authentication protocol, Sect. 4 provides possible security threats from attacks and loopholes of the protocol under observation followed by proposed protocol and its security analysis along with proposed comparison of results of existing protocols and proposed protocol in Sects. 5 and 6. Section 7 concludes the paper.

2 Related Work

Most of the protocols developed for mutual authentication between the Tag and the Reader in RFID system are based on hash functions. Despite of being not very secure and containing loopholes, these protocols are easily exploited by the attackers. Many schemes have been proposed to counter the privacy issues in RFID System. Few of them are as follows.

Okubhu et al. [4] proposed a hash-based protocol, which was the improved version of "*Kill Command Feature*"-based protocols introduced by Auto-ID Center [5]. However, the high cost of searching tags remained the major flaw of this protocol. Weis et al. [6] proposed "*Hash Lock Technique*" which was the improved version of this protocol. Meanwhile, many protocols have been proposed such as randomized hash lock scheme by MIT, anonymous ID scheme by NTT, external re-encryption scheme by RSA Laboratory and XOR based onetime pad scheme by RSA Laboratory. Tsudik [7] proposed an improved authentication protocol, namely, Yet-Another Trivial RFID Authentication Protocol (YATRAP), in order to provide tracing resistance Tag authentication through monotonically increasing time-stamps on the Tag. Furthermore, they proposed a new protocol [8] which was found vulnerable to replay attack due to lack of Reader's authenticity.

Similarly, Chatmon et al. proposed an anonymous RFID authentication protocol in [9]. Sun et al. [10] proposed a RFID technology to guard inpatient medication safety. In 2009, Huang and Ku [11] proposed the RFID protocol for medication safety of inpatient, which was found to be vulnerable to denial-of-service and replay attacks.

Chien et al. [12] proposed improved version of [11] which was still remained vulnerable to the impersonation and replay attacks. Peris-Lopez et al. [13] introduced a new concept of IS-RFID system; however, it is vulnerable to easy manipulation [14–16]. Chen et al. [17] proposed tamper resistant protocol which fails to guarantee the safety against impersonation, desynchronization attacks and traceability. To overcome the desynchronization problem, Cho et al. [18] followed by Kim [15, 16] proposed hash-based RFID Tag mutual authentication protocols. Suja et al. proposed an RFID authentication protocol based on cyclic redundancy check (CRC) and hamming distance calculation between Reader to Tag. They claimed that their protocol resists against tracing and cloning attacks in the most efficient way [19].

Considering the great success of RFID technology in Telecare sector, Srivastava et al. [3] proposed a hash-based mutual RFID Tag authentication protocol in telecare medicine information system. We have reviewed and cross-examined their protocol against all possible attacks, and we have shown possible loopholes of this protocol in subsequent section (Table 1).

Table 1 Preliminaries

Symbol details	
I_k	Identity of the kth Tag
$\|\|$	Concatenation operation
N	Random number
N_r	Random number of Reader
N_t	Random number of Tag
S	Secret value
S_j	Secret value used in the jth session
$H(\cdot)$	Hash function
$\oplus$	Bitwise XOR
DB	Database

3 State-of-the-Arts

This section presents state-of-the-arts of hash-based mutual RFID Tag authentication protocol and its functioning. The underlying work is the basis of our proposed work in this paper. The hash-based Tag authentication protocol constitutes three phases that are as follows.

3.1 *Pre-phase*

Initially, the Tag and Data server share the Tag identity I_k, hash function $H(\cdot)$, and the secret value S_j. The Tag and the Reader have their own random number generators.

3.2 *Reader's Request*

The Reader generates the random number N_{r} and sends a request to Tag with this random number.

3.3 *Tag's Response*

On receiving request from the Reader, Tag is invoked and Tag generates a random number N_{t} and subsequently perform the following computations:

- $X = H(S_j \parallel I_k) \oplus N_{\mathrm{t}}$
- Calculate $Y = X \oplus H(I_k \parallel N_{\mathrm{r}} \parallel N_{\mathrm{t}})$, and then
- $Z = H(Y \oplus T_1 \oplus N_{\mathrm{t}})$.

Tag sends the response message $(X; Z; T_1)$ to the Reader and Reader sends the response message $(X; Z; T_1; N_{\mathrm{r}})$ to the Data server after adding N_{r} to it.

3.4 *Data Servers Response and Tag Authentication*

Data server after extracting the Tag identity (I_k) and shared Secret value (S_j) from the database performs the following computations:

- If the expected legitimate time interval for transmission delay, $\Delta T < (T_2 - T_1)$ the server rejects the login request, where T_2 is current time-stamp at server.
- Computes $N_{\mathrm{t}}^* = X \oplus H(S_j \parallel I_K)$.
- Check $Z^* = H(X \oplus H(I_k \parallel N_{\mathrm{r}} \parallel N_{\mathrm{t}}^*) \oplus N^* \oplus T_1) \approx Z$.

- Checks until Z^* is equal to Z as extracted from response message sent by Reader.
- Computes $W = H(X \oplus H(I_k \parallel N_r \parallel N_t) \oplus T_2 \oplus S_j)$.

When $Z^* \approx Z$, the Data server sends the Tag data to the Reader. It sends the message $U = \text{DATA} \parallel W$ to the Reader. DATA is the information of Tag that needs to be transmitted.

3.5 *On Receiving* U, *Reader's Response*

On receiving U from Data server, Reader extracts the DATA from U and sends the remaining part to the Tag for further communication.

3.6 *Data Server Authentication and Secret Value Updation*

If $\Delta T < (T_3 - T_2)$, where T_3 is the current time-stamp and ΔT is the expected legitimate time interval for transmission delay then Tag rejects the request.

- Tag computes $W^* = H(Y \oplus S_j \oplus T_2)$.
- If $W^* \approx W$, Tag authenticates the Data server.
- Updates the S_j to $S_{j+1} = H(S_j \oplus N_r \oplus N_t)$ on both Tag and server side.

4 Security Analysis against Possible Attacks and Loopholes

We perform security analysis of the protocol which emphasizes that the protocol is not secure and is vulnerable to various attacks such as any attacker can invoke the Tag by sending the random number N_r. The Tag performs following operations:

- $X = H(S_j \parallel I_k) \oplus N_t$
- Calculates $Y = X \oplus H(I_k \parallel N_r \parallel N_t)$
- $Z = H(Y \oplus T_1 \oplus N_t)$.

Furthermore, the Tag sends the response message (X; Z; T_1) to the Attacker assuming him to be the correct Reader. The attacker sends the response message (X; Z; T_1; N_r) to the Data server adding N_r to it. Now with these values, the attacker can easily fetch the DATA from the message $U = \text{DATA} \parallel W$ sent by the Data Server assuming the attacker to be the correct Reader.

4.1 Impersonation Attack

In the first step of protocol, the Reader sends the request to the Tag along with a random number N_r. At Tag's side, there is no verification mechanism to ensure that request is made by the correct Reader.

4.2 Server-Masquerading Attack

Since, the Tag identity (I_k) and the Secret value (S_j) are stored as plaintext in the database. Any privileged insider having access to database can easily get these values and behave as the Data server by intercepting the response message ($X; Z; T_1; N_r$) send by the Reader. The Attacker with values ($I_k; S_j; X; Z; T_1; N_r$) computes:

- $N_t^* = X \oplus H(S_j \parallel I_k)$
- $W = H(X \oplus H(I_k \parallel N_r \parallel N_t) \oplus T_2 \oplus S_j)$
- $U = \text{InvalidDATA} \parallel W.$

In this way, the Attacker can send the invalid data to the Reader. The Reader will accept the message U from the Attacker, assuming him to be the Data server. Since, there is no mechanism for the verification of the correct Data server at the Reader's side.

4.3 Denial-of-Service Attack

The protocol is not fail-safe against the computation exhaustive attacks. At Data Server, there is an authentication step to verify the Reader.

- Check $Z^* = H(X \oplus H(I_k \parallel N_r \parallel N_t^*) \oplus N_t^* \oplus T_1) \approx Z.$

But there is no limit on the number of wrong messages can be sent to the Data server. The Attacker can send number of fake messages to the server which leads to the excessive computation on the server side keeping the server busy and unable to process any request.

4.4 Man-in-Middle Attack

All the messages between the Tag, Reader, and Data server are transmitted as plaintext on the communication channels. These messages can be intercepted and used by the Attacker as done in server-masquerading attack. Also the attacker can change these messages on the communication channel in order to invoke the denial-of-service attack by sending the fake messages to the Data server.

4.5 Insider Attack

Insider attacker is one who is having the administrative access to the server. Tag identity (I_k) and secret value (S_j) are stored as the plaintext to the server. The privileged insider, who has direct access to the server, can get these parameters and use the secret information for personal benefit as discussed in server-masquerading attack.

5 Encrypted Tag Identity and Secret Value Protocol: Proposed Work

We propose Encrypted Tag Identity and Secret Value Protocol (ETISVP) with the objective to fill the loopholes of the existing protocol. Our protocol uses the encryption scheme to safely communicate the messages between the Tag, Reader, and Data server. Encryption makes it impossible for the Attacker to intercept or modify the message during the transmission on communication channel. Table 1 shows the terminologies which are frequent in ETISVP.

The Tag identity (I_k) and Secret value (S_j) are stored as the hash function cipher in the database. So, the any privileged insider or attacker cannot hack and use these values from the database. The functioning of ETISVP is shown in Table 2.

5.1 Pre-phase

Initially, Tag, Reader, and Data server have their private keys for encryption and share their public keys with each other for decryption.

Hash function $H(\cdot)$, Tag identity (T) and secret value (V_j) are shared by Tag and Server. T and V_j are stored as H(T) and H(V_j) in the database. V_j is the secret value of Tag in jth session.

5.2 Readers Request

The Reader generates the random number N_{r} and performs the following computations.

- Encrypt N_{r} using its private key, $K_{\mathrm{R[PR]}}\{N_{\mathrm{r}}\}$.
- Sends a request to Tag with $K_{\mathrm{R[PR]}}\{N_{\mathrm{r}}\}$ on secure communication channel.

5.3 Tags Response

Decrypt the message received from Reader using Reader's public key: $K_{R[PU]}\{N_r\}$. Generates a random number N_t and performs the following computations:

- $X = (H(V_j) \parallel H(T)) \oplus N_t$
- Calculates $Y = X \oplus H(H(T) \parallel N_r \parallel N_t)$, and then
- $Z = H(Y \oplus T_1 \oplus N_t)$ where T_1 is the time-stamp at the Tag.

Table 2 Encrypted tag identity and secret value protocol

DATA SERVER	READER	TAG
	Generate random no. N_r and Encrypt it using Private Key. → $K_{R[PR]}(N_r)$ Send a request to Tag.	Decrypt the received message $K_{R[PU]}(N_r)$ Generates Random Number N_t and performs following computations: • $X = (H(V_j) \parallel H(T)) \oplus N_t$ • Calculates $Y = X \oplus H(H(T) \parallel N_r \parallel N_t)$ • And then, $Z = H(Y \oplus T_1 \oplus N_t)$ where T_1 is the timestamp at the tag. Encrypt message (X, Z, T_1) and send to the Reader. $K_{T[PR]}(X, Z, T_1)$ ←
	Decrypt message: $K_{T[PU]}(X, Z, T_1)$ Add N_r to it and encrypt it again: $K_{R[PR]}(X, Z, T_1, N_r)$ ←	
Decrypt Message: $K_{R[PU]}(X, Z, T_1, N_r)$ Extract $H(T)$ and $H(V_j)$ from database and computes: • Check if, $\Delta T < (T_2 - T_1)$ the server rejects the login request, where T_2 is current time-stamp at server. • Computes $N^*_t = X \oplus (H(V_j) \parallel H(T))$ • Check $Z^* = H(X \oplus H(H(T) \parallel N_r \parallel N^*_t) \oplus N^*_t \oplus T_1) \approx Z$. • Computes $W = H(X \oplus H(H(T) \parallel N_r \parallel N_t) \oplus T_2 \oplus V_j)$ • $DATA \parallel W$ and encrypt it. → $K_{D[PR]}(DATA \parallel W)$		
	Decrypt Message: $K_{D[PR]}(DATA \parallel W)$ Extract DATA and Encrypt W to send it to Tag. → $K_{R[PR]}(W)$	Decrypt Message: $K_{R[PU]}(W)$ • Tag rejects the request if, $\Delta T < (T_3 - T_2)$, where T_3 is the current time-stamp. • Computes $W^* = H(Y \oplus H(V_j) \oplus T_2) \approx W$ • Updates the V_j to $V_{j+1} = H(H(S_j) \oplus N_r \oplus N_t)$ on both tag and server side

Tag generates the response message (X, Z, T_1) and encrypts it using its private key: $K_{T[PR]}\{X, Z, T_1\}$. Tag sends $K_{T[PR]}\{X, Z, T_1\}$ to the Reader on secure communication channel.

5.4 Reader's Side

On receiving response of Tag on secure communication channel, the following operations take place on Reader's side.

- Decrypt the message received from Tag using Tag's public key: $K_{T[PU]}\{X, Z, T_1\}$.
- Add N_r and encrypt the message using its private key: $K_{R[PR]}\{X, Z, T_1, N_r\}$.
- Reader sends $K_{R[PR]}\{X, Z, T_1, N_r\}$ to the Data server on secure communication channel.

5.5 Data Server Response and Tag Authentication

Decrypt message received from Reader using Reader's public key: $K_{R[PU]}\{X, Z, T_1, N_r\}$. Extracts $H(T)$ and $H(V_j)$ from database and performs the following computations:

If the expected legitimate time interval for transmission delay, $\Delta T < (T_2 - T_1)$

- The server rejects the login request, where T_2 is current time-stamp at server.
- Computes $N_t^* = X \oplus (H(V_j) \parallel H(T))$.
- Check $Z^* = H(X \oplus H(H(T) \parallel N_r \parallel N_t^*) \oplus N_t \oplus T_1) \approx Z$.
- Checks until Z^* is equal to Z.
- Computes $W = H(X \oplus H(H(T) \parallel N_r \parallel N_t) \oplus T_2 \oplus V_j)$
- If $Z^* \approx Z$ the Data server generates the message: DATA $\parallel W$.
- Encrypt it using its private key: $K_{D[PR]}\{\text{DATA} \parallel W\}$.

5.6 Reader's Response

On receiving the encrypted data from Data server, Reader performs the following operations.

- Decrypt the message using Server's public key: $K_{D[PU]}\{\text{DATA} \parallel W\}$.
- Extract DATA from $\{\text{DATA} \parallel W\}$.
- Encrypt W using its private key: $K_{R[PR]}\{W\}$.

And sends $K_{R[PR]}\{W\}$ to the Tag on secure communication channel.

5.7 Data Server Authentication and Secret Value Updation

On receiving encrypted data from Tag, the following operations take place on Data server side.

- Decrypt the message using Reader's public key: $K_{\mathrm{R[PU]}}\{W\}$.
- Tag rejects the request if, $\Delta T < (T_3 T_2)$, where T_3 is the current time-stamp and ΔT is the expected legitimate time interval for transmission delay.
- Tag computes $W^* = H(Y \oplus H(V_j) \oplus T_2)$.
- If $W^* \approx W$, Tag authenticates the Data server.
- Updates the V_j to $V_{j+1} = H(H(S_j) \oplus N_{\mathrm{r}} \oplus N_{\mathrm{t}})$ on both Tag and server side.

6 Security Analysis

Proposed protocol ETISVP overcomes the loopholes of the existing protocol hash-based Tag authentication protocol, and hence, is highly safe against most common attacks like man-in-middle attack, eavesdropping attack, impersonation attack, server-masquerading attack and Insider Attack. The analysis of these attacks is provided in the following subsections.

6.1 Man-in-Middle Attack

In the proposed ETISVP protocol, during the message exchange, sender encrypts the message using the private key before transmitting it on communication channel. This message can only be decrypted by the corresponding public key. Therefore, any Attacker cannot modify the message without having the public key. Since, the attacker do not have the public key, he cannot use the message to harm the system.

6.2 Impersonation Attack

The Reader in the first step sends the random number to the Tag by encrypting it with the private key as $K_{\mathrm{R[PR]}}\{N_{\mathrm{r}}\}$. The Tag can decrypt the message by the corresponding public key. Therefore, the Attacker cannot send the random number to the Tag without knowing the private key of Reader to encrypt the message. Hence, the Attacker cannot fool the Tag by behaving as Reader.

Table 3 Performance Evaluation

Protocol	Computation cost (Tag)	Computation cost (Reader)	Communication rounds
YA-TRAP [13]	2H + 3RNG	RNG	4
Suja and Arivarasi [19]	3H + 2MOD + RNG	RNG	5
Existing Protocol [3]	2H + RNG	RNG	5
ETISVP	4H + RNG + Encryption	RNG + Encryption	5

6.3 *Server-Masquerading Attack*

According to the proposed protocol, Tag identity (T) and secret value (V_j) are stored in the encrypted form in database. Therefore, any privileged insider cannot fetch these values from database. Also, the message sent by the Reader to the Data server is encrypted by private key of Reader. Hence, cannot be decrypt by the Attacker without the corresponding public key. So the Attacker cannot get the required information for server masquerading.

6.4 *Eavesdropping Attack*

The proposed protocol is safe against the eavesdropping attack. Since, all the messages on the communication channel are in encrypted form. Therefore, any attacker cannot get the valuable information without knowing the corresponding public key for decryption. Hence, the proposed protocol meets the required challenges (Table 3).

7 Conclusion

In this paper, we have analyzed the hash-based mutual Tag authentication protocol based on random numbers and synchronized secret value. The protocol proved to be insecure against the eavesdropping attack, man-in-middle attack, impersonation attack, denial-of service attack, and insider attack. The protocol is vulnerable to these attacks due to the secret value, tag identity, and hash function stored in the database as plaintext. Also, the protocol requires safe communication mechanism for transferring intermediate computed values over the communication channel.

Proposed ETISVP protocol filled these loopholes by using encryption for message exchange and by storing the Tag identity and secret value in the database in encrypted form. The protocol introduced is secure against these attacks and hence, strengthen the security of RFID technology. The ETISVP provides immunity from several attacks; hence, it is safe to deploy in the industry. However, encryption of data delays the attack only rather than making it fool proof. In the modern world, we have several machines which are capable of holding huge computation power which poses threat to the proposed protocol for decrypting the encrypted information. Therefore, we are intended to work upon strengthening ETISVP as a future direction of work. Apart from this, we are also intended to work upon quantum computing environment as existing protocol are easily deciphered in quantum computing environment due to its immense computing power.

References

1. Shepard, S.: RFID: Radio Frequency Identification. McGraw Hill Professional (2005)
2. Landt, J.: The history of RFID. IEEE Potentials **24**(4), 8–11 (2005)
3. Srivastava, K., Awasthi, A.K., Kaul, S.D., Mittal, R.: A hash based mutual RFID tag authentication protocol in telecare medicine information system. J. Med. Syst. **39**(1), 153 (2015)
4. Ohkubo, M., Suzuki, K., Kinoshita, S., et al.: Cryptographic approach to privacy-friendly tags. In: RFID Privacy Workshop, Cambridge, USA, vol. 82 (2003)
5. Center, A.-I.: 860 MHz-960 MHz Class-I radio frequency identification tag radio frequency & logical communication interface specification proposed recommendation version 1.0. 0 (2002)
6. Weis, S.A., Sarma, S.E., Rivest, R.L., Engels, D.W.: Security and privacy aspects of low-cost radio frequency identification systems. In: Security in Pervasive Computing, pp. 201–212. Springer (2004)
7. Tsudik, G.: YA-TRAP: yet another trivial RFID authentication protocol. In: Pervasive Computing and Communications Workshops, 2006. PerCom Workshops 2006. Fourth Annual IEEE International Conference on, IEEE, pp. 4–7 (2006)
8. Tsudik, G.: A family of dunces: trivial RFID identification and authentication protocols. In: Privacy Enhancing Technologies, pp. 45–61. Springer (2007)
9. Chatmon, C., van Le, T., Burmester, M.: Secure anonymous RFID authentication protocols. Florida State University, Department of Computer Science, Tech. Rep
10. Sun, P.R., Wang, B.H., Wu, F.: A new method to guard inpatient medication safety by the implementation of RFID. J. Med. Syst. **32**(4), 327–332 (2008)
11. Huang, H.-H., Ku, C.-Y.: A RFID grouping proof protocol for medication safety of inpatient. J. Med. Syst. **33**(6), 467 (2009)
12. Chien, H.-Y., Yang, C.-C., Wu, T.-C., Lee, C.-F.: Two RFID-based solutions to enhance inpatient medication safety. J. Med. Syst. **35**(3), 369–375 (2011)
13. Peris-Lopez, P., Orfila, A., Mitrokotsa, A., Van der Lubbe, J.C.: A comprehensive RFID solution to enhance inpatient medication safety. Int. J. Med. Inform. **80**(1), 13–24 (2011)
14. Yen, Y.-C., Lo, N.-W., Wu, T.-C.: Two RFID-based solutions for secure inpatient medication administration. J. Med. Syst. **36**(5), 2769–2778 (2012)
15. Kim, H.: Enhanced hash-based RFID mutual authentication protocol. In: Computer Applications for Security, Control and System Engineering, pp. 70–77. Springer (2012)
16. Kim, H.: RFID mutual authentication protocol based on synchronized secret. Int. J. Secur. Appl. **7**(4), 37–50 (2013)
17. Chen, Y.-Y., Huang, D.-C., Tsai, M.-L., Jan, J.-K.: A design of tamper resistant prescription RFID access control system. J. Med. Syst. **36**(5), 2795–2801 (2012)
18. Cho, J.-S., Yeo, S.-S., Kim, S.K.: Securing against brute-force attack: a hash-based RFID mutual authentication protocol using a secret value. Comput. Commun. **34**(3), 391–397 (2011)
19. Suja, S., Arivarasi, A.: An RFID Authentication Protocol for Security and Privacy. In: International Conference on Computing and Control Engineering (2012)

Analysis of Binary PSK Modulations Over the Line-of-Sight Plus Scatter Fading Model

Veenu Kansal and Simranjit Singh

Abstract In this paper, the average bit error rate (ABER) performance of binary coherent and non-coherent modulation is investigated over a new fading channel which is well known as Beaulieu-Xie fading channel. This fading model consists of multiple no. of line-of-sight (LOS) components with some diffuse power. Specifically, the closed-form expressions are derived for the exact ABER of binary phase-shift keying (BPSK) and differential phase-shift keying (DPSK) by using the probability density function (PDF) approach. The final exact expression for BPSK is obtained in the form of infinite series which needs a reasonable number of terms to converge and the final expression for DPSK is expressed in terms of elementary functions. It is observed that ABER decreases with increase in specular power and/or with decrease in fading severity. The obtained results are applied to general scenarios for various values of fading parameters by adjusting the specular power and fading severity. The numerical results plotted by using the derived expressions show a close agreement with the results obtained by Monte Carlo simulation.

Keywords Average BER (ABER) · Binary phase-shift keying (BPSK)
Femtocells · High-speed trains

1 Introduction

In wireless communication system, bit error rate (BER) is the important performance metric to measure the quality of transmission over the fading channels. Over the past years, several mathematical models such as Rayleigh, Rician, Nakagami-m, k–μ, and two-wave with diffuse power (TWDP) [1–7] have been used to analyze the small-scale fading. However, none of the above-mentioned distributions are suitable

V. Kansal · S. Singh (✉)
Department of ECE, Punjabi University Patiala, Patiala, India
e-mail: simranjit@live.com

V. Kansal
e-mail: veenukansal@outlook.com

L. C. Jain et al. (eds.), *Data and Communication Networks*, Advances in Intelligent Systems and Computing 847, https://doi.org/10.1007/978-981-13-2254-9_2

for modeling the fading channel where multiple dominant specular components are present in addition to the diffuse scatter power. To model such a scenario, a fading model was recently proposed by Beaulieu and Jiandong [8]. This fading is based on non-central chi distribution (just like generalized Rician and k–μ fading); however, it is much more flexible than aforementioned distributions for modeling practical scenarios with multiple dominant specular waves. This model can be obtained from the non-central chi distribution as in similar way by which the Nakagami-m fading is obtained from the central chi distribution. While the Nakagami-m fading is ideally suitable to model scatter fading wireless channels, the Beaulieu-Xie fading is uniquely efficient to characterize practical LOS channels. It is reported that this new fading distribution is especially suitable for modeling high-speed trains and future femtocells channels. One major difference between typical fading channels and the fading channels of the train station for the high-speed railway is that the K-factor of the later is considerably smaller [9]. The existing fading models such as Rayleigh and Rician can be obtained from the Beaulieu-Xie fading channel by adjusting the parameters K and m.

In the literature, the ABER analysis of wireless fading channels such as Rayleigh, Rician, and TWDP, etc. can be found in [1–7]. In [1], the authors provide a generalized approach to analyze BER performance for different types of modulation techniques, with or without diversity reception over additive white Gaussian noise (AWGN) and also, same analysis is presented for generalized fading models. An analysis with DPSK modulation scheme over Rician fading channel is presented in [2]. Further, error probability for uncoded BPSK is investigated in TWDP fading in [5]. By using moment-generating function (MGF) approach, the performance of an average symbol error rate of quadrature amplitude modulation (QAM) is studied over TWDP fading by deriving the analytical expressions in [6]. The ABER analysis of M-ary phase-shift keying (MPSK) in the presence of TWDP fading is provided in [7]. In [10], the second order statistics for Beaulieu-Xie fading is investigated. Furthermore, the analysis of this fading channel is done with selection combining and equal combining techniques in [11, 12]. To the best of our knowledge, the exact ABER analysis of Beaulieu-Xie fading with BPSK and DPSK modulations is not reported in the literature yet. The exact BER analysis of BPSK and binary DPSK is presented in this paper with an aim to gain a better understanding about this fading channel.

The remaining paper is organized as follows: The description of PDF of fading model is provided in Sect. 2. The exact analytical expressions for ABER of BPSK and DPSK are derived in Sect. 3. Section 4 presents the discussion of analytical and simulation results. Section 5 contains the concluding remarks of this paper.

2 System Model

A novel fading model has been proposed by Beaulieu and Jiandong [8] which models multiple dominant specular components with some diffused power. The PDF of the

SNR can be evaluated by applying a square transformation [13] in (1). The resulting expression of PDF of SNR can be given by:

$$p_\gamma(\gamma) = \sqrt{\eta^{m+1}} \exp(-mK - \eta\gamma)\left(\frac{\gamma}{mK}\right)^{(m-1)/2} I_{m-1}\left(2\sqrt{mK\eta\gamma}\right), \quad (1)$$

where $\eta = L_1^{m-1}(-mK)/\bar{\gamma}$, $K = \lambda^2/\Omega$, $L_n^\alpha(\cdot)$ is the associated nth-order Laguerre polynomial [11], and $\bar{\gamma}$ is the average SNR, $I_{m-1}(\cdot)$ is the modified Bessel function of the first kind and $(m-1)$th order, m is the shape factor, Ω defines the spread of the model, and λ represents the height and location of the PDF. The Beaulieu-Xie fading behaves as Rician fading, when the fading parameter 'm' is equal to 1 for any value of 'K'. Also, it can be reduced to Rayleigh fading when $m = 1$, $K = 0$.

3 Performance Evaluation

Average Bit Error Rate The exact closed-form expressions for BPSK and DPSK modulation techniques over Beaulieu-Xie fading are derived in this section. The ABER for various modulation techniques over a fading model can be derived by integrating the conditional BER of any modulation techniques and the PDF of a fading model. The ABER, P_e of a modulation technique can be computed mathematically as:

$$P_e = \int_0^\infty P_e(\gamma) p_\gamma(\gamma) d\gamma, \quad (2)$$

3.1 ABER of BPSK Modulation Scheme Over the Beaulieu-Xie Fading Channel

The conditional BER of BPSK over an AWGN channel [5] is given by:

$$P_e(\gamma) = \frac{1}{2}\mathrm{erfc}\left(\sqrt{\gamma}\right), \quad (3)$$

To calculate the exact expressions for ABER of BPSK over the Beaulieu-Xie fading channel, substitute the values of $P_e(\gamma)$ and $p_\gamma(\gamma)$ from (3) and (1), respectively, in (2). By using ([14], 3.381.9) in (3), we obtain the expression:

$$P_e = \int_0^\infty \frac{1}{2\sqrt{\pi}} \Gamma(0.5, \gamma)\sqrt{(\eta)^{m+1}} \exp(-mK - \eta\gamma) \times \left(\frac{\gamma}{mK}\right)^{(m-1)/2} I_{m-1}\left(2\sqrt{mK\eta\gamma}\right) \mathrm{d}\gamma, \tag{4}$$

By expressing the modified Bessel function of the first kind and $(m-1)$th order in the form of an infinite series, and by using identity ([14], 6.455.1), the resulting expression for unconditional ABER of BPSK is given as:

$$P_e = \frac{1}{2\sqrt{(1+\eta)\pi}} \exp(-mK) \sum_{g=0}^{\infty} \frac{\Gamma(m+g+0.5)}{g!\,\Gamma(m+g+1)} \frac{(mK)^g}{\left(1+\frac{1}{\eta}\right)^{m+g}} {}_2F_1\left(1, m+g+0.5, m+g+1, \frac{1}{1+\frac{1}{\eta}}\right), \tag{5}$$

where ${}_2F_1(q; x; y; z)$ is a Gauss hypergeometric function.

3.2 ABER of DPSK Modulation Over the Beaulieu-Xie Fading Channel

The ABER for the DPSK modulation can be derived by integrating the conditional BER of the DPSK modulation with the PDF of the Beaulieu-Xie fading model as in (2). The conditional BER of DPSK over an AWGN channel [10] is given by:

$$P_e(\gamma) = \frac{1}{2} \exp(-\gamma), \tag{6}$$

To obtain the ABER of the DPSK modulation scheme over the Beaulieu-Xie fading channel, substitute the values of $P_e(\gamma)$ and $p_\gamma(\gamma)$ from (6) and (1), respectively, into (2), the expression is obtained as:

$$P_e = \frac{1}{2} \int_0^\infty \exp(-\gamma)(\eta)^{(m+1)/2} \exp(-mK - \eta\gamma) \left(\frac{\gamma}{mK}\right)^{(m-1)/2} I_{m-1}\left(2\sqrt{mK\eta\gamma}\right) \mathrm{d}\gamma, \tag{7}$$

Now by applying the identity ([14], 6.643.2), integral in (7) can be solved as:

$$P_e = \frac{\exp(-mK)}{2} \left(\frac{\eta}{mK(1+\eta)}\right)^m \exp\left(\frac{-mK\eta}{2(1+\eta)}\right) M_{-m,0}\left(\frac{-mK\eta}{1+\eta}\right), \tag{8}$$

With the help of ([14], 9.220.2) and $M_{u,s}(z) = \exp\left(\frac{-z}{2}\right)z^{s+1/2}M\left(s-u+\frac{1}{2};1+2s;z\right)$, the expression in (8) can be simplified as:

$$P_e = \frac{\exp(-mK)}{2\left(1+\frac{1}{\eta}\right)^m} M\left(m, m, \frac{mK\eta}{1+\eta}\right), \tag{9}$$

where $M(a, b, c)$ is a confluent hypergeometric function.

Further, by using the property $M(a, a, c) = \exp(c)$ in (9), the final expression for the unconditional ABER of DPSK as:

$$P_e = \frac{\eta^m}{2(1+\eta)^m} \exp\left(-\frac{mK}{1+\eta}\right). \tag{10}$$

4 Numerical Results and Discussion

The final ABER expressions for BPSK and DPSK in (5) and (10) are calculated analytically and curves are drawn for various combinations of fading parameter, 'K' and 'm', respectively. To validate the accuracy of analytical results, they have been compared with simulation results. The simulation values at very low BER have been eliminated since their accurate simulation requires a large amount of time.

Figures 1 and 2 show the results for ABER of BPSK over the Beaulieu-Xie fading channel with constant total power of 6.98 dB. Figure 1 depicts the ABER of BPSK against SNR per bit, $\bar{\gamma}$ for a constant value of $K = 2$ and various values of m. It can be demonstrated that for a constant K, the performance of system improves as m increases. It is observed from the results that $m = 5$ gives a SNR advantage of 1 dB over $m = 3$ and $m = 3$ gives a SNR advantage of 13 dB over $m = 1$ case to achieve an ABER of 10^{-3}. With increase in the value of 'm', the SNR advantage decreases because of decrease in the fading severity. The above observation comes from system, because fading severity decreases with increase in m. The numerical results are closely matched to the simulation results. Figure 2 shows similar results as Fig. 1 but for constant $m = 3$ and by varying K. On comparing the results for $K = 0$ with curves for $K = 2$, it is observed that $K = 2$ has a SNR advantage of 2 dB over $K = 0$ to achieve an ABER of 10^{-3}. This effect is because of the increase in specular power with increase in K.

Figure 3 shows the ABER of DPSK for several values of K and m. The simulation results obtained for the special case of Beaulieu-Xie fading as Rayleigh and Rician fading are also provided in Fig. 3. The obtained results are in accordance with the numerical results plotted by using final closed-form expressions of ABER of DPSK. It is clear that for an ABER of 10^{-2}, the SNR improvement for curves shown in Fig. 3, corresponding to $m = 1$, $K = 2$ over curves for $m = 1$, $K = 0$ is approximately

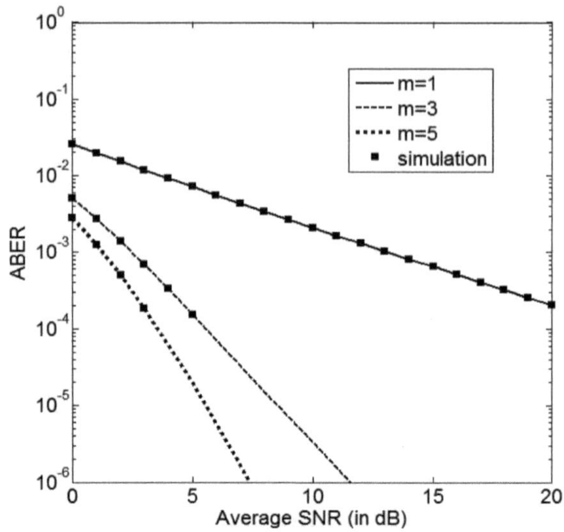

Fig. 1 ABER of BPSK over the Beaulieu-Xie fading for fixed $K = 2$ and various values of m

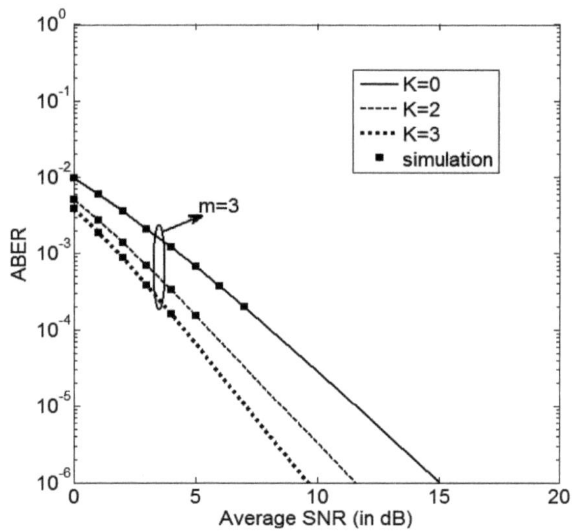

Fig. 2 ABER of BPSK over the Beaulieu-Xie fading for fixed $m = 3$ and various values of K

3 dB. The results presented in this section are useful to analyze the effects of fading parameters 'K' and 'm' on the ABER performance.

5 Conclusion

This paper presents exact analytical expressions for the ABER of BPSK and DPSK that are obtained over the Beaulieu-Xie fading. In particular, the final expressions

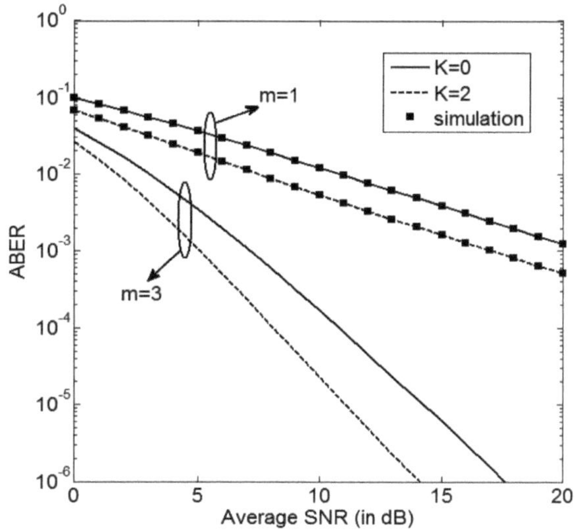

Fig. 3 ABER of DPSK over the Beaulieu-Xie fading for various combinations of K, m with constant total power of 6.02 dB

for BPSK and DPSK are in the form of infinite series and elementary functions, respectively. Various scenarios for different combination of 'K' and 'm' have been examined and discussed by which it is found that the ABER decreases as specular power, 'K' and/or fading severity parameter, 'm' increases. The analytically obtained expressions in this paper are verified through simulations.

References

1. Simon, M.K., Alouini, M.S.: A unified approach to the probability of error for noncoherent and differentially coherent modulations over generalized fading channels. IEEE Trans. Commun. **46**(12), 1625–1638 (1998)
2. Kam, P.Y., Zhong, C.: Tight bounds on the bit-error-probabilities of 2DPSK and 4DPSK in nonselective Rician fading. IEEE Trans. Commun. **46**(7), 860–862 (1998)
3. Yacoub, M.D.: The k-μ distribution and η-μ distribution. IEEE Antennas Propag. Mag. **49**(1), 68–81 (2007)
4. Durgin, G.D., Rappaport, T.S., de Wolf, D.A.: New analytical models and probability density functions for fading in wireless communications. IEEE Trans. Commun. **50**(6), 1005–1015 (2002)
5. Oh, S.H., Li, K.H.: BER performance of BPSK receivers over two-wave with diffuse power fading channels. IEEE Trans. Wirel. Commun. **4**(4), 1448–1454 (2005)
6. Suraweera, H.A., Lee, W.S., Oh, S.H.: Performance analysis of QAM in a two-wave with diffuse power fading environment. IEEE Commun. Lett. **12**(2), 109–111 (2008)
7. Singh, S., Kansal, V.: Performance of M-ary PSK over TWDP fading channels. Int. J. Electron. Lett. **4**(4), 433–437 (2015). Taylor and Francis
8. Beauliu, N.C., Jiandong, X.: A novel fading model for channels with multiple dominant specular components. IEEE Wirel. Commun. Lett. **4**(1), 54–57 (2015)

9. Guan, K., Zhong, D., Ai, B., Kuerner, T.: Propagation measurements and analysis for train stations of high-speed railway at 930 MHz. IEEE Trans. Veh. Technol. **63**(8), 3499–3516 (2014)
10. Olutayo, A., Ma, H., Cheng, J., Holzman, J.F.: Level crossing rate and average fade duration for the Beaulieu-Xie fading model. IEEE Wirel. Commun. Lett. **6**(3), 326–329 (2017)
11. Olutayo, A., Cheng, J., Holzman, J.F.: Asymptotically tight performance bounds for selection diversity over Beaulieu-Xie fading channels with arbitrary correlation. In: IEEE International Conference on Communication Proceedings, Paris, France, pp. 1–6 (2017)
12. Olutayo, A., Cheng, J., Holzman, J.F.: Asymptotically tight performance bounds for equal-gain combining over a new correlated fading channel. In: IEEE 15th Canadian Workshop on Information Theory Proceedings, Quebec, Canada, pp. 1–5 (2017)
13. Kansal, V., Singh, S.: Analysis of effective capacity over Beaulieu-Xie fading model. in Proc. IEEE International Women in Engineering Conference on Electrical and Computer Engineering, Dehradun, India, pp. 207–210 (2017)
14. Gradshteyn, I.S., Ryzhik, I.M.: Table of Integrals, Series and Products, 5th edn. Academic, San Diego, CA (1994)

Approximate Bit Error Rate of DPSK with Imperfect Phase Noise in TWDP Fading

Veenu Kansal and Simranjit Singh

Abstract This study provides closed-form expressions for average bit error rate (ABER) of differential phase shift keying (DPSK) with phase error in two-wave with diffuse power (TWDP) fading. It is considered that the envelope of phase error is Gaussian distributed. The effect of phase synchronization on wireless system is studied for different values of TWDP fading parameters and phase error. The analytical results are evaluated to study the impact of phase error on the system performance. Also, the results are compared with the case of perfectly synchronized.

Keywords BER · DPSK · TWDP

1 Introduction

The performance of PSK with noisy phase reference has been a topic of interest for many researchers since the 1960s. Initial work on modeling the phase error was done by Tikhonov and Viterbi. Since then a lot of work has been done to study the effect of phase error on different digital modulation schemes over various fading models such as Rayleigh, Nakagami-m and Rician [1–7]. In all these papers, the effects of imperfect carrier synchronization have been studied by assuming the phase distribution either Gaussian or Tikhonov. However, phase synchronization errors can have a significant effect on the receiver's capability to make the correct decisions. The problem gets more severe as data rates increase due to phase error. In 2002, a fading model was proposed by Durgin et al. which could be used to model fading scenarios having two specular paths in addition to the diffusely propagating power [8]. This fading model is known as TWDP and can be used to characterize the fading scenarios which are more severe than Rayleigh fading. The practical significance

V. Kansal · S. Singh (✉)
Punjabi University, Patiala 147002, India
e-mail: simranjit@live.com

V. Kansal
e-mail: veenukansal@outlook.com

L. C. Jain et al. (eds.), *Data and Communication Networks*, Advances in Intelligent Systems and Computing 847, https://doi.org/10.1007/978-981-13-2254-9_3

of this model comes from modern uses of wireless channels in applications such as data transmission in dynamic environments such as mobile personal devices (such as smartphones or smart/connected automobiles) and Internet service in aircraft that have become pervasive in recent years.

The existing fading models such as Rayleigh and Rician fading are included in TWDP fading as special cases. The error performance analysis of various digital modulation schemes in TWDP fading is reported in the literature [9–12]. However, till date no work is available on the analysis of modulation schemes with phase noise error in TWDP fading model. In this paper, we derive the approximate closed-form expression for average BEP of DPSK with noisy phase reference in TWDP fading channel. In Sect. 2, the PDF of TWDP fading model is presented. In Sect. 3, the PDF approach is used to derive the ABER of DPSK with noisy phase reference over TWDP fading model. The numerical results are evaluated and discussed in Sect. 4. Section 5 holds the conclusion of the paper.

2 TWDP Fading Model

The TWDP fading model is having two specular components with some Rayleigh distributed diffuse power. The instantaneous signal-to-noise ratio (SNR) is defined as $\gamma = r^2 E_b/N_0$, and the average SNR is defined as $\bar{\gamma} = E\left(r^2\right)\frac{E_b}{N_0}$. The PDF of SNR ($\gamma$) for TWDP fading channel can be evaluated as [9]:

$$p_\gamma(\gamma) = \frac{\eta}{2}\sum_{i=1}^{L} a_i \sum_{j=0}^{1} \exp\left(-P_{2i-j} - \eta\gamma\right) I_0\left(2\sqrt{P_{2i-j}\eta\gamma}\right), \tag{1}$$

where $\eta = K + 1/\bar{\gamma}$, $P_{2i-j} = K\left[1 + (-1)^j \Delta \cos\left(\pi\left(\frac{i-1}{2L-1}\right)\right)\right]$, Δ denotes the relative strength of two specular components, K is defined as ratio of specular power to diffuse power, and L is the order of PDF. The TWDP fading can be reduced to special cases [9] of fading such as Rayleigh fading ($K=0$) and Rician fading ($K>0$, $\Delta=0$).

3 Performance Analysis

The conditional BER of DPSK with noisy phase reference over AWGN channel can be expressed as [13]:

$$P(e|\gamma, \phi) = \frac{1}{2}\exp\left(-\gamma\, \cos^2\phi\right), \tag{2}$$

where ϕ represents the phase error. It is considered that ϕ is Gaussian distributed with zero mean and $2\sigma^2$ variance, i.e., $C\ (0, 2\sigma^2)$. The PDF of ϕ can be given by:

$$p_{\phi}(\phi) = \frac{1}{\sqrt{2\pi\sigma^2}}\exp\left(\frac{-\phi^2}{2\sigma^2}\right), \quad (3)$$

The ABER of a modulation scheme with phase noise error over a fading channel can be obtained by performing a twofold integration, first over the PDF of ϕ and second over the PDF of fading model [2],

$$P_e = \int_{0}^{\infty}\int_{-\pi}^{\pi} P(e|\gamma,\phi)p_{\phi}(\phi)p_{\gamma}(\gamma)\mathrm{d}\phi\,\mathrm{d}\gamma, \quad (4)$$

To derive the ABER of DPSK with Gaussian distributed phase error over TWDP fading model, first solve the inner integral of (4) as:

$$P(e|\gamma) = \int_{-\pi}^{\pi} P(e|\gamma,\phi)p_{\phi}(\phi)\mathrm{d}\phi, \quad (5)$$

Substitute $P(e|\gamma,\phi)$ and $p_{\phi}(\phi)$ from (2) and (3), respectively, into (5), and we get:

$$P(e|\gamma) = \int_{-\pi}^{\pi} \frac{1}{2}\exp(-\gamma\,\cos^2\phi) \times \frac{1}{\sqrt{2\pi\sigma^2}}\exp\left(\frac{-\phi^2}{2\sigma^2}\right)\mathrm{d}\phi, \quad (6)$$

By using the small angle approximation, $\cos(\Delta\phi) \cong 1-\frac{(\Delta\phi)^2}{2}$, (6) can be rewritten as:

$$P(e|\gamma) = \frac{\exp(-\gamma)}{2\sqrt{2\pi\sigma^2}}\int_{-\pi}^{\pi}\exp\left(-\phi^2\left(\frac{1}{2\sigma^2}-\gamma\right)\right)\mathrm{d}\phi, \quad (7)$$

Now by employing the general formula, i.e., $\int_{-\pi}^{\pi}\exp(-x^2)\mathrm{d}x \simeq \sqrt{\pi}$, the conditional BER of DPSK with Gaussian phase error over AWGN channel can be expressed as:

$$P(e|\gamma) = \frac{\exp(-\gamma)}{2\sqrt{1-2\sigma^2\gamma}}, \quad (8)$$

Now solving the outer integral of (4) over PDF of TWDP fading, we get

$$P_e = \int_{0}^{\infty} P(e|\gamma)p_{\gamma}(\gamma)\mathrm{d}\gamma, \quad (9)$$

By substituting $P(e|\gamma)$ and $p_{\gamma}(\gamma)$ from (8) and (1) into (9), we get:

$$P_e = \int_0^{\infty} \frac{\exp(-\gamma)}{2\sqrt{1-2\sigma^2\gamma}} \times \frac{\eta}{2}\sum_{i=1}^{L} a_i \sum_{j=0}^{1} \exp(-P_{2i-j} - \eta\gamma) I_0\left(2\sqrt{P_{2i-j}\eta\gamma}\right) \mathrm{d}\gamma, \tag{10}$$

Put $\sigma_\phi^2 = (C\gamma)^{-1}$ in (10), and rearrange (10) as:

$$P_e = \frac{\eta}{4}\sum_{i=1}^{L} a_i \sum_{j=0}^{1} \exp(-P_{2i-j}) \frac{1}{\sqrt{1-\frac{2}{C}}} \int_0^{\infty} \exp(-\gamma(1+\eta)) I_0\left(2\sqrt{P_{2i-j}\eta\gamma}\right) \mathrm{d}\gamma, \tag{11}$$

By using the formula given in the appendix, the final ABER expression is calculated as:

$$P_e = \frac{\eta}{4}\sum_{i=1}^{L} a_i \sum_{j=0}^{1} \exp(-P_{2i-j}) \left(1-\frac{2}{C}\right)^{-1/2} \times \frac{1}{1+\eta} M\left(1;1;\frac{P_{2i-j}\eta}{1+\eta}\right), \quad \mathrm{C} > 2. \tag{12}$$

where $M(a;b;c)$ is confluent hypergeometric function.

4 Numerical Results and Discussion

The numerical results are provided for the ABER of partially coherent differential PSK over TWDP fading model. By using (12), the results are plotted in Figs. 1, 2, and 3. As mentioned in (12), these results are valid only for value of C greater than 2. The results are taken for different parameters such as C, Δ, and K. The value of imperfect synchronization factor (C) is taken to be 3, 10, and ∞. When $C = \infty$, the DPSK modulation is considered to be having no phase error. For low values of C, the phase error is more, and as expected, the degradation is severe. As value of C increases, the degradation becomes less severe and closer to the perfectly synchronization case.

In Fig. 1, results are plotted for a combination of different values of C and Δ and for fixed value of K. It is observed that the performance of system deteriorates as Δ approaches 1.

Similarly, in Fig. 2, results are plotted for fixed value of Δ with different values of C and K. The results demonstrated that the ABER decreases as the value of specular power (K) increases.

Figure 3 shows the results for ABER versus C for fixed values of Δ, K, and SNR (in dB). The plot depicts that the performance of system improves as the value of C increases; i.e., as the phase error decreases, the ABER also decreases.

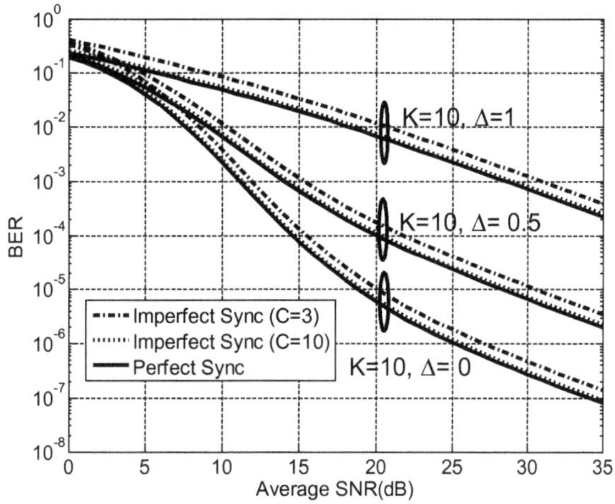

Fig. 1 ABER versus SNR for DPSK with phase error with fixed K and varying C and Δ

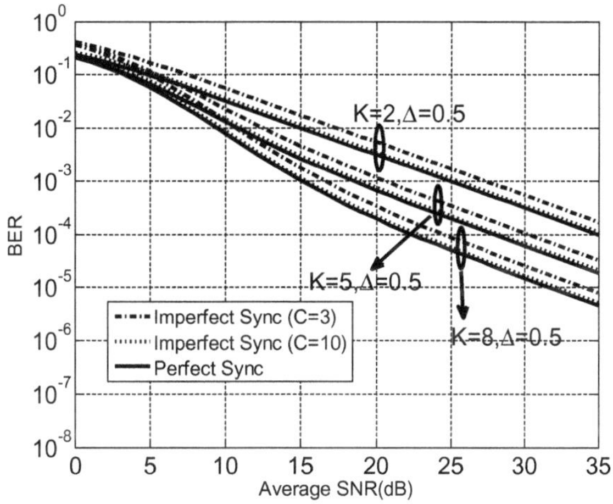

Fig. 2 ABER versus SNR for DPSK with phase error with fixed Δ and varying C and K

5 Conclusion

This paper analyzed the performance of ABER of partially coherent DPSK in TWDP fading channels. The closed-form expressions are derived for ABER which is a convenient tool for error performance analysis. The final expression of unconditional ABER of DPSK along with Gaussian phase error over TWDP fading is valid only for value of C greater than 2. The analysis depicted that if the imperfect synchronization factor, 'C' increases; the system performance is improved and is closed to perfectly synchronization case.

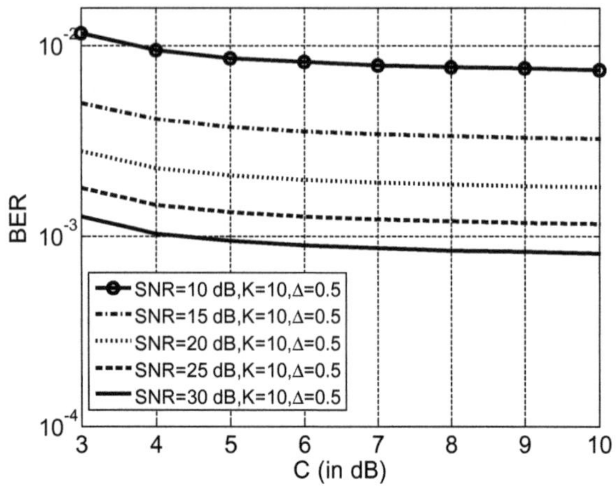

Fig. 3 ABER versus C (in dB) for DPSK modulation for fixed K, SNR, and Δ

Appendix

The expression in (11) can be further solved by using identity given in ([14], 6.6.14.3) as:

$$\int_0^\infty \exp(-\alpha x) I_{2v}\left(2\sqrt{\beta x}\right)\mathrm{d}x = \frac{1}{\sqrt{\alpha\beta}}\exp\left(\frac{\beta}{2\alpha}\right)\frac{\Gamma(v+1)}{\Gamma(2v+1)}M_{-1/2,v}\left(\frac{\beta}{\alpha}\right)$$

Now, (11) can be simplified as:

$$P_e = \frac{\eta}{4}\sum_{i=1}^{L} a_i \sum_{j=0}^{1}\exp\left(-P_{2i-j}\right)\left(1-\frac{2}{C}\right)^{-1/2}\exp\left(\frac{P_{2i-j}\eta}{2(1+\eta)}\right) \times \frac{1}{\sqrt{(1+\eta)P_{2i-j}\eta}}M_{-1/2,0}\left(\frac{P_{2i-j}\eta}{1+\eta}\right) \quad (13)$$

Use $M_{k,s}(z) = \exp\left(\frac{-z}{2}\right)z^{s+1/2}M\left(s-k+\frac{1}{2}; 1+2s; z\right)$ [14] in (13), and the final equation can be obtained as given in (12).

References

1. Viterbi, A.J.: Phase-locked loop dynamics in the presence of noise by Fokker-Planck techniques. Proc. IEEE **51**, 1737–1753 (1963)
2. Simon, M.K., Alouini, M.S.: Simplified noisy reference loss evaluation for digital communication in the presence of slow fading and carrier phase error. IEEE Trans. Veh. Technol. **50**(2), 480–486 (2001)

3. Lo, C.M., Lam, W.H.: Error probability of binary phase shift keying in Nakagami-m fading channel with phase noise. Electron. Lett. **36**(21), 1773–1774 (2000)
4. Lindsey, W.C.: Phase-shift-keyed signal detection with noisy reference signals. IEEE Trans. Aerosp. Electron. Syst. **2**, 393–401 (1966)
5. Prabhu, V.K.: PSK performance with imperfect carrier phase recovery. IEEE Trans. Aerosp. Electron. Syst. **12**, 275–285 (1976)
6. Chandra, A., Patra, A., Bose, C.: Performance analysis of PSK systems with phase error in fading channels: a survey. Phys. Commun. **4**, 63–82 (2011)
7. Smadi, M.A., Aljazar, S.O., Ghaeb, J.A.: Simplified bit error rate evaluation of Nakagami-m PSK systems with phase error recovery. Wirel. Commun. Mob. Comput. **12**, 248–256 (2012)
8. Durgin, G.D., Rappaport, T.S., de Wolf, D.A.: New analytical models and probability density functions for fading in wireless communications. IEEE Trans. Commun. **50**(6), 1005–1015 (2002)
9. Subadar, R., Singh, A.D.: Performance of SC receiver over TWDP fading channels. IEEE Wirel. Commun. Lett. **2**(3), 267–270 (2013)
10. Singh, S., Kansal, V.: Performance of M-ary PSK over TWDP fading channels. Int. J. Electron. Lett. **4**(4), 433–437 (2015)
11. Das, P., Subadar, R.: Performance of M-EGC receiver over TWDP fading channels. IET Commun. **11**(12), 1853–1856 (2017)
12. Singh, S., Sharma, S.: Performance analysis of spectrum sensing techniques over TWDP fading channels for CR based IoTs. Int. J. Electron. Commun. **80**, 210–217 (2017)
13. Lindsey, W.C., Simon, M.K.: Telecommunication Systems Engineering. Prentice-Hall, Englewood Cliffs, NJ (1973)
14. Gradshteyn, I.S., Ryzhik, I.M.: Table of Integrals, Series, and Products, 7th edn. Academic Press (2000)

Analysis of Workloads for Cloud Infrastructure Capacity Planning

Eva Patel and Dharmender Singh Kushwaha

Abstract Workload analysis and characterization are the first steps toward effective cloud infrastructure capacity planning. Identifying workload patterns based on resource utilization not only enables informed decisions about mapping of current request to available capacity, but also serves as a meaningful indicator for future resource requirements. Of paramount concern is the optimal utilization of data center server capacity, i.e., the CPU, I/O, and memory. The compute capacity of modern servers can be further harnessed by optimal utilization of individual CPU cores. A precise CPU core-level usage monitoring and provisioning can lead to cumulative benefits of optimal CPU utilization, efficient VM placement, reduced VM migrations, and energy efficiency through lower power consumption. In this paper, we make a preliminary analysis of usage patterns of CPU cores in the case of CPU- and memory-intensive workloads on an experimental cloud setup in our laboratory. The aim is to make a comparative analysis of the utilization of individual CPU cores with that of aggregated CPU usage to explore the feasibility of incorporating a fine-grained usage detail for resource scheduling and VM provisioning. Initial experiments reveal observable differences between the utilization of individual CPU cores and that reported by aggregate CPU usage. Usage difference ranges from 1 to 29% below and between 4 and 20% above the aggregate. Incorporating such finer details can leverage the vast compute capacity of multicore servers and effective power usage.

Keywords Workload analysis · Workload characterization · Capacity planning
Server capacity planning · Workload intensity · Job inter-arrival patterns
Multicore scheduling

E. Patel (✉) · D. S. Kushwaha
Department of Computer Science and Engineering, MNNIT Allahabad, Allahabad, UP 211004, India
e-mail: evapatel08@gmail.com

D. S. Kushwaha
e-mail: dsk@mnnit.ac.in

L. C. Jain et al. (eds.), *Data and Communication Networks*, Advances in Intelligent Systems and Computing 847, https://doi.org/10.1007/978-981-13-2254-9_4

1 Introduction

Scalability, agility, guaranteed quality of service, the illusion of infinite computational resources, and access to high-end computing infrastructure on a pay-as-per-usage basis, without the painstaking exercise of setting up a private IT infrastructure, have led to the relentless adoption of cloud computing solutions for a range of computing requirements—business intelligence, engineering design, scientific computing, social networking, and content delivery. These services are hosted on data centers that house hundreds of thousands of servers and supporting infrastructure. For instance, Amazon, which made the IT infrastructure available via the Internet in 2006, a technology termed cloud computing [1], has around 450,000 servers spread across seven different data centers around the world [2] that provide the infrastructure services—compute, storage, and networking. Google data center caters to the diverse requests of its cloud users through 900,000 plus servers hosted in data centers in 16 different geographical locations [2]. The envisioned growth in computing and storage requirements, diversity of services, and emergence of innovative server technologies that also entail growing power consumption and increasing carbon footprints calls for adoption of capacity planning of cloud data centers as a continual process.

Data center capacity planning aims to determine a system configuration for given service requests that comply with the service-level agreement (SLA) [3] with minimal over- or under-provisioning of resources. The first step to effective capacity planning is the analysis of resource usage patterns of cloud workloads to identify associated challenges and their implications on resource provisioning and overall system performance. A vast body of literature records design of scheduling algorithms with CPU utilization as the primary performance metric [4–10]. With several CPU cores crammed into a single socket, and multisocket servers with massive compute capacity, the use of individual core usage as decision parameter for VM scheduling is one possible approach to effective utilization of server compute capacity.

Driven by the observation that aggregate CPU usage is not always reflective of the usage of individual CPU cores, we conduct a preliminary analysis of utilization of CPU cores in case of CPU-intensive and memory-intensive workloads. Through this study, the aim is to achieve following objectives:

1. Effective utilization of compute capacity of all the cores of a multicore server by considering individual CPU core capacity rather than the aggregate CPU usage.
2. Identify and segregate cores that are underutilized, to schedule to incoming workload thereby reducing energy wastage.
3. Mitigate situations of overloaded CPU cores to maintain desired performance and sustainable energy consumption levels.

The rest of this paper is organized as follows. Section 2 gives a brief overview of related work on workload analysis and characterization. Section 3 discusses some background concepts and findings from past research that reveal suboptimal use of CPU cores. Experimental setup and results are discussed in Sect. 4. Section 5 concludes with directions for future work.

2 Related Work

Resource requirements of applications deployed on the cloud depend on their processing needs. Consequently, each workload has different challenges which guide the analysis and characterization process.

Based on behavioral characteristics, Mahambre et al. [4] identify five cloud workload patterns, i.e., periodicity, threshold, relationship, variability, and image similarity, for capacity planning decisions such as migration, re-provisioning, load balancing, and initial placement of the workloads. Moreno et al. [5] present a comprehensive analysis of workload characteristics to study heterogeneity due to user behavior and task resource usage. Their study shows that users are responsible the most for introducing heterogeneity in the cloud as compared to task diversity which is a consequence of diverse service requests due to the illusion of infinite resources to the users. Authors in [6] characterize the interactive behavior of Web applications in terms of number of instructions executed per second, and CPU, memory and disk utilization. Peng et al. [7] develop a classification scheme based on computational needs of the workloads for compute-intensive, I/O-intensive, and network-intensive applications. Zhang et al. [8] consider both the heterogeneity of machine hardware and workload diversity for dynamic capacity provisioning. Cloud infrastructure is heterogeneous due to the presence of machines from multiple generations, heterogeneous processor architectures and speed, and different memory and disk capacities. Singh and Chana [9] characterize workloads based on QoS requirements for different workload types such as scientific computing, online transaction processing, performance testing, and storage and backup services. Authors in [10] classify cloud workloads based on their functional characteristics, into six categories—scientific processing, Big Data application, OLTP, caching, streaming, and Web serving. The aim is to customize the resource requirements with the actual utilization of a specific workload for effective cloud monitoring.

From our study of the above work and the conclusions drawn by Lozi et al. [11], we infer that compute capacity of data center servers can be harnessed more effectively by considering per-core CPU usage rather than the aggregated utilization.

3 Related Concepts

3.1 Capacity Planning

Capacity planning is the process of determining computing infrastructure (hardware, software, and connection interface) and the floor area to house these components to cater to services for a future time period [12]. Data center capacity planning involves server capacity planning as well as network capacity planning. This work focusses on *server capacity planning* in which an IT department determines the amount of server hardware resources required to provide the desired levels of service for a given workload mix for the least cost [12]. The central objective is to provision resources in compliance with SLA.

The notion of capacity planning as perceived by cloud actors depends on their corresponding role. National Institute of Standards and Technology (NIST) [13] defines five cloud actors: consumer, provider, auditor, broker, and carrier. Of interest for server capacity planning are the roles of cloud consumer and cloud provider. The main aim of a cloud provider is to provision resources in a way that maximizes returns on investment, whereas a cloud user is concerned with quality of service that would justify rental cost. These benefits can be fully realized through optimal utilization of compute capacity of server.

3.2 *Workload Analysis and Characterization*

Workload analysis involves a detailed investigation of workload features of interest, to identify their behavioral characteristics, intra- and inter-dependencies, and impact on system performance. *Workload characterization* is the process of precisely describing the system's global workload in terms of its main components. Workload analysis measures cloud services along different dimensions including infrastructure capacity planning, energy efficiency, performance, reliability, and security. Two fundamental issues with workload analysis specific to infrastructure capacity planning [3] are:

1. Workload arrival pattern
2. Workload intensity

Workload arrival pattern measures the rate at which jobs arrive for service. *Workload intensity* refers to the amount of work done over a unit of time. Job arrival patterns are highly unpredictable and exhibit different behavioral patterns—diurnal, seasonal, and flash crowd [14]. Additionally, workloads may have either steady resource demands or may manifest temporal patterns such as periodic, bursting, growing, and on-and-off. An effective server capacity planning scheme should ensure resource availability on demand and provisioning for required time duration.

Job Arrival Patterns as Non-homogeneous Poisson Process. Non-homogeneous Poisson process is one way to model real-world job arrival patterns. Formally, given the occurrence of events at a constant rate λ, over a period T, the job inter-arrival time can be modeled as a Poisson process with exponential distribution which is governed by Eq. (1) below:

$$P(N(t) = k) = \frac{(\lambda t)^k}{k!}e^{-\lambda t}, \quad k = 0, 1, 2, \ldots \tag{1}$$

where the random variable $N(t)$ denotes the number of events that occur in t time units, P is the probability of occurrence of k events, and λ, known as the rate parameter, measures the number of events that occur per unit of time. Variability in job arrival can be captured using non-homogeneous Poisson process in which the rate parameter λ changes with time.

3.3 Multicore Scheduling

Multicore processors are the consequence of unsustainable power consumption and heat dissipation resulting from increased CPU clock frequency for performance gains [15]. The layout of a typical quad-core processor chip is shown in Fig. 1. Each core has a separate Level 1 (L1) cache for data and instructions and a Level 2 (L2) cache that holds both instruction and data, which are private. L2 cache is either united or distributed. L2 cache is usually physically distributed with a united Level 3 (L3) cache. Some multicores have an off-chip Level 4 (L4) cache. Two central issues when scheduling on multicore servers are contentions due to access to shared resources and efficient utilization of individual cores. The vast processing capability of modern multicore processors can be tapped, by keeping all the cores busy to the maximum. This has been achieved by adapting the operating system scheduler for single-core systems to that of scheduling large number of cores.

The Linux Scheduler. Completely Fair Scheduler (CFS), introduced in October 2007, is the default process scheduler in Linux since kernel version 2.6.23 and supersedes the $O(n)$ scheduler and $O(1)$ scheduler. $O(n)$ scheduler was used in kernel versions from 2.4 to 2.6 and was replaced by $O(1)$ scheduler in 2001, due to non-scalability issues. The $O(1)$ scheduler maintains a constant scheduling time irrespective of the number of jobs in the system and uses average sleep time of a process to distinguish between interactive and non-interactive jobs. However, this

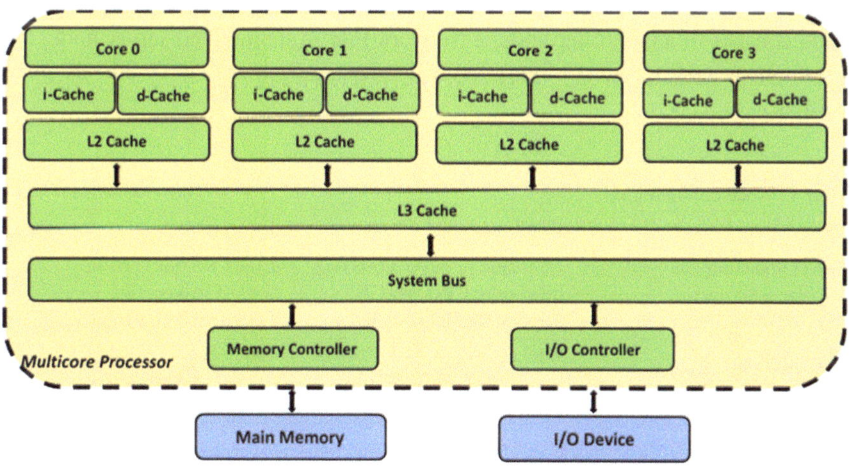

Fig. 1 A multicore chip

heuristic makes the scheduler complex with incidental cases of non-interactive jobs being identified as interactive jobs. CFS uses the concept of execution time and alleviates such miscalculations by organizing the jobs as red–black tree. The job with lowest execution time is the leftmost node of the tree and is picked up by CFS when invoked to schedule a process. CFS is a work-conserving scheduler which implies that the cores are prevented from being idle if there are processes ready for execution. However, experimental studies conducted by Lozi et al. [11] reveal that the Linux scheduler leaves cores idle while threads are waiting in run queues. This is revealed by performance degradation.

4 Results and Discussion

To make a comparative analysis of individual core usage with that of aggregate CPU utilization, we monitor per-core CPU utilization for workload types given below:

- A *CPU-intensive workload* is one that involves massive computation such as in financial modeling and scientific applications.
- *Memory-intensive workloads* involve high memory activity such as pushing of large volumes of data into and out of memory and frequent and long durations of memory read and write operations. Such workloads need not necessarily bloat memory usage and involve minimal CPU processing.

4.1 Experimental Cloud Setup

Usage of CPU cores is collected on a cloud setup in the laboratory. The experimental cloud consists of a controller node, a Network File System (NFS) server node, and two compute nodes. The configuration and software environment details are given in Table 1.

Table 1 Node configuration

Node	Configuration	Environment
Server and compute	Intel core i7 4790 3.60 GHz quad-core CPU, 16-GB 1600 MHz DDR3 RAM	Linux Kernel Version 4.4.0-116-generic, QEMU-KVM Version 2.5.0
Controller	Intel core i7 4770 3.40 GHz quad-core CPU, 16-GB 1600 MHz DDR3 RAM	Linux Kernel Version 4.4.0-116-generic

4.2 Design of Experiments

Applications used to generate different workload types are listed in Table 2, and different test cases are given in Table 3. Experiments are conducted on VMs running on a single host that uses full virtualization with Kernel-based Virtual Machine (KVM). The core virtualization infrastructure is implemented as a loadable kernel module (kvm.io). The modules, kvm-intel.ko and kvm-amd.ko, correspond to processors specific to Intel and AMD, respectively. Virtualization with KVM leverages basic mechanisms of CPU scheduling, memory management, and I/O access, provided by the Linux kernel.

To emulate real-world service request scenario, we generate variable job arrival times using a non-homogeneous Poisson process. Workloads are executed through a Bash script according to the generated job arrival times. Resource utilization log is collected for 1 hour at 3 seconds interval using System Activity Report (sar) system monitoring tool for both the guest and the host. Usage statistics extracted from the collected log are analyzed using R programming environment.

Baseline Resource Usage. To obtain CPU utilization for running basic system services and virtualization overheads, we run two VMs with no other workload on the guest or the host. A plot of the aggregated CPU usage is shown in Fig. 2, and usage statistics are given in Table 4. Spikes in the graph indicate some high CPU activity for a short duration. Mean value indicates that only 1–3% of the core usage accounts for running operating system as well as virtualization software.

CPU-Intensive Workload on Single VM. Per-CPU core usage of host and guest machines when running CPU-intensive workload on a single machine is depicted in Fig. 3. Summary statistics for host and guest are given in Tables 5 and 6, respectively.

Table 2 Workload generation

Workload type	Real-world application
CPU intensive	Prime factorization, Fibonacci series generation, factorial calculation
Memory intensive	Swap two integers, array copy

Table 3 Test cases for usage monitoring

Test case	Measurement objective
Baseline resource usage	Overhead due to basic system services and virtualization
CPU-intensive workload on single VM	Individual CPU core usage
Memory-intensive workload on single VM	Individual CPU core usage
CPU-intensive workload on one VM, memory-intensive workload on second VM	CPU core usage for workloads with complementary computational requirements
CPU-intensive workloads on three VMs	CPU core usage for workloads having similar computational requirements

Fig. 2 Baseline aggregate host usage

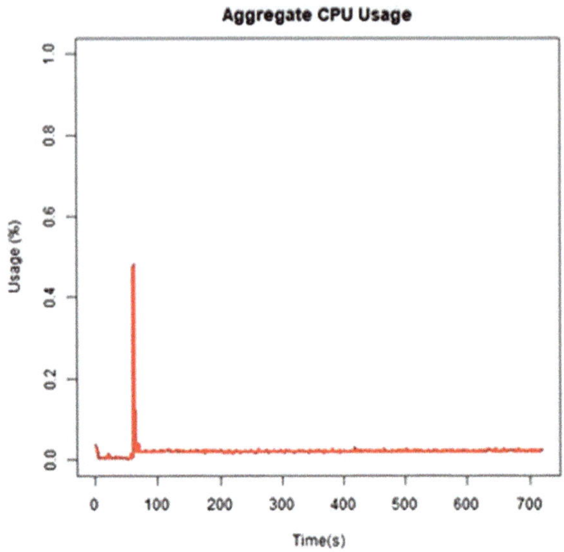

Table 4 Summary statistics of baseline host CPU core usage

CPU usage	Min.	Median	Mean	Max.	Standard deviation
Aggregate	0.15	2.05	2.15	48.34	2.53
Core 0	0.00	2.40	2.35	39.28	2.55
Core 1	0.00	2.20	1.99	64.87	3.13
Core 2	0.00	2.40	1.80	80.49	3.91
Core 3	0.00	1.00	2.24	44.42	2.51

Table 5 Summary statistics for host CPU core usage with CPU-intensive workload

CPU usage	Min.	Median	Mean	Max.	Standard deviation
Aggregate	1.96	75.48	70.60	80.53	13.01
Core 0	0.00	04.00	41.57	100.00	44.97
Core 1	0.34	100.00	80.73	100.00	33.48
Core 2	0.00	100.00	90.54	100.00	20.21
Core 3	0.34	100.00	68.12	100.00	41.44

Mean utilization of the cores lies between 41 and 90%, whereas aggregate usage is 70% with core 2 being highest utilized and core 0 being the least utilized than what is given by the aggregate usage. High values for standard deviation show the widely varying usage levels of individual cores. Core usage by the VM as given in Table 8 reveals near equal usage of all the vCPUs for executing the workload.

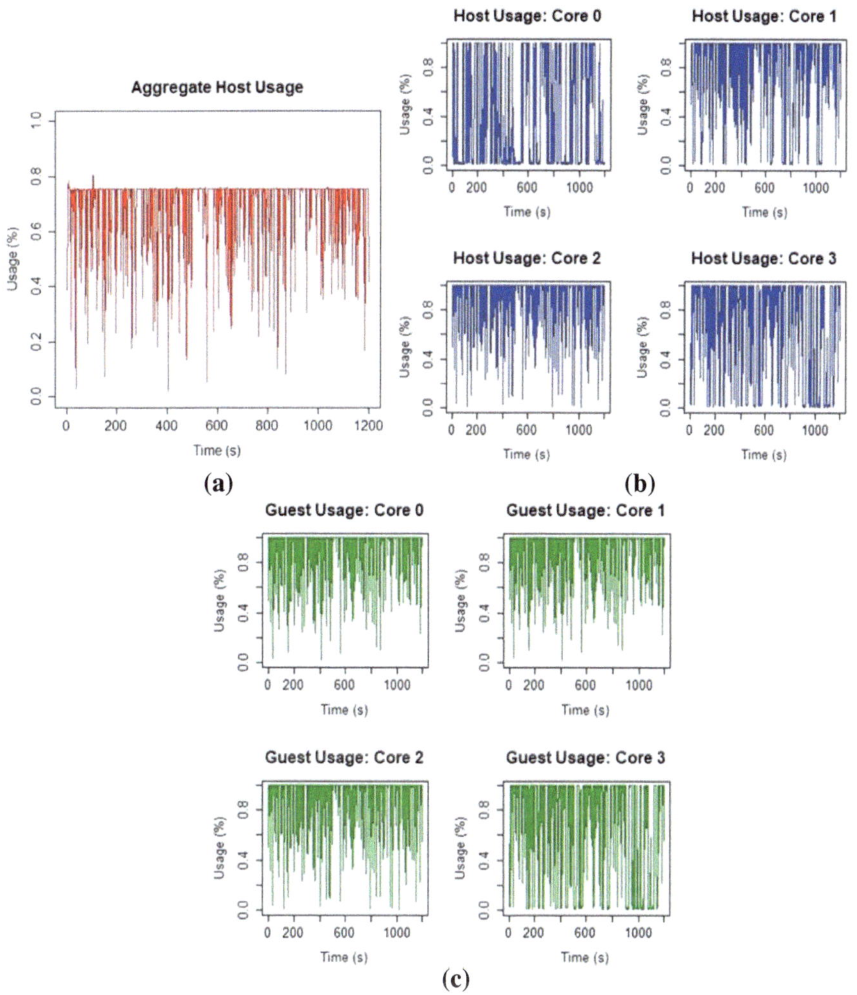

Fig. 3 CPU utilization for CPU-intensive workload. **a** Host aggregate, **b** per-core host, and **c** per-core guest

Memory-Intensive Workloads. Figure 4 shows a plot of per-core CPU usage for host and guest, running memory-intensive workloads on a single VM. The plots reveal that CPU activity is low with memory-intensive workloads as compared to CPU-intensive workloads, as the case should be. The cores either are mostly idle or have very low utilization as the median values in Table 7 reveal.

In this case, utilization of core 1 is 2% lesser than the aggregate and utilization of core 2 is 5% higher as given in Table 6.

Memory- and CPU-Intensive Workloads. The usage plots for this case are shown in Fig. 5, and Table 8 contains the usage statistics. While mean aggregated CPU

Table 6 Summary statistics for guest CPU core usage with CPU-intensive workload

CPU usage	Min.	Median	Mean	Max.	Standard deviation
Aggregate	2.11	100.00	92.11	100.00	17.28
Core 0	3.01	100.00	92.22	100.00	17.06
Core 1	1.67	100.00	92.09	100.00	17.32
Core 2	1.67	100.00	92.11	100.00	17.27
Core 3	1.33	100.00	92.08	100.00	17.34

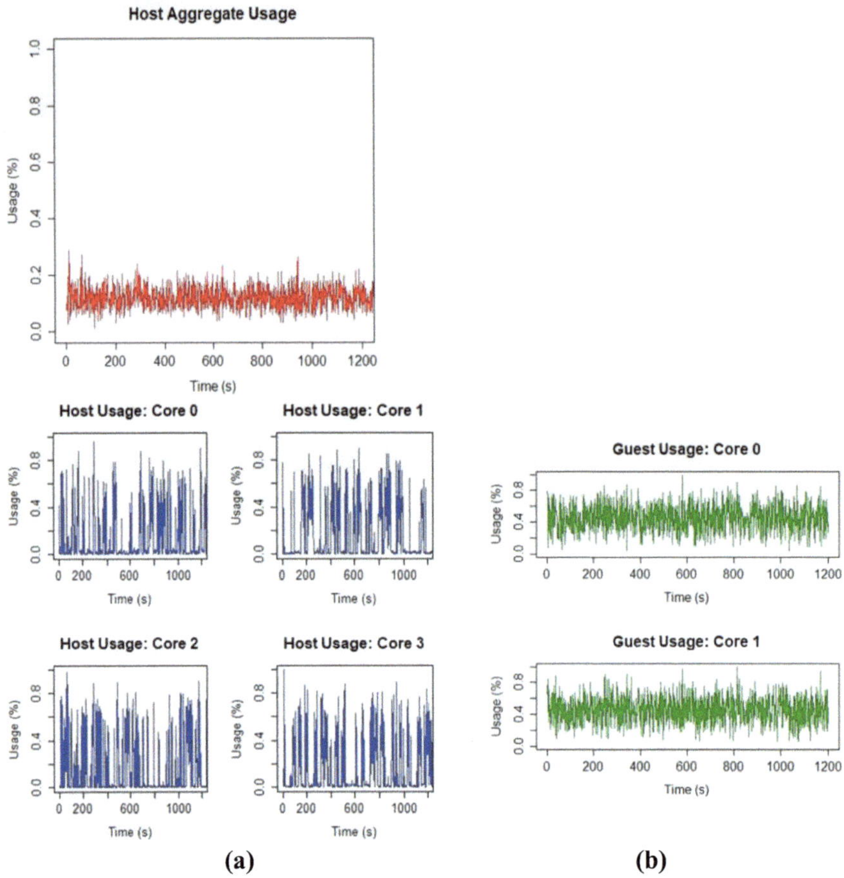

Fig. 4 Memory-intensive workloads **a** Aggregate CPU usage, **b** per-CPU core host usage, and **c** per-core guest usage

usage is close to 21%, core 0 and core 1 have higher mean usage. Such a usage detail can be detrimental when scheduling pCPUs to vCPUs in a virtualized environment. Also, since computational requirements of CPU- and memory-intensive workloads

Table 7 Summary statistics of host and guest CPU core usage for memory-intensive workload

CPU usage	Min.	Median	Mean	Max.	Standard deviation
Host usage					
Aggregate	1.40	11.59	11.84	28.85	03.76
Core 0	0.00	1.97	13.135	96.20	20.81
Core 1	0.00	0.60	08.389	98.00	21.59
Core 2	0.00	1.60	15.485	100.00	22.10
Core 3	0.00	0.200	10.446	77.110	08.22
VM usage					
Aggregate	5.51	45.05	45.76	100.00	14.74
Core 0	4.19	45.80	45.92	100.00	16.33
Core 1	1.60	45.20	45.59	100.00	16.23

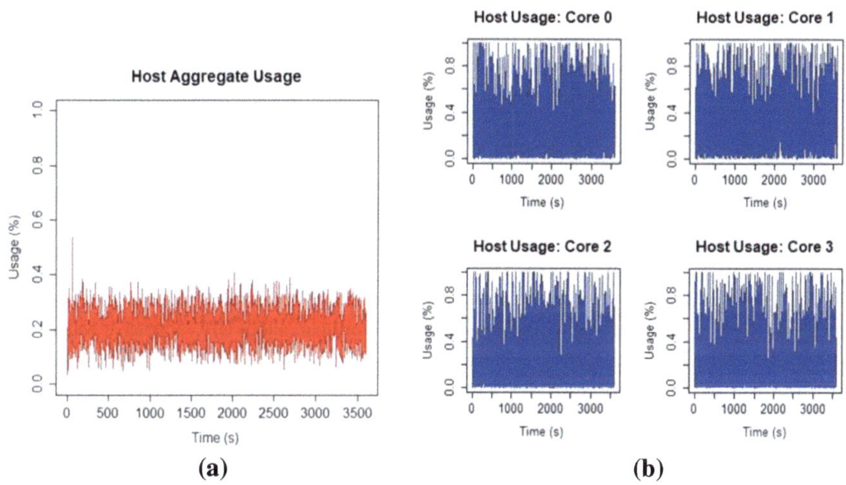

Fig. 5 Compute- and memory-intensive workloads: **a** host aggregate usage and **b** host per-core utilization

are complementary, the workloads do not contend for compute capacity of cores as can be seen from Fig. 6.

Compute-Intensive Workloads on three VMs. To study CPU core usage with workloads having similar computational requirements, we run two VMs, each with compute-intensive workloads, on a single host. Figure 7 contains the usage plots, and utilization values are given in Table 9.

From Table 9, we make similar observations as in the previous cases that in the case of multiple VMs running workloads in terms of similar resource requirements, core utilization is not uniform. Mean utilization of Core 1 is higher than that reported by the aggregate usage while that of Core 3 is below the aggregate value.

Table 8 Summary statistics of host CPU Core usage for memory- and CPU-intensive workloads

CPU usage	Min.	Median	Mean	Max.	Standard deviation
Aggregate	3.23	20.62	20.95	53.79	6.23
Core 0	0.00	17.14	24.63	100.00	22.34
Core 1	0.00	18.30	25.02	100.00	24.93
Core 2	0.00	25.27	19.95	100.00	24.43
Core 3	0.00	13.21	19.17	100.00	21.94

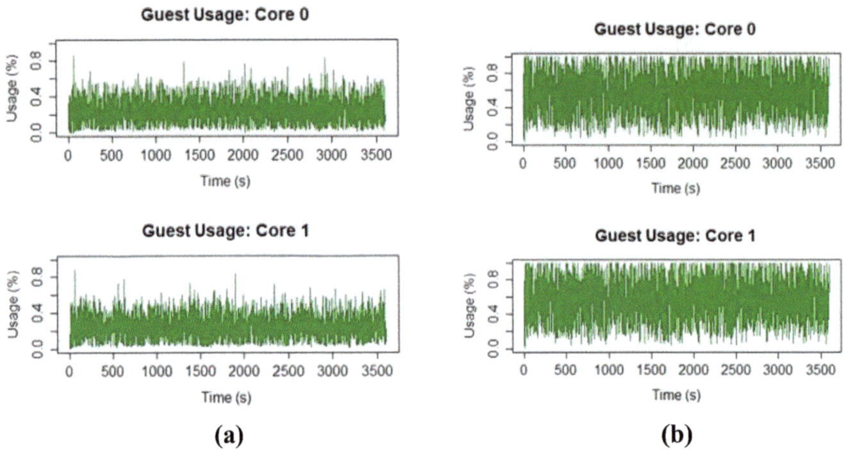

Fig. 6 Guest per-core utilization. **a** Memory-intensive workload and **b** CPU-intensive workload

Table 9 Summary statistics of host with CPU-intensive workloads on three VMs

CPU usage	Min.	Median	Mean	Max.	Standard deviation
Aggregate	53.36	73.85	75.08	100.00	10.51
Core 0	07.12	94.67	74.70	100.00	25.18
Core 1	07.38	72.65	82.02	100.00	26.16
Core 2	08.45	96.67	75.52	100.00	23.61
Core 3	10.03	94.64	67.95	100.00	22.37

Table 10 summarizes the percentage difference in utilization of individual cores from that of the aggregated usage. With compute-intensive workloads on single VM (Case II), a core usage of 29% below and 20% above the aggregate usage is recorded. When running only memory-intensive workloads (Case III), individual core usage is 3% above and below the aggregate usage. Similarly, in case of CPU- and memory-intensive workloads (Case IV) and CPU-intensive workloads on multiple VMs (Case V), the core usage varies between 1% below and 4% above the aggregate usage for Case IV and 7% above and below the aggregate usage in Case V. Knowledge

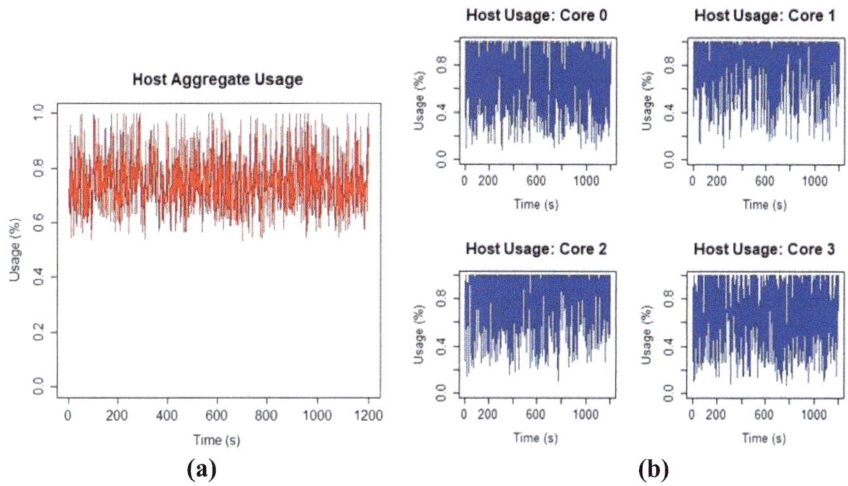

Fig. 7 CPU-intensive workloads on three VMs. **a** Host aggregate usage and **b** per-core host usage

Table 10 Usage comparison of CPU cores with aggregate usage

	Aggregate	Core 0	Core 1	Core 2	Core 3
Case II	70.60	41.57	80.73	90.54	68.12
Case III	11.84	13.14	8.40	15.48	10.45
Case IV	20.95	24.63	25.02	19.95	19.17
Case V	75.08	74.70	82.02	75.52	67.95

of individual core usage can be instrumental in effective scheduling and resource utilization.

5 Conclusion and Future Scope

In this work, we conducted an empirical study of utilization of CPU cores of a multicore server to analyze core-level CPU usage to verify our observations that aggregated CPU usage is not reflective of utilization at the core level. We collected CPU usage for five test cases—baseline usage and CPU core usage in case of CPU-intensive workload on single VM, memory-intensive workload on single VM, CPU-intensive workload on one VM and memory-intensive workload on the other, and CPU-intensive workloads on two VMs. We observe that basic system services and virtualization software incur some overhead in terms of CPU usage, as the case should be. The remaining four cases clearly bring out differences between aggregated CPU usage and individual cores which ranges from 1 to 29% below and between 4 and 20% above the aggregate usage. A precise knowledge of individual core usage will help in identifying cores that are underutilized and making informed scheduling decisions. Additionally, it will also be a vital input for reducing the load of overloaded cores and devising policies to prevent aggravating of the overload situation.

As future research, we would study core usage for I/O- and network-intensive workloads. A model based on the usage measurements of the four workload types—CPU, memory, I/O, and network intensive—would then be developed for workload characterization and prediction at the level of individual CPU cores. We would delve deeper into the varied usage levels of the CPU cores to study their criticality and impact on overall performance on different workload types. Scheduling policies that consider usage of CPU cores and account for inferences derived from the investigations would be developed, the impact of such a fine-grained scheduling on overall system performance would be analyzed, and approaches to mitigate performance loss would be investigated.

References

1. Varia, J., Mathew, S.: Overview of Amazon Web Services. Amaz. Web Serv. (2014)
2. https://creative-brackets.com/business/interesting-facts-statistics-largest-data-centers-world/. Last Accessed 2018/06/10
3. Feitelson, D.G.: Workload modeling for computer systems performance evaluation. Cambridge University Press, Cambridge (2015)
4. Mahambre, S., Kulkarni, P., Bellur, U., Chafle, G., Deshpande, D.: Workload characterization for capacity planning and performance management in IaaS cloud. In: 2012 IEEE International Conference on Cloud Computing in Emerging Markets (CCEM), pp. 1–7 (2012)
5. Moreno, I.S., Garraghan, P., Townend, P., Xu, J.: Analysis, modeling and simulation of workload patterns in a large-scale utility cloud. IEEE Trans. Cloud Comput. **2**, 208–221 (2014)
6. Magalhães, D., Calheiros, R.N., Buyya, R., Gomes, D.G.: Workload modeling for resource usage analysis and simulation in cloud computing. Comput. Electr. Eng. **47**, 69–81 (2015)
7. Peng, J., Chen, J., Zhi, X., Qiu, M., Xie, X.: Research on application classification method in cloud computing environment. J. Supercomput. **73**, 3488–3507 (2017)
8. Zhang, Q., Zhani, M.F., Boutaba, R., Hellerstein, J.L.: Dynamic heterogeneity-aware resource provisioning in the cloud. IEEE Trans. Cloud Comput. **2**, 14–28 (2014)
9. Singh, S., Chana, I.: QRSF: QoS-aware resource scheduling framework in cloud computing. J. Supercomput. **71**, 241–292 (2015)
10. Orzechowski, P., Proficz, J., Krawczyk, H., Szymański, J.: Categorization of cloud workload types with clustering. In: Proceedings of the International Conference on Signal, Networks, Computing, and Systems, pp. 303–313 (2017)
11. Lozi, J.P., Lepers, B., Funston, J., Gaud, F., Quéma, V., Fedorova, A.: The Linux scheduler: a decade of wasted cores. In: Proceedings of the Eleventh European Conference on Computer Systems, p. 1 (2016)
12. Capacity Planning—Discipline for Data center decisions, https://www.teamquest.com/files/6814/2049/9759/tqeb01.pdf. Last Accessed 2018/06/10
13. Sokol, A.W., Hogan, M.D.: NIST Cloud Computing Standards Roadmap (2013)
14. Calzarossa, M.C., Della Vedova, M.L., Massari, L., Petcu, D., Tabash, M.I.M., Tessera, D.: Workloads in the clouds. In: Principles of Performance and Reliability Modeling and Evaluation, pp. 525–550. Springer, Berlin (2016)
15. Vajda, A.: Programming many-core chips. Springer Science & Business Media, Berlin (2011)

Trust-Based Scheme for Location Finding in VANETs Using Trustworthiness of Node

Sanjoy Das, Indrani Das, Rishi Pal Singh, Prashant Johri and Ashwini Kumar

Abstract The purpose of vehicular ad hoc networks is to improve the traffic conditions and mitigate the problems occurred during ply. Any wrong decision taken by the vehicle may lead to hazardous problems and may create various serious issues. Thus, they need to be handled very carefully. Here, we have used trust-based methodology to find location with the help of trustworthiness of the node. In our proposed trust-based scheme, firstly calculation of trustworthiness of each vehicle exists in the network. After using trustworthiness, the location information of a vehicle is calculated. The purpose of calculating trustworthy is that higher the value of trustworthy of a node, higher the chances of providing the most accurate response. We have implemented the scheme in Java. The result shows that proposed scheme solves problem in the previously existing scheme.

Keywords Trust · Trustworthiness · Location · RSU

S. Das (✉)
Indira Gandhi National Tribal University, Regional Campus Manipur, Manipur, India
e-mail: sdas.jnu@gmail.com

I. Das
Assam University, Silchar, India
e-mail: indranidas2000@gmail.com

R. P. Singh
Guru Jambheshwar University of Science and Technology, Hisar, India
e-mail: pal_rishi@yahoo.com

P. Johri · A. Kumar
Galgotias University, Greater Noida, India
e-mail: johri.prashant@gmail.com

A. Kumar
e-mail: ashwinipaul@gmail.com

L. C. Jain et al. (eds.), *Data and Communication Networks*, Advances in Intelligent Systems and Computing 847, https://doi.org/10.1007/978-981-13-2254-9_5

1 Introduction

The vehicular ad hoc networks (VANETs) is used to provide information that is critical in the decision-making process of vehicles at the time of hazardous situations like congestion, earthquake, floods, and jams to the users. This network helps minimize road accidents, fuel wastage, time, and provide much safer, economical, and more fruitful environment for driving [1]. In VANET, there are various applications that will provide security to information that is being exchanged between the vehicles. This helps in overcoming problems related to accidents, traffic condition, etc. The trust management scheme is a very important and basic requirement to maintain a safe and trustworthy exchange of communication in VANET. This helps maintain a reliable and secured communication among vehicles during their journey [2]. The reputation and trust issues are very important aspect in this environment for the decision-making process. The reputation is basically the opinion of vehicles on each others [3]. We have introduced the concept of evaluation the trustworthiness of the vehicle on the basis of their responses. Trust is basically the confidence on the messages that are passed to establish communication in VANETs [4].

The paper is organized as follows: In Sect. 2, we have discussed the work-related trustworthiness proposed by researchers and academicians. In Sect. 3, various algorithms for computing the trust percentage and trusted location using trustworthiness are discussed. In Sect. 4, computation of trustworthiness of node is presented. Section 5 discusses the simulation environments and result analysis. Finally, we have concluded the paper with future direction for further research in Sect. 6.

2 Related Work

The geographical constraint restricts the movement of the vehicles in VANETs. Furthermore, the vehicles, due to dependent velocities, are spatially dependent on each other. As a result, the quality of connectivity largely depends upon the quality of communication link [5–8]. When a new node (i.e., vehicle) joins the network, its validation is essential to maintain security of the network. A trust-based evaluation is proposed in [6, 9, 10] based on security authentication mechanisms. The authors have classified them as direct authentication and indirect authentication. The direct method is based on the historical security event records. Whenever nodes and authority unit communicate each other, the events relating to security are recorded in their database. The authentication units (AUs) that belong to a particular organization can easily share the same database. The recorded data is very much useful in determining credibility of a newly joined node. When a group of nodes are communicating, it solely depends on their decision to accept a new node or not. The acceptance of a new node depends on trust value and recommendation given by other nodes. If selfish nodes exist in the network, it is possible that they may deny the new vehicle node. This condition of the network is undesirable, and this degrades the overall performance of the network. So, the presence of malicious nodes needs to be identified before computing the indirect trust value in the indirect method. In this scheme, correlation coefficient of the recommendation trust value is used for distinguishing malicious nodes. This helps in discarding the recommendation of trust values received from malicious nodes. Trust evaluation based on the historical security events and recommendation is discussed in [11].

There is a huge scope of work in trust management for point-to-point networks, wireless sensor networks (WSNs), VANETs, mobile ad hoc networks (MANETs), etc. The trust establishments for VANETs discussed in [12–14] are very less. In [12], authors discussed a trust computation method based on Perron–Frobenius theorem for VANETs. The trust management based on AHP [13] provides a secure alternative to replace the current security mechanism. A trust and reputational infrastructure-aided approach TRIP for VANETs is discussed in [14]. The trust value is computed with the help of reputation score given by other vehicles. In [15], authors discussed the intrusion detection system by using watchdog scheme for evaluating them. In [16], authors devise two techniques that improve throughput for ad hoc network. In [5] introduce an intrusion detection scheme, which produces a more realistic system, this system is more reliable in handling malicious nodes. The currently available trust management schemes and works focus on evaluating and protecting the privacy of messages.

3 Methodology

We have proposed the following schemes for finding the trusted location of vehicle. Firstly, trust percentage of information is computed by using the responses received from vehicles. Based on this, we have calculated the trust percentage of the information on the basis of the number of request and number of positive responses. In this particular case, the trust percentage of the information is calculated. Each vehicle giving the positive response about the information will be rewarded as points for its true information providing, thus calculating the trustworthiness of each node present in the network. In the second case when the trust is below 50%, then instead of going to RSU or TA the vehicle will ask the trustworthiness of the each node in the network and take the response of the most trustworthy node.

3.1 Computation of Trust Percentage of Vehicles

In this case, vehicle wants to find that whether the location information is present or not. It could be about anything like nearby hospitals and hotels. So the source vehicle broadcast the request (REQ) message to its neighbor vehicles. The neighboring vehicles of the source vehicle reply to REQ with reply (REP) messages. The source node received REP messages from its neighbour nodes, depending upon true responses vehicle will be rewarded a (+1) point and its number of true response will be updated. The trust value is computed by source node based on REP messages. In our work, we have proposed four cases for calculating the trust percentage of whether the location is present or not.

Case 1 (Computation of trust percentage of location present using vehicle): The source node received REP messages from its neighbour nodes and compute ratio of trusted information about a location. If source vehicle receives REP messages from 10 nodes and 6 nodes replied that the location information is present and 4 nodes replied with no information is available i.e. false reply. Every true response received, and the vehicle will be rewarded one point thus increasing its trustworthiness. In this scenario, the ratio is [(6/10) * 100] = 60% of the trusted information about location. The source vehicle may trust broadcast location till it is above 50%; otherwise, a vehicle can compute location using case 2.

Case 2 (Computation of trusted location using trustworthiness of the most trustworthy node): In this case, we have assumed that at every true response of the vehicle, we will count the total number of true responses of the vehicles and the total number of the responses of the vehicle. Thus, we will calculate the trustworthiness of each node present in the network by dividing the total number of true responses by the total number of responses multiplied by 100. Now, thus, when the trust percentage of the information is below 50%, source node will directly ask the most trustworthy node and the response from the most trustworthy node will be more reliable.

Case 3 (Computation of trusted location using RSU): In sparsely populated network scenario, sometimes it is possible that source vehicle is alone plying. In this scenario, the particular vehicle collects information from Road-Side Units (RSUs). The location information received from RSU is fully trusted. In case there is no neighboring node and RSU available to the source node, then the location is computed with the help of Case 4.

Case 4 (Computation of Trusted Location Using Trust Authority (TA)): The source vehicle asks for the location information from trusted authority (TA) (e.g., police vans, ambulance, and post office vehicles).

4 Computation of Trustworthiness of Node

For the calculation of the trustworthiness of the node, a buffer is maintained at every node. This buffer will maintain the information regarding the trustworthiness of the node. On the basis of the buffer information, the source vehicle will choose the most trustworthy node among the various vehicles in the range of the source vehicle. Source vehicle, when found the trust percentage of the information and ask the most trustworthy node and the reply from the most trustworthy node, will be final. The proposed working of the scheme involves these steps.

Table 1 Trustworthiness of nodes

Total number of responses by the vehicle	Total number of true responses by the vehicles	Trustworthiness of the vehicles in percentage
50	50	100
50	45	90
50	40	80
50	35	70
50	30	60

Step 1

Source vehicle broadcast REQ message to its neighboring vehicles and vehicles responding with positive reply will be awarded one point. A buffer is used for storing these awarded points. Also, two different counters are used to keep record of responses received from vehicle. One of the counters stored the total number of true responses received from the vehicle, and the other keeps the record of total number of responses received.

The trustworthiness of a node computed according to Eq. 1

$$\text{Trustworthiness} = \left(\frac{\text{Total Number of True Response}}{\text{Total Number of Response}}\right) \times 100 \quad (1)$$

Step 2

At each response, there is always update among the counters of the nodes regarding their total true response and total responses, thus updating the trustworthiness of each node.

Step 3

If at a certain point of time trust percentage of information is below 50%, then the vehicle will ask the trustworthiness of the nodes within the communication range, thus sharing the buffer data stored at each node among all the nodes present in the communication range (Table 1).

Table 2 Simulation parameters

Parameter	Value
Number of nodes	10, 20, 30, 40, 50
Simulated area	1000 × 1000
Antenna	Omnidirectional
Transmission range of node	150 m
Simulation time (in seconds)	1000
IDE)	NetBeans 8.0.2 J2SE

Table 3 Number of node exists within the transmission range of different source node

Node density	Number of node within the range		
	Source node: 2	Source node: 3	Source node: 7
10	6	9	9
20	13	19	19
30	20	29	29
40	27	39	39
50	34	47	46

Table 4 Trust percentage of source nodes with variable node density

Number of nodes	Trust percentage source nodes		
	2	3	7
10	33.33	67	55
20	46.15	57	42
30	40.00	51	41
40	44.44	43	48
50	44.11	38	43

5 Simulation and Result Analysis

In this paper, we are implementing our scheme on NetBeans IDE 8.0.2 J2SE. VANET scenario composed of 10–50 vehicles distributed in an area of 1000 × 1000 units, and the maximum range of vehicles is 150 m. Vehicles are considered to be at a traffic point situation. Here, we have arbitrarily considered nodes 2, 3, and 7 as source nodes and with a different number of nodes present in the range. The network size gradually increases from 10, 20, 30, 40, and 50. The source nodes (i.e. node 2, 3, 7) are broadcast messages to know whether a location is present or not. The number of nodes within the transmission range of source node for a particular time shown in Table 3. The trust percentage of the information computed after getting the response from other nodes fall within the transmission range. In different node densities, the trust of source nodes 2, 3, and 7 is computed as shown in Table 4. In Table 5, we have computed the trustworthiness of the each node present in the simulation scenario with various source nodes as 2, 3, and 7 (Table 2).

5.1 Scenario 1 (Source Node Is 2)

Figure 1 shows the node deployment of various nodes considered for simulation. Figure 2 shows the trust percentage of source node 2 under different node densities. The trust percentage obtain is always below 50% in different node density. For source node 2 in Table 5 and Fig. 9, nine nodes present in the network whose trustworthiness

Table 5 Trustworthiness of each node

Node ID	Source node			Node ID	Source node		
	2	3	7		2	3	7
1	41	47	66	26	30	66	77
2	0	47	50	27	40	46	55
3	41	0	41	28	0	40	80
4	66	47	58	29	0	46	77
5	66	52	33	30	0	60	55
6	50	41	58	31	44	57	37
7	33	58	0	32	55	42	62
8	0	41	58	33	55	60	62
9	0	64	58	34	44	66	50
10	0	29	41	35	40	50	75
11	36	62	63	36	66	42	87
12	69	56	45	37	44	64	25
13	63	56	45	38	0	*71*	50
14	63	43	63	39	0	57	50
15	45	43	45	40	0	35	62
16	36	43	45	41	**75**	46	*100*
17	45	37	18	42	37	38	50
18	0	51	81	43	37	69	50
19	0	68	45	44	66	46	0
20	0	31	45	45	37	38	50
21	40	46	66	46	37	53	*100*
22	30	68	55	47	37	53	0
23	30	66	55	48	0	46	0
24	30	60	66	49	0	0	50
25	50	60	33	50	0	0	50

is above 50%. It can ask the node 41 directly with the trustworthiness of 75%. The answer by this node would be more reliable in comparison to the first technique which was just asking to the nodes for the information without any prior knowledge of the nodes participating for the trust calculation (Fig. 3).

5.2 *Scenario 2 (Source Node Is 3)*

Figure 4 shows the deployment of nodes for source node 3. Now, in Fig. 5, it observed that twice trust percentage goes below 50. To overcome such situation, Table 5 and Fig. 6, there are 23 nodes present in the network whose trustworthiness is above 50% and one being 71% as the highest trustworthy node. Thus in such a situation, we have a chance to test the information on a better scale and have a better results. It can ask the node 38 directly with the trustworthiness of 71%. The answer by this node would be more reliable in comparison to the first technique which was just asking to the nodes for the information without any prior knowledge of the nodes participating for the trust calculation. In this particular scenario, we can also notice that time to take decision will be much less if we have to take request and response from one vehicle.

5.3 *Scenario 3 (Source Node Is 7)*

Figure 7 shows the node deployment for source node 7. In Fig. 8, trust percentage is above 50% only when the number of node in the network is 10 and for other node density the values goes below 50. To overcome this problem, from Table 5 and Fig. 9, we can see that there are 24 nodes present in the network whose trustworthiness is above 50% and two being 100% as the highest trustworthy nodes. Thus in such a situation, we have a chance to test the information on a better scale and have better results. It can ask the node 41 or 46 directly with the trustworthiness of 100%. The answer provided by this node would be more reliable in comparison to the first technique which was just asking to the nodes for the information without any prior knowledge of the nodes participating for the trust calculation. Here we can notice that the source node 7 has an option of two nodes in the range that too with 100% of trustworthiness. Thus in this particular case would be like it is better to go for such vehicle which has its trustworthiness 100% instead of trust percentage below 50%.

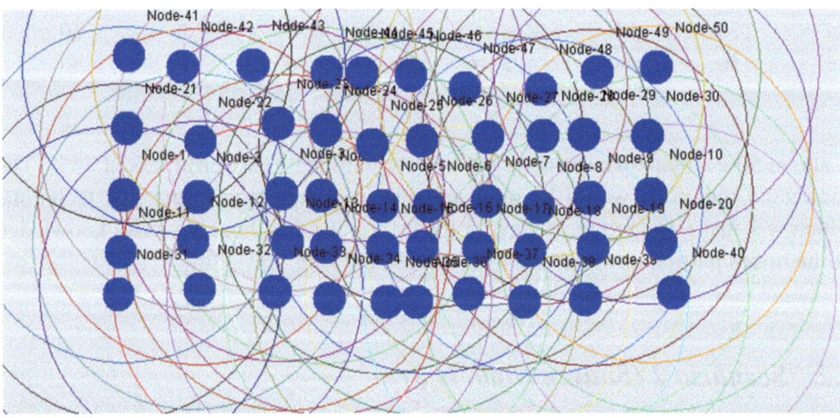

Fig. 1 Node placement

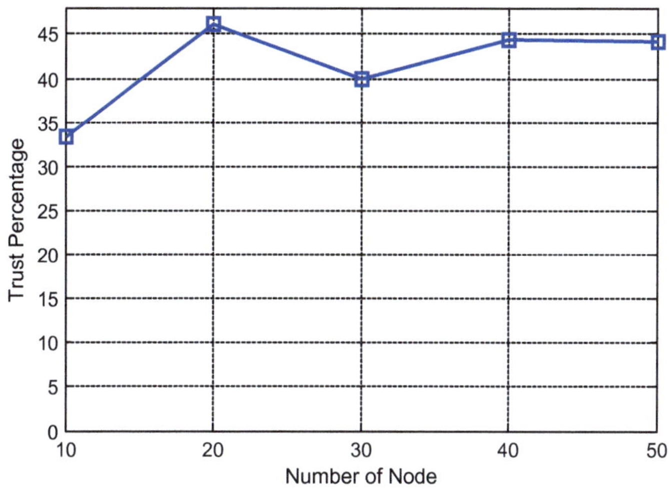

Fig. 2 Trust percentage of node 2 with varying node density

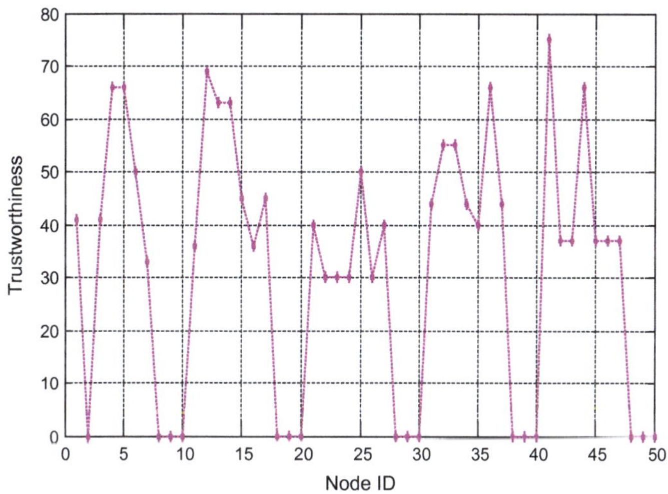

Fig. 3 Trustworthiness of each node w.r.t node 2

6 Conclusion and Future Direction

We can see the variation in trust percentage of information when we increase the number of nodes and thus increasing the number of responses. Now we can use the trustworthiness of the each node and can find the most trustworthy node. Request the most trustworthy node, and the response of the most trustworthy node will be more reliable when the trust percentage of the information is below 50%. Thus we

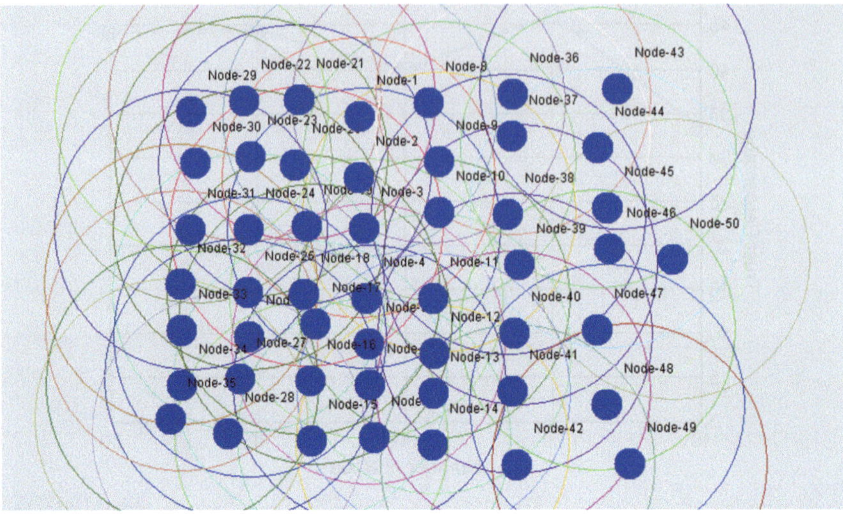

Fig. 4 Node placement during simulation (source node = 3)

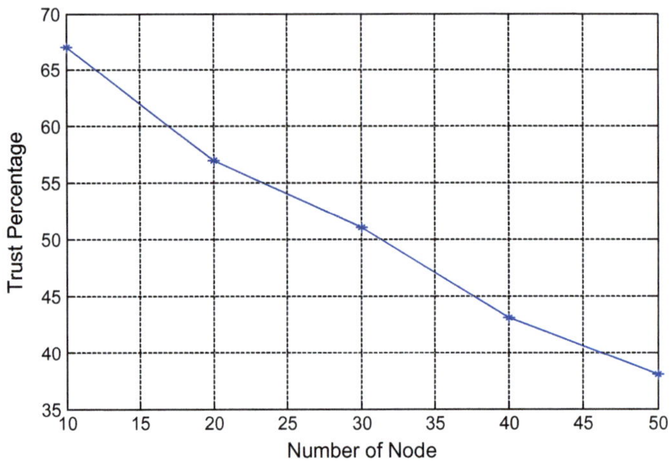

Fig. 5 Trust percentage of node 3 with varying node density

found that our proposed scheme is more reliable in finding the genuiness of requested information and gives a more satisfactory response to the situation. Thus, we find that the above technique is more worthy for the current situation proposed in this paper.

There is a lot to emphasize on in this area such as reducing the response time of the vehicles in serious conditions. We must focus on bringing those schemes which could handle the openness in the environment of the VANETs. There could be various ways of handling the situation if we can come up with it may result in much

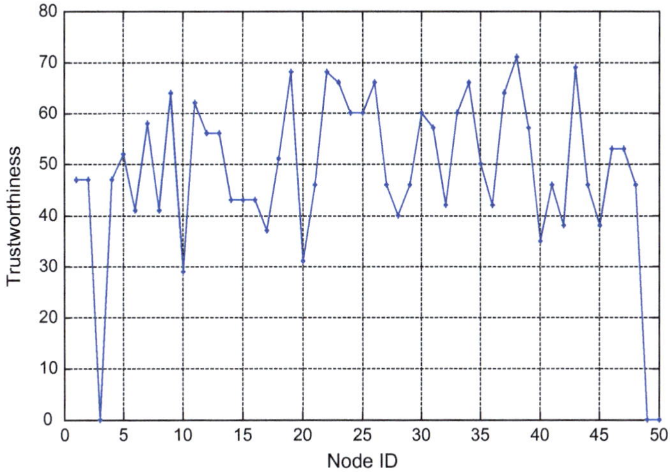

Fig. 6 Trustworthiness of each node (source node is 3)

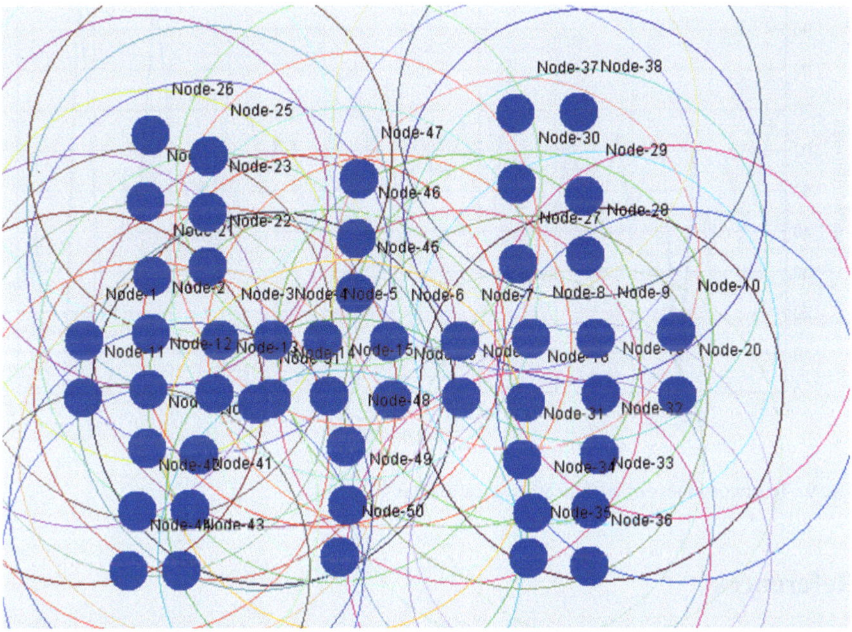

Fig. 7 Node placement during simulation (source node = 7)

more accuracy. There are many issues in which VANET scenario could be handled in better ways. The whole concept revolves around if we could find a way to check whether the response is trustworthy or not. So considering such scenario, we must try to progress our work for further improvement.

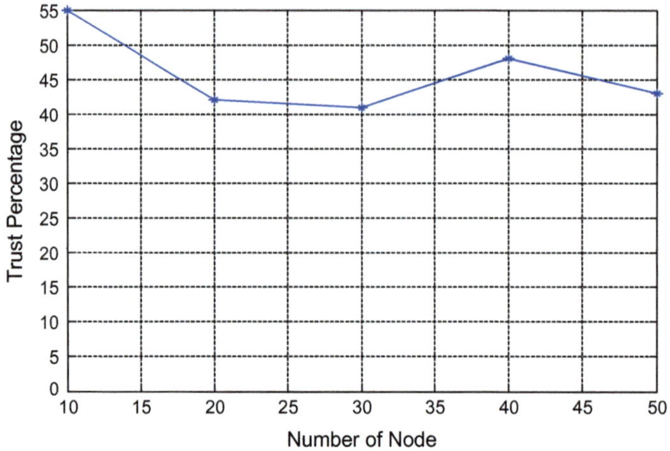

Fig. 8 Trust percentage of node 7 with varying node density

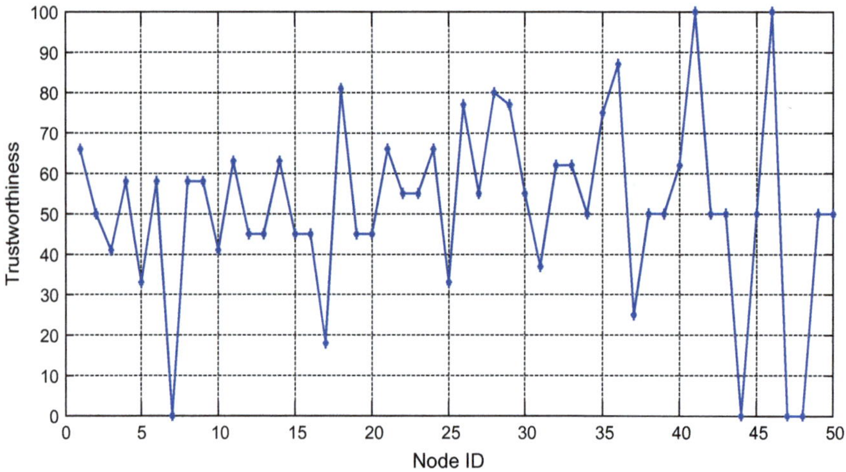

Fig. 9 Trustworthiness of each node (source node is 7)

References

1. Toor, Y., Muhlethaler, P., Laouti, A.: Vehicle ad hoc networks: applications and related technical issues. IEEE Commun. Survey Tutor. **10**(3), 74–88 (2008)
2. Sharma, K., Soni, S., Chaurasia, B.K.: Reputation and trust computation in VANETs. In: Proceedings of the IEMCON-14 Conference on Electronics Engineering and Computer Science, pp. 118–122 (2014)
3. Ding, Q., Jiang, M., Li, X., Zhou, XH.: Reputation-based trust model in vehicular ad hoc networks. In: International Conference on Wireless Communications and Signal Processing (WCSP), pp. 1–6 (2010)

4. Sumra, I.A., Hasbullah, H., Lail, J., Rehman, M.: Trust and trusted computing in VANET. Comput. Sci. J. **1**(1), 29–51 (2011)
5. Hasswa, A., Zulkernine, M., Hassanein, H.: Routeguard: an intrusion detection and response system for mobile ad hoc networks. In: Proceedings of the IEEE International Conference on Wireless and Mobile Computing, Networking and Communications, vol. 3, pp. 336–343(2005)
6. Tyagi, P., Dembla, D.: Investigating the security threats in vehicular ad hoc networks (VANETs): towards security engineering for safer on-road transportation. In: International Conference on Advances in Computing, Communications and Informatics, pp. 2084–2090 (2014)
7. Zhang, J.: A survey on trust management for VANET. In: International Conference on Advanced Information Networking and Applications, IEEE Computer Society, pp. 105–112 (2011)
8. Yang, X., Liu, L., Vaidya, N.: A vehicle-to-vehicle communication protocol for cooperative collision warning. In: 1st Annual International Conference on Mobile and Ubiquitous Systems: Networking & Services, pp. 114–123 (2004)
9. Shuai, Z., Jun, T., Yijian, Y.: Anonymous authentication—oriented vehicular privacy protection technology research in VANET. In: IEEE International Conference on Electrical and Control Engineering, pp. 4365–4368 (2011)
10. Zhang, C.: An efficient message authentication scheme for vehicular communication. IEEE Trans. Veh. Technol. **57**(6), 3357–3368 (2008)
11. Tong, Z.: P2DAP—Sybil attacks detection in vehicular ad hoc networks. IEEE J. Sel. Areas Commun. **29**(3), 582–594 (2011)
12. Chaurasia, B.K., Verma, S., Tomar, G.S.: Trust computation in VANETs. In: Proceedings of the International Conference on Communication Systems and Network Technologies (CSNT), pp. 468–471 (2013)
13. Saraswat, D., Chaurasia, B.K.: AHP based trust model in VANETs. In: Proceedings of the IEEE 5th International Conference on Computational Intelligence and Communication Networks (CICN), pp. 391–393 (2013)
14. Marmol, F.G., Perez, G.M.: TRIP a trust and reputation infrastructure based proposal for vehicular ad hoc networks. J. Netw. Comput. Appl. **35**(3), 934–941 (2012)
15. Obimbo, C., Arboleda, L.M., Chen, Y.: A watchdog enhancement to IDS in MANET. In: Proceedings of the IASTED Conference on Wireless Networks, pp. 510–515 (2006)
16. Marti, S., Giuli, T.J., Lai, K., Baker, M.: Mitigating routing misbehavior in mobile ad hoc networks. In: Proceeding of the 6th ACM Annual International Conference on Mobile Computing and Networking, pp. 255–265(2000)

Implementation and Analysis of Energy Efficiency of M-ary Modulation Schemes for Wireless Sensor Network

Samir Ahmad Sheikh and Sindhu Hak Gupta

Abstract In a wireless sensor network because of the path loss, there is a considerable difference in the signal strength from the transmitted node to receiving node. Path loss is a crucial factor for a WSN, and it may be evaluated using stochastic, deterministic, or empirical methods. For a WSN, it is challenging task to optimize the transmission power, reliability, and data rate in the presence of path loss. Choice of appropriate modulation scheme is the most important aspect of the physical layer. As optimum modulation scheme is capable of minimizing the error and making WSN more reliable. In this paper, a new approach is considered to relate path loss of the WSN to M-ary modulation schemes. Critical comparative analysis for M-ary FSK and M-ary PSK is done for the case scenario. Performance is analyzed for free space earth model and plane earth model. Results indicate that PSK is more reliable than FSK, whereas the data transmission rate for FSK is greater than PSK. Power required to transmit the data in FSK is less than less.

Keywords Path loss · Data rate · Power transmitted · PSK · FSK · Reliability

1 Introduction

A wireless sensor network (WSN) is the combination of autonomous devices. These autonomous devices are spatially scattered and are used for monitoring and recording physical parameters like pressure, temperature, humidity, EEG, and ECG for environment or living body, respectively [1]. Based on this fact, WSN has got various applications in upcoming fields like wireless body area network (WBAN), emergency relief, and precision agriculture (PA). Wireless sensor networks are used in

S. A. Sheikh (✉) · S. H. Gupta
Department of Electronic and Communication Engineering, ASET,
Amity University, 125, Noida, UP, India
e-mail: samirkashmir@gmail.com

S. H. Gupta
e-mail: shak@amity.edu

L. C. Jain et al. (eds.), *Data and Communication Networks*, Advances in Intelligent Systems and Computing 847, https://doi.org/10.1007/978-981-13-2254-9_6

different fields to gather the information, and thus collecting the information, the sensor nodes communicate with each other and pass the information further to cluster head. Cluster heads transmit it to the base station and from that it is processed and examined. But while communicating with each other they face attenuation, fading, noise and path loss [2, 3]. The prime focus is energy efficiency between the sensor nodes, as low battery power and limited computation capability are inherent features of sensors. This may cause data loss evaluation of transmitting signal. This leads for advanced design methodologies to address the energy content in any wireless sensor network. For this research paper, main focus would be path loss, power transmitted, data rate, and reliability between two sensor networks.

Path loss models are used to abstract the actual characteristic of the sensor nodes in the wireless sensor network. The appropriate modeling of the propagation and path loss is of predominant in the WSN system analysis and design [4]. Most important parameters of any WSN are energy dissipation, reliability, route optimization, decision support system, data rate, power transmitted, reliability, and connectivity. The sensor nodes which are used in the precision agriculture communicate with each other and have low power efficiency. Power to the nodes is usually provided by the batteries. But batteries have a finite energy and need replacement when depleted, which increase the maintenance cost. Path loss is one of the important parameters used to evaluate the performance of WSN. This factor depicts the difference between which exists between the transmitted power and the received power. There are different types of path loss models like free space path loss model, plane earth path loss model, and COST 231 [2, 3]. These all models are all in terms of the frequency. But new models correlate path loss to different modulation schemes which are developed for a wireless sensor network.

In every WSN modulation scheme, it plays an important role in transmitting the information in an efficient way. There are different types of M-ary modulation schemes that exist in the literature which govern the transmission of data because of the three fundamental parameters as frequency, time, and phase. These digital modulation schemes include basic schemes such as M-ary frequency shift keying, M-ary phase shifting keying, and amplitude shifting keying (ASK). But our focus is on M-ary phase shift keying and M-ary frequency keying. The choice of efficient modulation schemes is important for reliable communication in wireless sensor network [5]. Using M-ary modulation schemes in transmission of data can reduce the transmission time by sending multiple bits per symbol, and it results in the tangled circuitry and increased radio power consumption. The trade-off parameters are formulated in and it is concluded that under start-up power most influential conditions, the binary modulation scheme is more efficient. So in any wireless sensor network, crucial point is the choice of modulation scheme by which the life time, efficiency, and reliability of the sensor nodes can be increased. Several factors like data rate and symbol rate, reliability, power transmitted, data rate, and path loss have been analyzed in order to determine the efficiency of different M-ary modulation schemes [5]. This will lead to improve the performance, life time, and reliability of the sensors in a wireless sensor network. The main contribution of this paper is:

- To develop a mathematical model that correlates path loss to M-ary modulation schemes;
- With the help of simulation results analyze the performance of M-ary PSK and M-ary FSK for considered schemes.

2 System Model

The system model is defined for a wireless sensor network model which is linear and the nodes are linearly spaced and are equidistant to each other with a fixed distance of 10 m. The total numbers of nodes are fixed, and each node is homogenous. In this scenario, receivers are assumed to be synchronized for simulation. The communication between the sensor and the sink is wireless. To evaluate the path loss plane earth path loss (PE) and free space path loss model (FS) can be used.

Free space path loss model is derived from path loss equation and which is expressed in dBm [2, 3]. The application of the model can be conceivably applied to development as the minimum level for the computation of the path loss as this Eq. (1) represents free space plane path loss model.

$$\mathrm{PL(dBm)} = -27.56 + 20\log_{10}(f) + 20\log_{10}(d) \quad (1)$$

where d is the distance between the sender and the receiver node in meters, and f is the frequency in Mhz.

For frequency of 2.4 Ghz, Eq. (1) becomes:

$$\begin{aligned} \mathrm{PL(dBm)} &= -27.56 + 20\log_{10}(2400) + 20\log_{10}(d) \\ \mathrm{PL(dBm)} &= 40.40 + 20\log_{10}(d) \\ 20\log_{10}(d) &= \mathrm{PL} - 40.40 \\ d &= 10^{(\mathrm{PL}-40.40)/20}. \end{aligned} \quad (2)$$

Plane earth path loss model is derived from path loss equation and which is expressed in dBm [2, 3]. The estimated path loss is expressed as follows:

$$\mathrm{PL(dBm)} = 40\log_{10}(d) - 20\log_{10}(h_{\mathrm{t}}) - 20\log_{10}(h_{\mathrm{r}}) \quad (3)$$

where h_{t} and h_{r} are height of transmitter and receiver antenna in cms, and d is the distance between the transmitter and the receiver node in meters. For $h_{\mathrm{t}} = h_{\mathrm{r}} =$ 13 cm, Eq. (3) can be written as follows:

$$\begin{aligned}\text{PL} &= 40\log_{10}(d) - 20\log_{10}(0.13) - 20\log_{10}(0.13)\\ \text{PL} &= 40\log_{10}(d) + 35.44\\ 40\log_{10}(d) &= \text{PL} - 35.44\\ d &= 10^{(\text{PL}-35.44)/40}\end{aligned} \tag{4}$$

In a WSN, it has been stated in [6] that energy per bit E_b in M-ary modulation schemes may be represented as given in Eq. (5)

$$\begin{aligned}\frac{E_b}{N_o} &= \text{SNR}\cdot\frac{1}{R} = \frac{P_{\text{revd}}}{N_o}\cdot\frac{1}{R}\\ &= \frac{1}{N_o\cdot R}\frac{P_t\cdot G_r\cdot G_t\cdot\lambda^2}{(4\pi)^2\cdot d_o^y\cdot L}\left(\frac{d_o}{d}\right)^{\gamma}\end{aligned} \tag{5}$$

where $G_t \cdot G_r$ and γ are gain of the transmitter, gain of receiver antenna, and path loss exponent, respectively, and R is data rate, P_t is the power transmitted, and $\frac{E_a}{N_o}$ is the ratio between the energy per bit and the noise level which is fix for different M-ary modulation schemes.

A novel mathematical relation is proposed using Eqs. (2) and (4) in Eq. (5) which correlates path loss to different modulation schemes and is developed for a WSN. Path loss can be related to the power transmitted and the data rate by keeping one as the constant. For M-ary PSK and FSK modulation schemes using Free Space Path Loss model which correlates power transmitted, data rate with the Path Loss:

$$P_t = \frac{E_b}{N_o} \times \frac{N_0\cdot R\cdot(4\pi)^2\cdot L}{G_t\cdot G_r\cdot\lambda^2} \times 10^{\frac{\gamma(\text{PL}-40.04)}{20}} \tag{6}$$

For M-ary PSK and FSKmodulation schemes using Plane Earth Path Loss model which correlates power transmitted, data rate with the Path Loss:

$$P_t = \frac{E_b}{N_o} \times \frac{N_o\cdot R\cdot(4\pi)^2\cdot L}{G_t\cdot G_r\cdot\lambda^2} \times 10^{\frac{\gamma(\text{PL}-35.44)}{40}} \tag{7}$$

where γ is path loss exponent$=3.5$, $d_o = 1\text{m}, G_r = G_t = L = 1$, R (data rate) and N_o(noise level)$= -180$ dB.

Using Eq. (6) and (7) in Eq. (8), reliability [1] can be expressed in terms of path loss for both free space path loss model and earth space path loss model.

$$\text{Reliablity} = W\log_2\left(1 + \frac{P_t * |h|^2}{N_o * W}\right) \tag{8}$$

where W is the bandwidth, and since in this scenario ISM band is assumed, the value of W will be 2.4 GHz.

Table 1 $\frac{E_b}{N_0}$ [dB] required for M-ary FSK and M-ary PSK [6]

m	2	4	8	16	32	64
M-ary PSK: $\frac{E_b}{N_0}$	10.5	11	14	18.5	23.4	28.5
M-ary FSK: $\frac{E_b}{N_0}$	13.5	10.8	9.3	8.2	7.5	6.9

From the data base, different M-ary modulation schemes can be analyzed. The table below shows the variation of $\frac{E_b}{N_0}$ with changing the value of M for PSK and FSK modulation schemes [1] (Table 1).

The proposed mathematical system model operates over M-ary modulation under different conditions. These conditions include increasing number of modulation levels and $\frac{E_b}{N_0}$ which is different for different modulation schemes [6, 4].

3 Results and Discussion

In order to determine the performance of the system, we compare different parameters like power transmitted, data rate, and reliability with the path loss in a wireless sensor network using free space earth path loss model and plane earth path loss model.

Figures 1 and 2 illustrate the effect of path loss on the power transmitted for M-ary modulation schemes (PSK and FSK) using free space path loss model. From the above figures, it is observed that as the path loss increases, there is an increase in the power transmitted. With increase in the number of bits, the power required to transmitted increases but for M-ary FSK less power to transmit the data as compared to the M-ary PSK.

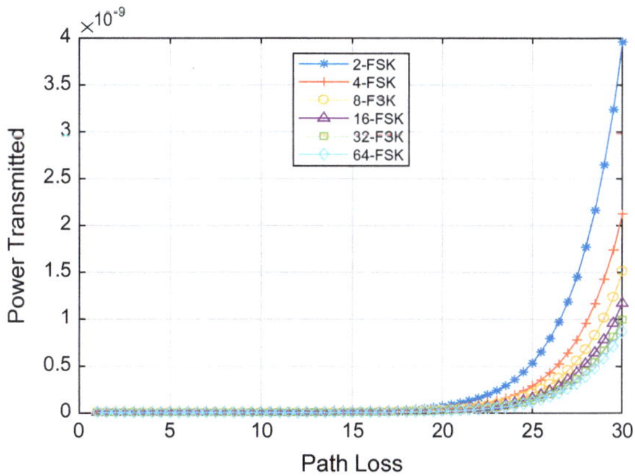

Fig. 1 Power transmitted versus path loss using free space earth path loss model

Fig. 2 Power transmitted versus path loss using free space earth path loss model

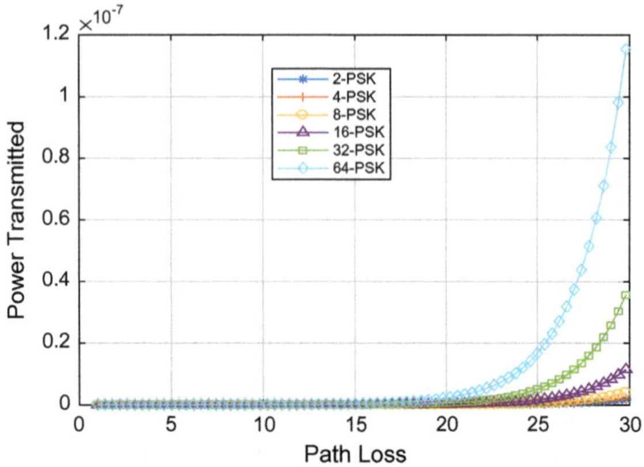

Figures 3 and 4 illustrate the effect of path loss on the data rate for M-ary modulation schemes (PSK and FSK) using free space path loss model. It is observed as the path loss increases, the data rate decreases, and with increasing, the number of bits the data rate to transmitted information increases but for M-ary FSK shows better data rate as compared to the M-ary PSK.

Figures 5 and 6 illustrate the effect of path loss on the reliability for M-ary modulation schemes (PSK and FSK) using free space path loss model. It is observed that with increasing the number of bits, the reliability to transmitted information increases but for M-ary PSK shows better reliability as compared to the M-ary FSK.

Figures 7 and 8 illustrate the effect of path loss on the power transmitted for M-ary modulation schemes (PSK and FSK) using plane earth path loss model. It is observed that with increasing the number of bits, the power required to transmitted information increases but for M-ary FSK less power to transmit the data as compared to the M-ary PSK.

Fig. 3 Data rate versus path loss using free space earth path loss model

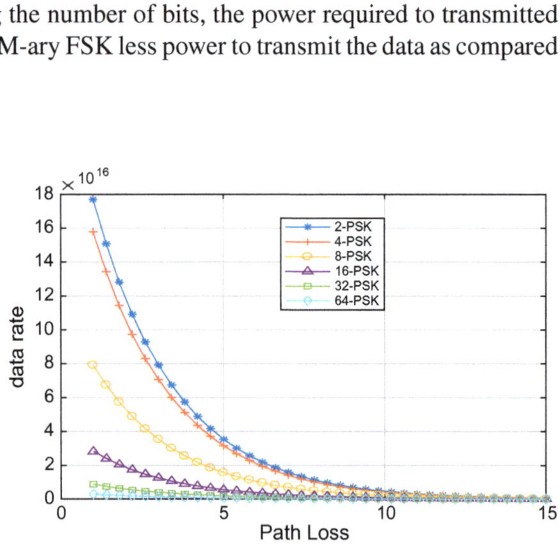

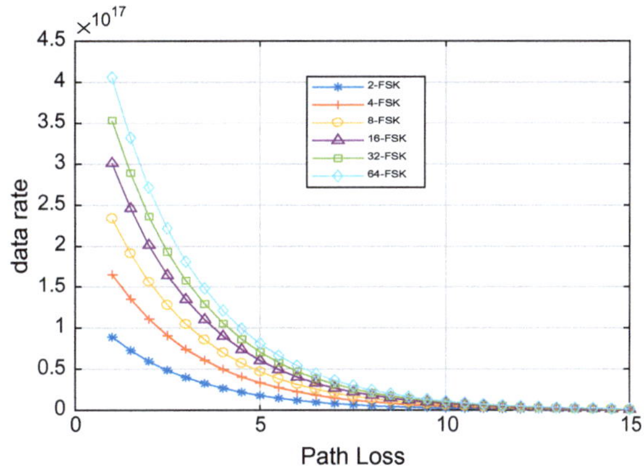

Fig. 4 Data rate versus path loss using free space earth path loss model

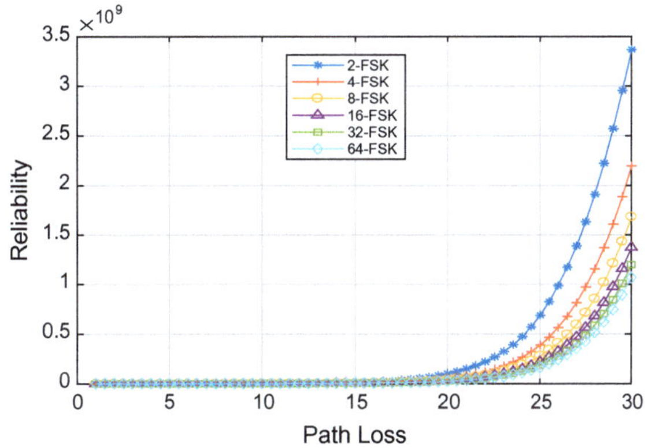

Fig. 5 Reliability versus path loss using free space earth path loss model

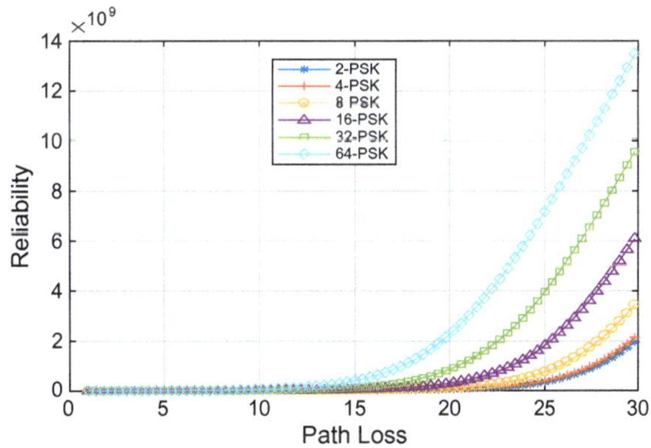

Fig. 6 Reliability versus path loss using free space earth path loss model

Figures 9 and 10 illustrate the effect of path loss on the data rate for M-ary modulation schemes (PSK and FSK) using plane earth path loss model. It is observed

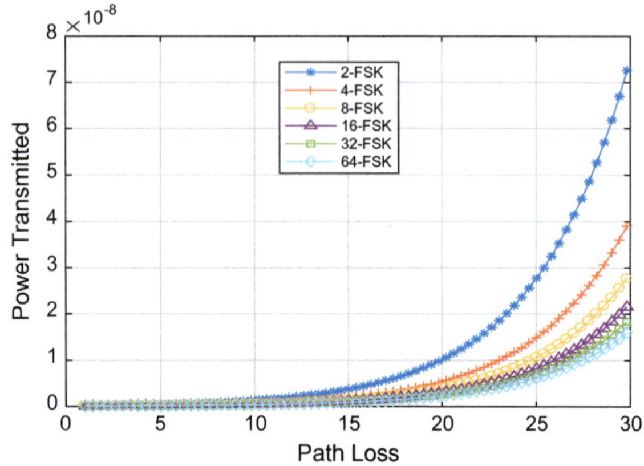

Fig. 7 Power transmitted versus path loss using plane earth path loss model

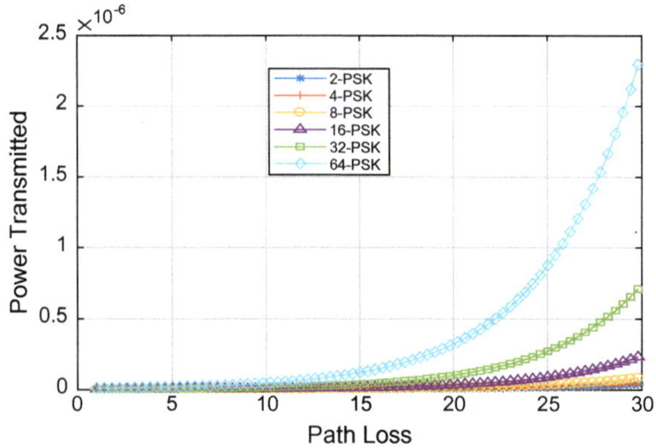

Fig. 8 Power transmitted versus path loss using plane earth path loss model

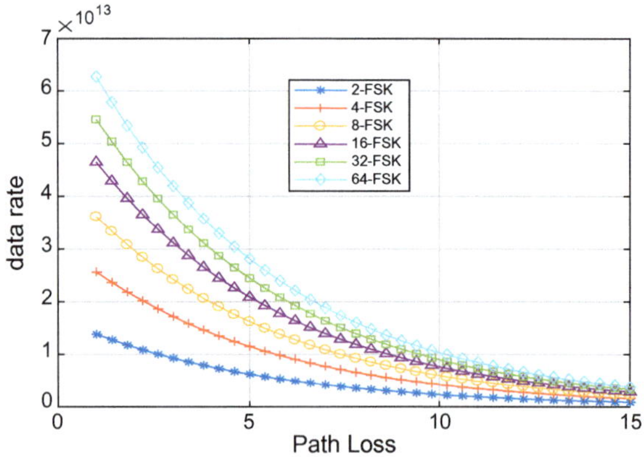

Fig. 9 Data rate versus path loss using plane earth path loss model

that with increasing the number of bits, the data rate to transmitted information increases but for M-ary FSK shows better data rate as compared to the M-ary PSK.

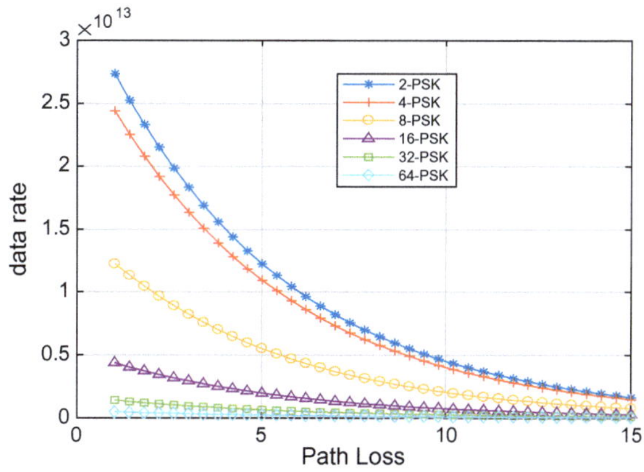

Fig. 10 Data rate versus path loss using plane earth path loss model

Figures 11 and 12 illustrate the effect of path loss on the reliability for M-ary modulation schemes (PSK and FSK) using plane earth path loss model. It is observed that with increasing the number of bits, the reliability to transmitted information increases but for M-ary PSK shows better reliability as compared to M-ary FSK.

Tables 2 and 3 illustrate the variation in the power transmitted, data rate, and reliability at path loss of 10 dBm. Clearly, it satisfies the results which are above mentioned. But when comparing the results of free space path loss model and the plane earth path loss model, the plane earth path loss model illustrates the better results in terms of power transmitted, reliability as compared to the free space earth path loss model. Data rate is less in plane earth path loss model as compared to free space earth path loss model.

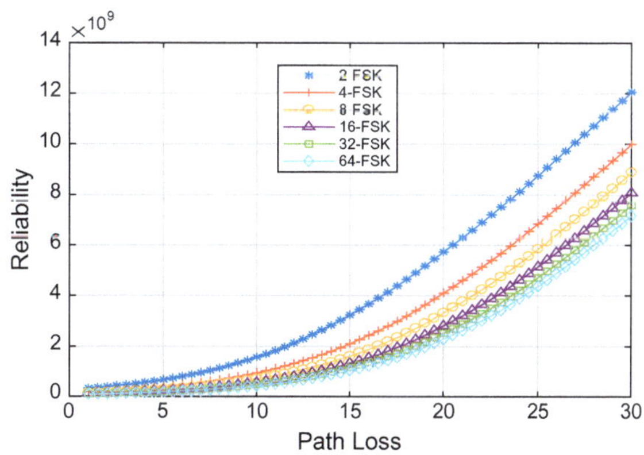

Fig. 11 Reliability versus path loss using plane earth path loss model

Table 2 Effect of path loss on M-ary modulation schemes in wireless sensor network at PL = 10 dBm using free space path loss model

M-ary modulation	Power transmitted (mW)	Data rate (kbps)	Reliability
2-FSK	12.50×10^{-10}	2.26×10^{15}	18.10×10^{5}
4-FSK	6.71×10^{-10}	4.39×10^{15}	9.66×10^{5}
8-FSK	4.75×10^{-10}	6.20×10^{15}	6.85×10^{5}
16-FSK	3.69×10^{-10}	7.99×10^{15}	5.32×10^{5}
32-FSK	3.14×10^{-10}	9.39×10^{15}	4.53×10^{5}
64-FSK	2.73×10^{-10}	10.7×10^{15}	3.94×10^{5}
2-PSK	6.28×10^{-10}	4.70×10^{15}	9.04×10^{5}
4-PSK	7.03×10^{-10}	4.19×10^{15}	1.01×10^{6}
8-PSK	1.40×10^{-9}	2.10×10^{15}	2.02×10^{6}
16-PSK	3.95×10^{-8}	7.63×10^{14}	2.71×10^{6}
32-PSK	1.22×10^{-8}	2.36×10^{14}	1.75×10^{7}
64-PSK	3.95×10^{-8}	7.60×10^{13}	5.65×10^{7}

Table 3 Effect of path loss on M-ary modulation schemes in WSN at PL = 10 dBm using plane earth path loss model

M-ary modulation	Power transmitted (mW)	Data rate (kbps)	Reliability
2-FSK	1.34×10^{-6}	2.23×10^{12}	15.39×10^{8}
4-FSK	7.21×10^{-7}	4.15×10^{12}	9.09×10^{8}
8-FSK	5.10×10^{-7}	5.87×10^{12}	6.67×10^{8}
16-FSK	3.96×10^{-7}	7.87×10^{12}	5.29×10^{8}
32-FSK	3.37×10^{-7}	8.89×10^{12}	4.55×10^{8}
64-FSK	2.93×10^{-7}	10.29×10^{12}	4.00×10^{8}
2-PSK	6.73×10^{-7}	4.45×10^{12}	8.56×10^{8}
4-PSK	7.53×10^{-7}	3.97×10^{12}	9.47×10^{8}
8-PSK	1.50×10^{-6}	1.99×10^{12}	1.68×10^{9}
16-PSK	4.24×10^{-6}	7.06×10^{11}	3.52×10^{9}
32-PSK	1.31×10^{-5}	2.23×10^{11}	6.46×10^{9}
64-PSK	4.24×10^{-5}	7.06×10^{10}	1.01×10^{10}

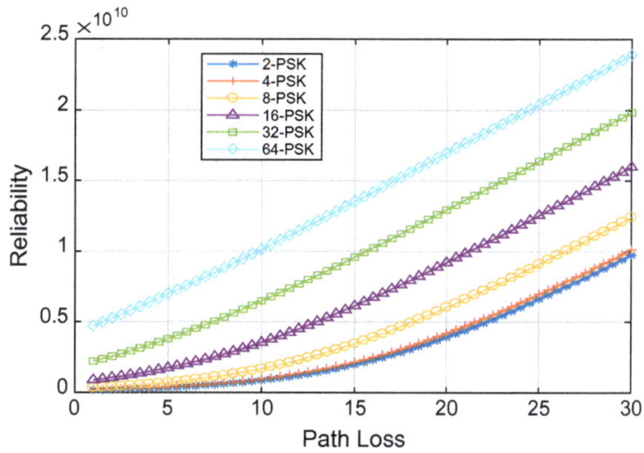

Fig. 12 Reliability versus path loss using plane earth path loss model

4 Conclusion

This paper analyzes the performance of different M-ary modulation schemes like PSK and FSK for wireless sensor networks. The aim is to analyze the energy efficiency, reliability, data rate, and lifetime of the sensor nodes. For the considered scenario, better modulation strategy to minimize the total power transmitted and data rate is required to send a given number of bits. Also evaluate reliability of free space earth model and plane earth model is analyzed, and the results are validated. Results indicate that PSK is more reliable than FSK, whereas the data transmission rate for FSK is greater than PSK. Power required to transmit the data in FSK is less than PSK. Plane earth path loss model illustrates the better results in terms of power transmitted and reliability as compared to the free space earth path loss model. Data rate is less in plane earth path loss model as compared to free space earth path loss model. With increasing the number of bits the power required to transmit the data increases but for M-ary PSK less power to transmit the data as compared to the M-ary FSK. The obtained results can be used as a design approach for modulation techniques in wireless sensor network.

References

1. Proakis, J.G.: Companders. In: Wiley Encyclopedia of Telecommunications (2003)
2. Vougioukas, S., Regen, C., Anastasiuamd, H., Zude, M.: Comparison of radio path loss model for wireless sensor networks in orchard environments. In: The CIGR-AgEng International Conference of Agricultural Engineering, Valencia, Spain, 2012
3. Song Meng, Y., Chong Ng, B., Hui Lee, Y.: Empirical near ground path loss modelling in forest at VHF and UHF bands. IEEE Trans. Antennas Propag. **57**(5), 1461–1468 (2009)
4. Al-Turjman, F.: Cognitive-node architecture and development strategy for the future WSNs. Mob. Netw. Appl. 1–19 (2017)
5. Shivaprkasha, K.S., Patkar, R., Kulkarni, M.: Performance analysis of energy efficient modulation and coding schemes for wireless sensor networks. Int. J. Parallel Emergent Distrib. Syst. **28**(6), 576–589 (2013)
6. Karl, H., Willing, A.: Protocols and Architecture for Wireless Sensor Networks. Wiley. ISBN 0-470-09510-5
7. Kulkami, R.V., Forster, A., Venayaganmoorthy, G.K.: Computational intelligence in wireless sensor networks: a survey. Commun. Surv. Tutor. **13**(1), 68–96 (2011)
8. Srbinovska, M., Dimcev, V., Borozan, V., Gavrovski, C.: Environmental parameters monitoring in precision agriculture using wireless sensor networks. J. Clean. Prod. **88**, 297–307 (2015)
9. Yuan, X., et al.: A genetic algorithm-based dynamic clustering method towards improved WSN longevity. J. Netw. Syst. Manage. **25**(1), 21–46 (2017)
10. Mafuta, M., et al.: Successful development of a wireless sensor network for precision agriculture in Malawi. Int. J. Distrib. Sens. Netw. **9**(5), 150703 (2013)

Protecting Multicast Sessions Against Single-Link Failure in Survivable WDM Mesh Networks

B. Mohapatra and Kunal Sain

Abstract Optical fiber network happens to be dominant transport medium because of its high data-carrying capacity. A single fiber failure in such networks can disrupt the information disseminated to several destination nodes, resulting in huge loss of data. Thus, it is imperative that multicast sessions in WDM networks should be protected from fiber failures. In this paper, we propose three novel protection schemes to protect multicast sessions against single fiber cut. The proposed schemes reduce the amount of network resource used by a session by reducing the number of branching in the multicast tree. As it is observed that more the number of leaf nodes in the multicast tree, more is the amount of network resource required to protect it. The performance of the proposed schemes is evaluated and compared with the reported state-of-the-art algorithm in the presence of dynamic traffic. An improvement of 7–15% is achieved using blocking probability as a performance metric.

Keywords Multicasting · Protection · Survivability
Wavelength-Division multiplexing

1 Introduction

Recent development in the field of optical networking, especially in WDM technology, high-bandwidth applications (both unicast and multicast) such as video conferencing, multiplayer gaming, high-definition television, movie broadcast, and other bandwidth-intensive real-time applications, has become very popular. Optical WDM network will perhaps also be used as the backbone for tomorrow's Internet. An efficient method of achieving multicast is through establishment of a multicast tree, with

B. Mohapatra (✉)
Department of Electronics and Communication Engineering, School of Electrical, Electronics and Communication Engineering, Galgotias University, Greater Noida 201310, India
e-mail: writetobm@gmail.com

K. Sain
Oracle India Pvt. Ltd., Bangalore 560025, India

L. C. Jain et al. (eds.), *Data and Communication Networks*, Advances in Intelligent Systems and Computing 847, https://doi.org/10.1007/978-981-13-2254-9_7

the destination nodes as the leaf or intermediate nodes and the source as the root node. In case of optical networks, such a multicast tree is often known as a light-tree.

In order to support multicasting, in optical WDM networks, the network nodes must be equipped with multicast-capable wavelength-routing switches (MWRSs) [1] or optical cross-connects (MC-OXC) [2]. Multicast-capable wavelength-routing switches (MWRSs) replicate a bit stream from one input port to multiple output ports. MWRS use opaque cross-connects to perform optical–electronic–optical (O/E/O) conversion. A signal arriving on the input fiber link is replicated into multiple copies in the electronic domain. One copy is dropped locally at the node, and the remaining copies are switched to different channels on outgoing fibers. Full wavelength conversion is inherent in such cross-connects and, hence, no explicit wavelength convertors are needed.

Fiber cuts in an optical network lead to huge loss of data, (generally in the order of terabytes). In the case of a single fiber failure in a unicast session, only one single destination is affected. However, even a single-link failure in a multicast session can be devastating as it causes communication failure to multiple destinations. Thus, provision of adequate backup is essential in optical network to provide survivable multicast-based applications. The problem of designing survivable WDM network to support multicast-based applications is more challenging than to provide unicast-based applications. The problem gets increased when the number of destinations gets increased (especially when capacity sharing is taken into consideration).

A. **Definition**

In this section, some of the terms that are defined have been used throughout the paper.

Light-path: In an optical network, it is a point-to-point path of light through which data is transferred.
Light-tree: In case of multicast sessions, the source and destinations are linked by a collection of light-paths. For a particular multicast session, the collection of light-paths connecting the source and the destinations form a light-tree.
Link-disjoint: Two paths/trees are said to be link-disjoint if they do not have any common link among them.
Spanning tree: A spanning tree T of a connected, graph G is a tree composed of all the vertices and some (or all) of the edges of G.

B. **Related work**

Significant work has been done to protect and restore light-paths for point-to-point communication [3–6]. Several schemes, such as path protection, sub-path protection and link protection with or without resource sharing, have been proposed. Since a single fiber cut may cause serious data loss in case of multicasting than an individual light-path, hence, protection of multicast session from a single-link failure has been considered in recent literature [7–11].

One of the simplest solutions to the problem of protecting multicast tree is by computing a link-disjoint backup tree. Two link-disjoint trees together can provide dedicated 1 + 1 protection to the multicast session. However, this scheme leads to usage of excessive network resources. Link protection scheme is proposed in [12]. In this scheme, for each link on the light-tree, a backup route is derived. This scheme suffers from the obvious disadvantage of enormous network resource usages. In [10], the authors proposed SDS (Shared Disjoint Segment), a segment protection scheme and OPP_SDP (Optimal Path Pair-based Shared Disjoint Paths), a path protection scheme. A segment in a multicast tree is a sequence of nodes from the root or a splitting point to a leaf or another downstream splitting point. In SDS, each segment in the primary light-tree is protected using link-disjoint backup path. In OPP_SDP, a link-disjoint path pair is found for each destination. There is also provision for sharing of a common link between different path pairs (i.e., between primary and backup; as well as, backup and backup of the same session). However, both the above algorithms (i.e., SDS and OPP_SDP) do not take into account the concept of spare capacity sharing among different multicast trees.

The author proposed two new schemes in [13] based on ADALINE (average number of destination nodes that are affected by a single-link failure). Assuming a multicast session has a source 'S' and four destinations 'A', 'B', 'C', 'E'. Costs of the links are shown adjacent to the links. Then the ADALINE is calculated as follows: failure of the links S–A, A–B, A–C and B–F affects 4, 2, 1 and 1 destination, respectively. Hence, the ADALINE is calculated as 8/4 = 2. On an average, a single-link failure will prevent 2 destination nodes.

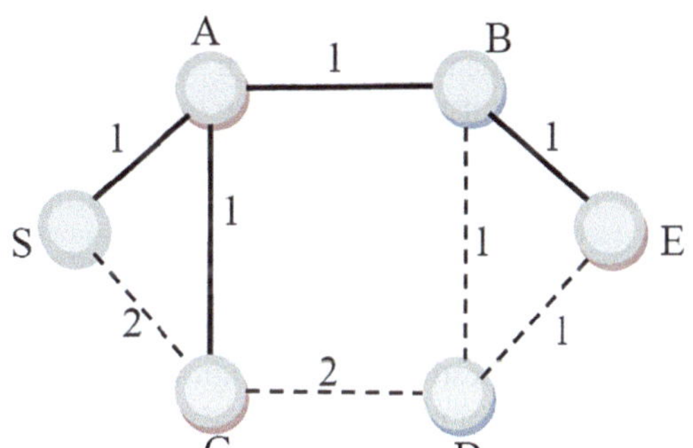

Example Network

A modified version of the state-of-the-art scheme OPP_SDP and named as multicast protection through spanning paths (MPSP) is presented in [14]. A spanning path is defined as a path in a multicast tree which extends from a leaf node to other leaf nodes. Here, the authors consider three kinds of backup path sharing, such as self-sharing, intra-request sharing, and inter-request sharing. Self-sharing scheme suggests sharing of path among primary and backup path in a primary light-tree. Intra-request sharing scheme suggests sharing of path among different backup paths to different destinations in a primary light-tree. Whereas, inter-request sharing schemes considers, sharing of backup paths of two different primary light-tree under single-link failure assumption. The result shows significant improvement in terms of average cost and blocking probability over the OPP_SDP scheme. The author assumes that wavelengths reserved in a fiber can be used in forward and in backward directions by reconfiguring the switches. However, this assumption cannot be achieved practically.

Cost and blocking probability can be reduced by removing the residual links from the backup paths of protected tree are presented in [15]. On the basis of the above theory, the author presented two algorithms, named as, OPP-RRL (Optimal Path Pair-based Removing Residual Links) and SLP-ARL (Source Leaf Path-based Avoiding Residual Links). The basic idea is presented here, using the example network. First derive the backup paths for all the individual links of the primary light-tree. S–C–A for S–A, S–C for A–C, C–D–B for A–B and B–D–F for B–F, respectively, are the backup paths for individual links. Instead of the above, one backup path may be derived as S–C–D–E to provide connection between source and all the destinations. Using this technique, the results claim marginal improvements over OPP_SDP in terms of wavelength link cost and blocking probability.

Segment-based protection scheme is proposed in [16]. A segment is a link which originates from the source or from intermediate nodes. Failure of a segment (link) adjacent to the source node disrupts many destinations; hence, a huge of revenue loss will occur. To overcome this problem, the author proposed segment-based protection tree (SPT) scheme and claims a significant reduction of cost over OPP_SDP scheme. A light-tree is constructed for a multicast session. For protection, another multicast tree is constructed. Each protection tree is originated from the source and reaches to every destination of the multicast session. Each segment of the primary tree is protected by the protection tree; also, the protection tree can share wavelength links with primary tree as well as another protection tree. Hence, it is a case of different sharing approaches, such as self-sharing, intra-request sharing, and inter-request sharing, already discussed.

P-cycle is another technology, which also applied to protect multicast sessions is presented in [17–19]. This technique aims to constructs a cycle which traverse through the source and all the destinations. Though this offers fast recovery time (which is an important criterion for restoration), however, it is less capacity efficient.

Recently, elastic optical network (EON) has attracted the attention among the researchers. To further increase the efficiency and data rate offered by the optical network, coherent optical OFDM technique is a viable option. Based on this, the author proposed both mathematical and heuristic models to protect multicast sessions in [20]. Here, a shared-based protection strategy is adopted. Link-disjoint backup path is constructed for each source–destination pairs in the primary tree. The backup path can be shared to protect several light trees under single-link failure assumption. As the OPP_SDP is a state-of-the-art and widely accepted scheme, our objective is to compare the proposed algorithm with the same. The future scope of this work is to implement the proposed algorithms in EON for better results may be obtained in terms of resource utilization.

C. **Our Contribution**

In this paper, we propose three routing algorithms for protection of multicast connection in WDM network. The Limited Splitting Spanning Tree (*LSST*) Scheme aims at minimizing the number of branching in the multicast tree with minimum increase in resource utilization. This is because more the number of branching more is the number of destination nodes that appear as the leaf nodes in the multicast tree. Thus, more network resource is required to protect the multicast tree. Then it aims at linking each leaf node to the source node by a path which is link-disjoint to the primary light-tree. Our second algorithm is an *enhanced version of LSST Scheme*, called enhanced LSST scheme (*ELSST*) which uses the same concept as the *LSST Scheme* in order to derive the primary light-tree. However, it aims at maximizing self-sharing between the backup path and the primary light-tree. Our third algorithm, the Circular Path Session Protection *(CPSP) Scheme* tries to put all the participating nodes of a particular session (source and destinations) into a cycle, thus providing effective use of backup resources to establish a protected multicast session. In each of the above-mentioned algorithms, we use two different concepts for evaluating the usage cost of a link:

- Length of the link or other physical costs (i.e., hop count, etc.) of getting data from one end to the other.
- Usage of the link. This means that the more the number of wavelengths in a link is used, the more costly it is.

We also evaluate the performance of an existing algorithm, OPP_SDP [10], which we believe to be a state-of-the-art algorithm, based on the above two cost parameters. However in [10], the authors have evaluated their algorithm only on the basis of physical link cost.

The organization of the paper is as follows. Section 2 describes the assumptions and calculation of link usage cost for the schemes. Section 3 illustrates the proposed algorithms in details. Results and discussions are presented in Sect. 4. Finally, Sect. 5 concludes the paper with future scope of the work.

2 Problem Formulation and Assumptions

A. **Problem Definition**

The problem of protecting multicast sessions in WDM Networks can be stated as follows:

- The network is given as $G = (V, E)$ where G is a weighted undirected graph 1. V is the set of vertices or network nodes. E is the set of edges or optical fiber links, each link has a capacity of carrying a maximum of W wavelengths. Each link $x \in E$ has a positive cost assigned to it representing the cost of moving traffic from one end to the other.

- We also make the assumption that each node is equipped with switches capable of performing full wavelength conversion (MWRSs) and light splitters.
- We represent a multicast session as *R*(*s*, *D*, *k*) where *s* is the source, *D* is the set of destinations, and k is the group size (including the source), i.e. (no of Destinations served + 1). The connections are considered to be unidirectional from the source to the destination. If the routing algorithm is able to discover sufficient resources to establish the protected multicast session, only then the session is established else it is considered to be a blocked session.

Our objective is to maximize the number of protected multicast sessions with minimum possible consumption of network resources.

B. **Link Usage Cost**

As mentioned earlier, we calculate the link usage cost in two different ways:

- Taking into consideration the physical length of the link or other physical costs (such as hop count, etc.) for getting the data from a node on one end of the link to the node on the other end.
- Taking into consideration, the fraction of the number of available wavelengths used up on the link by existing sessions.

While taking physical cost of the link as the link usage cost, the user might consider the actual physical cost of data transmission through the link. However, *hop count* is considered as the measure of distance between any two nodes in this work. Thus, we have taken the weight of every edge to be *one*. While evaluating the link usage cost on the basis of the number of used-up bandwidths of a link, we calculate the link usage cost as follows: We have used the above measure of link cost in order to stop the quick use-up of available wavelengths in a *busy* link. Even though this method leads to slightly costlier paths, it helps to spread the load almost evenly over the whole network. An optimal link cost measure would be to calculate the link cost based on both physical cost and availability of wavelengths on the link.

$$C_{\mathrm{ij}} = \frac{W_{\mathrm{ij}} - A_{\mathrm{ij}}}{W_{\mathrm{ij}}} \times 100$$

where C_{ij} is the link usage cost of the link (*i*, *j*). W_{ij} is the capacity of the link (*i*, *j*) in terms of number of wavelengths. A_{ij} is the number of available wavelengths on the link (*i*, *j*).

3 Proposed Schemes

The LSST (Limited Splitting Spanning Tree) Scheme

When a multicast request arrives, apart from allocating resource for establishing the session, some extra resources have to be reserved to protect it against single-link failure. The approach to the problem is achieved in two major steps. Firstly, establish a multicast tree while minimizing the number of leaves. Reducing the number of leaves generally tends to increase the cost of the tree. We try to keep this as minimum as possible. Then, the scheme protects the multicast tree against single-link failure by allocating some extra resources to the respective session.

Step 1 *Allocating Resources for the Primary Light-tree*—The objective while calculating the primary light-tree is to reduce the number of branching while ensuring that the excessive network resource is not used. In order to ensure this, we use a modified form of Dijkstra's algorithm for *single source single destination shortest path*. Here we modify the algorithm to find the shortest path between two nodes, one belonging to a set of source nodes and the other belonging to a set of destination nodes. However if there are two sets of path having equal length, the one with the source that was more recently found is considered. Once a path is found between two nodes a and b, where $a \in S$ (set of sources) and $b \in D$ (set of destinations), b is removed from the set D and added to the set S. By repetitively applying this, the *primary light-tree* is derived.

Step 2 *Allocating Resources for the Backup Light-tree*—In order to allocate resources for the *backup light-tree,* we use a traditional approach. Simply by joining the leaf nodes to the source node using paths that are *link-disjoint* to the *Primary Light-tree*.

The LSST Scheme
Finding Primary Light-tree

(1) $Q \leftarrow Source$
(2) $D \leftarrow Destinations$
(3) **While** $(D \neq \emptyset$ ***AND*** $test \neq 1)$

(a) $\forall x \in Q$, ***set*** $d[x] \leftarrow 0$
(b) $P \leftarrow \emptyset$
(c) **while** $(u \notin D$ ***AND*** $P \neq V)$
(i) $u \leftarrow Extract\_Min(d)$
(ii) $P \leftarrow P \cup \{u\}$
(iii) $\forall y \in Adj(u)$ ***AND*** $y \notin P$
If $d[y] > (d[u] + w(u, y))$
(A) $d[y] \leftarrow (d[u] + w(u, y))$
(B) $Pred[y] \leftarrow u$
(d) **If** $(u \in D)$
(i) $D \leftarrow D - \{u\}$
(ii) $Q \leftarrow Q \cup \{u\}$

(e) **If** $(u \notin D)$, ***set*** $test = 1$

Finding Back-path
findBackupTree_LSST(AdjMatrix, PrimaryTree, Source)

(1) $D \leftarrow leafNodes(Primary\,Tree)$
(2) $\forall x \in D$

  (a) $Path \leftarrow graphshortestpath(PrimaryTree, Source, x)$
  (b) $adj \leftarrow AdjMatrix$
  (c) $\forall \{y, z\} \in Path$
    (i) $adj[y, z] \leftarrow 0$
    (ii) $adj[z, y] \leftarrow 0$
  (d) $B\_Path \leftarrow cstgraphshortestpath(adj, Source, x)$

The Enhanced LSST (ELLST) Scheme
The *ELLST Scheme* deals with the problem of protecting multicast sessions in almost the same way as in the *LSST Scheme*. However this time we not only try to maximize the amount of self-sharing, we also try to find backup paths for the leaf nodes that use less network resource. We protect the session by connecting leaf nodes not only to the source, but to "already protected" leaf nodes too. This algorithm too uses a two-step approach to protect a multicast session.

Step 1 *Allocating Resources for the Primary Light-tree*—In order to allocate resources for the primary light-tree, the algorithm use the same approach that is used in the case of LSST scheme.

Step 2 *Allocating Resources for the Backup Light-tree*—In order to allocate reserve the resources for the backup light-tree, a leaf node l_l is randomly chosen from the set of leaf nodes L. Then the shortest path between the source s and the leaf node l_1 along the primary light-tree P_1 is calculated. All the links on the primary light-tree that are not in P_1 are set to negligible cost (in order to encourage *self-sharing* while guaranteeing that uselessly long paths are not formed). A backup path which is link-disjoint to P_1 joining s and l_1 is found. l_1 is removed from the set L, while the source node s and l_1 is added to the set S. Again a leaf node is randomly chosen from the set L, say l_2. Again the shortest path between the source s and this node l_2 along the primary light-path P_2 is calculated. Now the shortest path, link-disjoint to P_2 between l_2 and any node in the set S is calculated. Then l_2 is removed from the set L and added to the set S. This is continued until backup path for all leaf nodes are found.

The ELSST Scheme
Finding Primary Light-tree

1. $Q \leftarrow Source$
2. $D \leftarrow Destinations$
3. **While** $(D \neq \emptyset\ \boldsymbol{AND}\ test \neq 1)$

(a) $\forall x \in Q$, **set** $d[x] \leftarrow 0$
(b) $P \leftarrow \emptyset$
(c) **while** $(u \notin D \; \boldsymbol{AND} \; P \neq V)$
  (i) $u \leftarrow Extract\_Min(d)$
  (ii) $P \leftarrow P \cup \{u\}$
  (iii) $\forall y \in Adj(u) \; \boldsymbol{AND} \; y \notin P$
   If $d[y] > (d[u] + w(u, y))$
   (a) $d[y] \leftarrow (d[u] + w(u, y))$
   (b) $Pred[y] \leftarrow u$
(d) **If** $(u \in D)$
  (i) $D \leftarrow D - \{u\}$
  (ii) **If** $Q \leftarrow Q \cup \{u\}$
(e) **If** $(u \notin D)$, **set** $test = 1$

Finding Back-path
findBackupTree_ELSST(AdjMatrix, PrimaryTree, Source)

1. $D \leftarrow leafNodes(PrimaryTree)$
2. $S \leftarrow Source$
3. $\forall x \in D$

 (a) $Path \leftarrow graphshortestpath(PrimaryTree, Source, x)$
 (b) $adj \leftarrow AdjMatrix$
 (c) $\forall link \in PrimaryTree$, **set** $linkCost = \varepsilon$
 (d) $\forall \{y, z\} \in Path$
  (i) $adj[y, z] \leftarrow 0$
  (ii) $adj[z, y] \leftarrow 0$
 (e) $B\_Path \leftarrow cstgraphshortestpath(adj, Source, x)$
 (f) $S \leftarrow S \cup \{x\}$
 (g) $D \leftarrow D - \{x\}$

cstgraphshortestpth(adj, Sources, Destinations)

(1) $Q \leftarrow Source$
(2) $D \leftarrow Destinations$
(3) $\forall x \in Q$, **set** $d[x] \leftarrow 0$
(4) $P \leftarrow \emptyset$
(5) **while** $(u \neq D)$

 (a) $u \leftarrow Extract\_Min(d)$
 (b) $P \leftarrow P \cup \{u\}$
 (c) $\forall y \in Adj(u) \; \boldsymbol{AND} \; y \notin P$
  (i) **If** $d[y] > (d[u] + w(u, y))$
   (A) $d[y] \leftarrow (d[u] + w(u, y))$
   (B) $Pred[y] \leftarrow u$

The CPSP (Circular Path Session Protection) Scheme

The CPSP Scheme deals with the problem protecting multicast sessions by creating a cycle containing the source and all the destination nodes. However, an exact method of creating such a cycle is an NP-complete problem. Hence, a heuristic approach is proposed to create such a cycle.

The Algorithm

Initially, the algorithm starts with finding the shortest distance between the source node s and the destination node d_1 from the set of destination nodes participating in the multicast session. The nodes s and d_1 are added to a set S and are linked by the shortest path. The edges on this path are removed from the network. Then we try to find the node in the remaining set of destination nodes that is the shortest distance away from any of the two nodes in S. Let this destination node be d_2.

Then d_2 is added to the set S and the node in S that d_2 is found to be shortest distance away, is to be removed from the set S. Hence, a single path is derived by joining an unlinked destination node to any of the nodes at either end of the current path. Also new paths found must be link-disjoint to the currently existing path. Once all the destination nodes and the source node are linked by one path, the nodes at the two ends are linked, which completes the cycle. This guarantee that even with a single-link failure, data will continue to flow from the source to all the destination nodes.

The CPSP Scheme
FindSingleCycle(AdjArr, Source, Destinations)

(1) $Q \leftarrow Source$
(2) $D \leftarrow Destinations$
(3) **while** $(D \neq \emptyset$ **AND** $test \neq 1)$

 (a) $\forall x \in Q,$ ***set*** $d[x] \leftarrow 0$
 (b) $P \leftarrow \emptyset$
 (c) **while** $(u \notin D$ **AND** $P \neq V)$
  (i) $u \leftarrow Extract\_Min(d)$
  (ii) $P \leftarrow P \cup \{u\}$
  (iii) $\forall y \in Adj(u)$ **AND** $y \notin P$
  If $d[y] > (d[u] + w(u, y))$
   (A) $d[y] \leftarrow (d[u] + w(u, y))$
   (B) $Pred[y] \leftarrow u$
 (d) **if** $(u \in D)$
  (i) $D \leftarrow D - \{u\}$
  (ii) $Q \leftarrow Q \cup \{u\}$
  (iii) *If* $(length(Q) > 2) then\, Q \leftarrow Q - \{Pred[u]\}$
  (iv) *Remove the path from* $Pred[u]$ *to* u *from the* $AdjArr$

(4) *Find the shortest path connection* $Q[1]$ *and* $Q[2]$

4 Results and Discussions

In this section, we present a series of simulation results to evaluate the performance of the proposed schemes (*LSST*, *ELSST,* and *CPSP*). We also compare the results against the performance of a standard state-of-the-art algorithm *OPP_SDP*. The above schemes are simulated using two widely used networks such as USNet and NSFNet. It is assumed that all the nodes in the networks are capable of wavelength conversion and light splitting. Each link in the network is considered to have a maximum capacity of W wavelengths. Multicasts requests of size k are assumed to arrive with Poisson's distribution with arrival rate λ. Each request is assumed to have unit holding time. Thus, the average load of the network is λ. Source and destination nodes are randomly distributed across the whole network. The simulation results are as shown in below given figures. Figures 1, 2 show the results obtained for NSF Net with $k = 4$ and $k = 6$; and value of λ varied from 10 to 60 for link capacity $(W) = 16$; and value of λ varied from 10 to 100 for link capacity $(W) = 32$.

As is seen from all the above results, our algorithm performs quite better as compared to *OPP_SDP* in all the cases, especially under heavily loaded conditions. *CPSP* performs the best under heavily loaded conditions (though it gives slightly high blocking probability under very lightly loaded conditions). It is also seen that the blocking probability is slightly less.

While using *Link Usage* as the metric while allocating the light-paths as compared to while using *hop count* as the metric. Figures 3, 4 show the results obtained for USNet with $k = 6$ and $k = 8$; and value of λ varied from 10 to 60 for link capacity $(W) = 16$; and value of λ varied from 10 to 100 for link capacity $(W) = 32$. The results obtained for USNet are similar to those obtained for NSFNet.

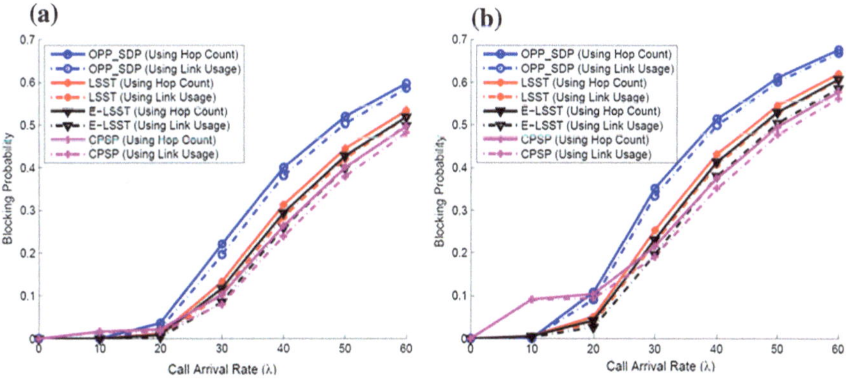

Fig. 1 **a** Session size $k = 4$ with link capacity $W = 16$ for NSFNet, **b** Session size $k = 6$ with link capacity $W = 16$ for NSFNet

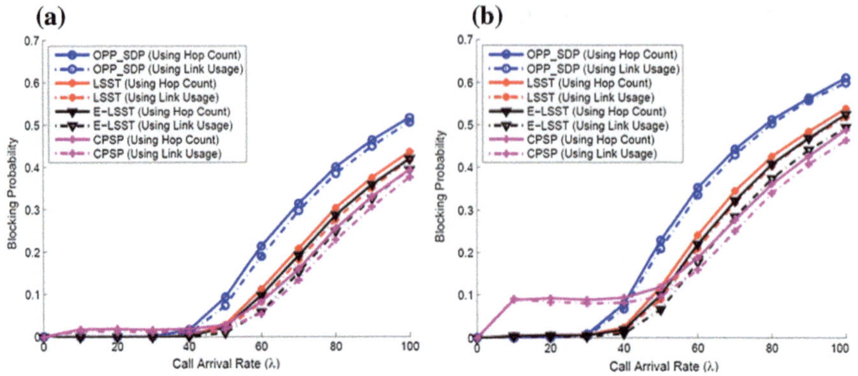

Fig. 2 **a** Session size $k=4$ with link capacity $W=32$ for NSFNet, **b** Session size $k=6$ with link capacity $W=32$ for NSFNet

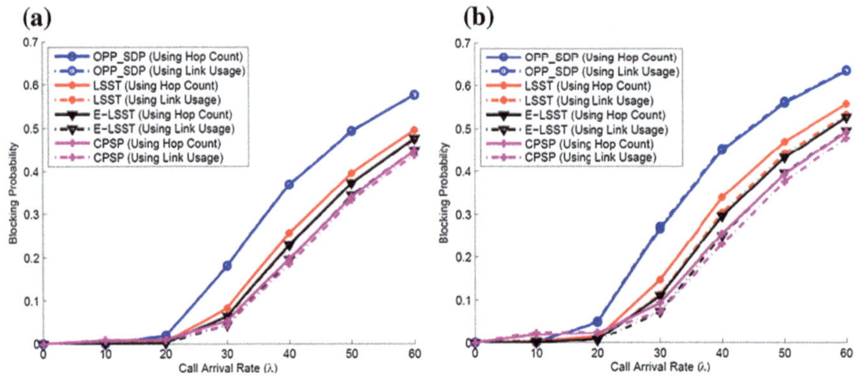

Fig. 3 **a** Session size $k=6$ with link capacity $W=16$ for USNet, **b** Session size $k=8$ with link capacity $W=16$ for USNet

5 Conclusions

In this paper, three schemes have been proposed to protect multicast session against single-link failure. The first scheme, *LSST* tries to reduce the network resource usages and in turn reduce blocking probability by reducing the number of splitting points in the primary light-tree that is to be protected. The second scheme, *ELSST* is an enhancement of the *LSST* scheme which tries to increase the degree of *self-sharing* among the primary light-tree and the backup paths for each of the failed link. The third scheme, *CPSP* approaches the problem of protecting multicast session using a single cyclic light-path. The results obtained from simulations performed on NSFNet and USNet indicate that the proposed algorithms provide significant improvement over a state-of-the-art existing algorithm *OPP_SDP* in terms of reduction in blocking probability.

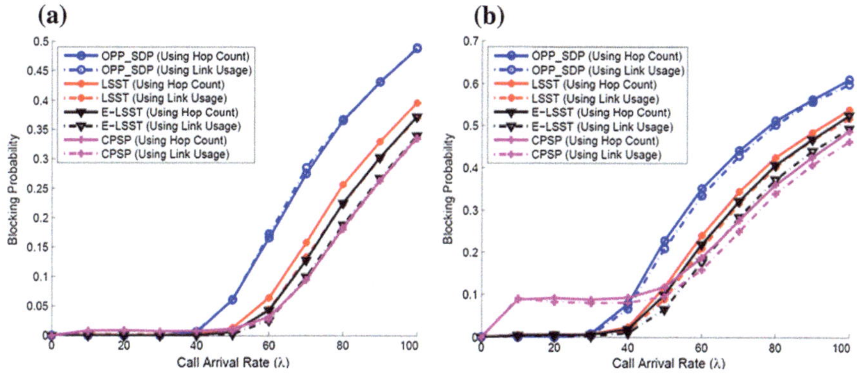

Fig. 4 **a** Session size $k = 6$ with link capacity $W = 32$ for USNet, **b** Session size $k = 8$ with link capacity $W = 32$ for USNet

In the case of NSFNet, when the link capacity $W = 16$, the improvement offered by the above schemes are 6–7, 8–9, and 10–11%, respectively. Whereas when the link capacity $W = 32$, the improvement appeared to be 8–10, 10–12, and 13–14%, respectively. On the other hand, in the case of USNet, when the link capacity $W = 16$, the improvement offered by the above schemes are 10–12, 12–15, and 15–16%, respectively. Whereas when the link capacity $W = 32$, the improvement appeared to be 11–13, 13–16, and 16–17%, respectively. It is observed that the performance of CPSP scheme is best scheme among the others and it is observed that this improvement is achieved at the cost of higher working capacity or to higher primary light-tree cost.

References

1. Sahasrabuddhe, L., Mukherjee, B.: Light trees: optical multicasting for improved performance in wavelength routed networks. Commun. Mag. IEEE **37**, 67–73 (1999)
2. Hu, W., Zeng, Q.: Multicasting optical cross connects employing splitter-and-delivery switch. Photonics Technol. Lett. IEEE **10**, 970–972 (1998)
3. Ramamurthy, S., Sahasrabuddhe, L., Mukherjee, B.: Survivable wdm mesh networks. J. Lightwave Technol. **21**, 870 (2003)
4. Ellinas, G., Hailemariam, A., Stern, T.: Protection cycles in mesh wdm networks. IEEE J. Sel. Areas Commun. **18**, 1924–1937 (2000)
5. Van Caenegem, B., Van Parys, W., De Turck, F., Demeester, P.: Dimensioning of survivable wdm networks. IEEE J. Sel. Areas Commun. **16**, 1146–1157 (1998)
6. Crochat, O., Le Boudec, J.-Y.: Design protection for wdm optical networks. IEEE J. Sel. Areas Commun. **16**, 1158–1165 (1998)
7. Singhal, N.K., Mukherjee, B.: Protecting multicast sessions in wdm optical mesh networks. J. Lightwave Technol. **21**, 884 (2003)
8. Rahman T., Ellinas, G.: Protection of multicast sessions in wdm mesh optical networks. In: Optical Fiber Communication Conference, 2005. Technical Digest. OFC/NFOEC, vol. 2, 3 pp, 2005

9. Khalil, A., Hadjiantonis, A., Ellinas, G., Ali, M.: Dynamic provisioning of survivable heterogeneous multicast and unicast traffic in wdm networks. In: IEEE International Conference on Communications, 2006. ICC '06, vol. 6, pp. 2465–2470, 2006
10. Singhal, N.K., Sahasrabuddhe, L.H., Mukherjee, B.: Provisioning of survivable multicast sessions against single-link failures in optical wdm mesh networks. J. Lightwave Technol. **21**, 2587 (2003)
11. Singhal, N.K., Ou, C., Mukherjee, B.: Cross-sharing versus self-sharing trees for protecting multicast sessions in mesh networks. Comput. Netw. **50**(2), 200–206 (2006). (Optical Networks)
12. Wu, C.-S., Lee, S.-W., Hou, Y.-T.: Backup vp preplanning strategies for survivable multicast atm networks. In: 1997 IEEE International Conference on Communications, ICC 97 Montreal, 'Towards the Knowledge Millennium', vol. 1, pp. 267–271, 1997
13. Luo, H., Li, L.: A new consideration for provisioning survivable multicast sessions in WDM mesh networks. In: Proceedings of International Conference on Communications, Circuits and Systems, Kokura, Japan, pp. 544-548, 2007
14. Luo, H., Li, L., Yu, H., Wang, S.: Achieving shared protection for dynamic multicast sessions in survivable mesh WDM networks. IEEE J. Sel. Areas Commun. **25**(9), 83–95 (2007)
15. Xiong Wang · Sheng Wang · Lemin Li · Yunji Song, "Achieving resource reduction for protecting multicast sessions in WDM mesh networks", Photon Netw Communication, Vol. 15, pp. 131–140, 2008
16. Long, L., Kamal, A.E.: Tree-based protection of multicast services in WDM mesh networks. In: Proceedings of 28th IEEE conference on Global telecommunications, GLOBECOM'09, pp. 1529–1534, 2009
17. Feng, T., Ruan, L., Zhang, W.: Intelligent *p*-cycle protection for multicast sessions in WDM networks. In: Proceedings of International Conference on Communications (ICC), pp. 5165–5169, 2008
18. Feng, T., Ruan, L., Zhang, W.: Intelligent *p*-cycle protection for dynamic multicast sessions in WDM networks. J. Opt. Commun. Netw. **2**(7), 389–399 (2010)
19. Frikha, A., Cousin, B., Lahoud, S.: Extending node protection concept of p-cycles for an efficient resource utilization in multicast traffic. In: Proceedings of IEEE 36th Conference on Local Computer Networks (LCN), Bonn, Germany, pp. 175–178, 2011
20. Cai, A., et al.: Multicast routing and distance-adaptive spectrum allocation in elastic optical networks with shared protection. J. Light. Technol. **34**(17), 4076–4088 (2016)

A Cross-Layer Routing Protocol for Wireless Sensor Networks

Pallavi Yarde, Sumit Srivastava and Kumkum Garg

Abstract In the era of Internet of things (IoT), the sensors play an important role and also face a challenge of energy consumption. Sensors in wireless sensor networks (WSNs) deals with accumulation and processing of data and forward that to the remote locations, generally considered as cloud. Generally, communication is done between the nodes which are placed at a far locations in the field. Hence, the energy consumption required to communicate the nodes plays an important role. In this paper, the proposed algorithm is based on low-energy adaptive clustering hierarchical (LEACH) routing algorithm named as multi-hop cluster LEACH (MC LEACH) algorithm. The proposed protocol is a cross-layer routing protocol that deals with physical, MAC, and network layers for the analysis of energy consumption at individual node as well as in whole network.

Keywords Wireless sensor networks · Cross-layer optimization · LEACH
MC LEACH

1 Introduction

There is a big resolution occurred with Internet of things (IoT) in the field of wireless sensor networks (WSNs). The aim of IoT is to communicate the data gathered into sensors network to remote location with one of wireless technologies. The data from network should reach to remote location, generally called sink with minimum delay and energy consumption. Delay and energy consumption are very important quality of service (QoS) measuring metrics. There are numerous delay measuring protocols available for WSN [1–4]. The proposed work focuses on optimizing energy dissipation in whole network. Each layer of communication network model plays a

P. Yarde (✉) · S. Srivastava
Manipal University Jaipur, Jaipur, Rajasthan, India
e-mail: pallavi.yarde@jaipur.manipal.edu

K. Garg
Bhartiya Skill Development University, Jaipur, Rajasthan, India

L. C. Jain et al. (eds.), *Data and Communication Networks*, Advances in Intelligent Systems and Computing 847, https://doi.org/10.1007/978-981-13-2254-9_8

vital role. When these different layers work in collaboration then it is considered as cross-layer optimization. The proposed protocol also implemented using cross-layer approach. Here, three layers, network, media access control (MAC), and physical layers, work in coordination to analyze energy consumption, number of dead nodes, and number of alive nodes. The proposed protocol is based on low-energy adaptive clustering hierarchy (LEACH) protocol with some add-ons. The function of the proposed protocol refers to the probability as well as distance for selecting cluster head. The cluster head selection process involves probability, distance among the nodes and sink, and also energy of nodes. The number of cluster heads makes a chain to propagate the processed data to remote sink, hence called multi-hop protocol. The proposed protocol MC LEACH involves functions like selection process for cluster heads, gathering data, processing of data, and designing a chain, i.e., choosing the best path to reach to the sink. These functions involves role of network layer, MAC layer, and physical layer in collaboration to as a cross-layer function.

2 Existing Work

There exist many protocols which are energy efficient and also focus on cross-layer optimization. These protocols use different QoS metrics as measures of performance like delay [5–8], throughput, network lifetime, residual energy [9–14]. Some of the protocols also implemented with cross-layer optimization. The proposed protocol is based on LEACH protocols with some modifications. Most of sensor nodes in WSN protocols directly transmit accumulated data to the sink instead; this protocol supports selecting a cluster head to each designed clusters in rotation. The proposed protocol also supports cross-layer optimization.

Packet size: The amount of data (raw data and supporting information) in bits used to travel between source and destination is considered as a packet. The size of packet may vary or of fixed size.

Traffic: Highly dense network has too many packets transmission and hence some of packets may have a chance to drop. This is because of collision or stop-and-wait situation at the node. This introduces a delay in communication.

Scalability: It is an ability to enhance number of nodes in the network. The number of nodes should be enhanced up to optimization level only, so that maximum delay cause can be avoided there.

An energy-efficient differentiated directed diffusion (EDDD) protocol was proposed in 2006 [15]. The data in network may be destined to the node from its surrounding or from a node to the other node. The first type of data is considered as real-time (RT) data, while other one is considered as best-effort (BE) data. This protocol has a capability to differentiate between these two data and work accordingly to avoid maximum delay.

A cross-layer priority-based congestion control protocol (PCCP) was proposed by Wang et al. [16]. It is a routing protocol. Here, there are two data queues generated: one at MAC layer and other at network layer. The node calculates the time consumed by a packet to wait and communicate and estimates the probability of congestion and actual congestion happened. Based on this, it notifies other nodes about congestion and avoids further delay in communication.

3 Proposed Work

The proposed protocol, multi-hop clusters LEACH (MC LEACH), is a cross-layer optimized routing protocol which focuses on media access control) (MAC) and network layers. The proposed protocol is a cross-layer routing protocol, which emphasizes minimized energy consumption for communication.

3.1 Network Structure

The different energy-level sensor nodes are considered to be randomly spread in bounded area and considered as heterogeneous network. Whole network nodes are divided into normal and advanced nodes. The complete network is virtually divided into number of clusters with each has a radius *R*. A sink is also placed at the center of the network. Sink received sensed data from network. Sending sensed data directly to the sink may exhaust sensor nodes easily. The proposed protocol follows LEACH, in which each cluster is appointed to of the node as a head called cluster head (CH). Here, the responsibility of CH is to gather the data from surrounding sensor nodes of its own cluster and to process that data and finally send it to the sink [17]. In MC LEACH protocol, each sensor node calculates the distance between all the remaining nodes and sinks. When a node is elected as CH, it already knows the distance between itself and the sink as well as between all available CHs of the network. The distance is calculated using the Euclidian distance formula.

In Fig. 1, each cluster head calculates the distance between all remaining cluster heads and the sink. Here, we have four clusters C1, C2, C3, and C4. Each cluster has a cluster head (black circles). The distances between all cluster heads are d23, d12, d13, d24, d34, and d14. The distances between clusters C1, C2, C3, C4 and the sink are d1, d2, d3, and d4, respectively.

3.2 Cluster Setup and Selection of Cluster Head

Each sensor node is initiated with a specific amount of energy. The selection of cluster head in a cluster depends on the probability like LEACH protocol and on energy of individual nodes; i.e., the node with highest probability and energy among the cluster nodes will be selected as CH.

Each node manages a routing table. The routing table contains information of its own and neighboring nodes of its cluster. The parameters stored with routing table are

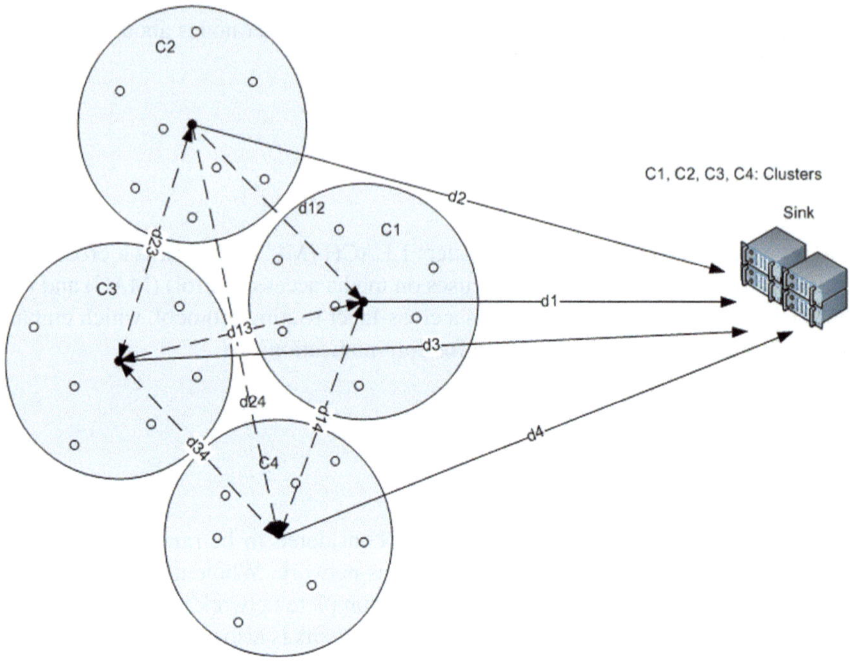

Fig. 1 Architecture of wireless sensor networks for MC LEACH protocol

probability to be cluster head, residual energy, and distance between the nodes. Like LEACH protocol, the node becomes cluster head on the basis of probability and also with an additional parameter residual energy. The node with maximum probability and residual energy among the nodes in a cluster will be selected as cluster head. Each node selects a random number where $0 < r < 1$. If, $r <$ threshold $T(n)$, then one of the criteria for selecting that node as cluster head fulfills. The threshold calculation is given in Eq. 1.

$$T(n) = \begin{cases} \frac{P}{1 - P*\left(r \bmod \frac{1}{P}\right)} & n \in G \\ 0 & \text{Otherwise} \end{cases} \quad (1)$$

where n is nth node for which threshold is calculated, G is the group of sensor nodes, and P is the probability of node to become cluster head.

Once the criteria of threshold fulfills, the energy of sorted nodes is compared and the node with maximum energy and threshold greater than value r is selected as cluster head for a round. This process keeps repeating for each round; i.e., for each round, number of clusters and cluster head are created.

3.3 Cluster Setup

Once the cluster is formed, the responsibility of sensor nodes is to introduce themselves with the other nodes of cluster. The nodes broadcast message signal consisting of node details like their node id and residual energy. Based on receive signal strength indicator (RSSI) [18], the node calculates the distance between the message sending node and itself.

Each cluster of MC LEACH protocol has a cluster head and sensor nodes. The cluster setup details for MC LEACH are shown in Fig. 2. Here, each node is initiated with a specific energy level. When a cluster is formed, each sensor node broadcasts about its presence in the network and calculates distance between itself and sink. The calculated distance is Euclidian distance. The setup phase further works in twofolds, intra-cluster and inter-cluster.

Intra-cluster Setup Phase Once the cluster is formed and CH is selected, the remaining nodes of cluster must know about the CH. Hence, CH broadcasts a message in cluster about its identification. When this message is received by remaining nodes in cluster, they calculate the distance between themselves and CH with the parameter receive signal strength indicator (RSSI).

$$P_r(d) = C\frac{P_t}{d^{\alpha}} \tag{2}$$

where $P_r(d)$ is the received power by receiver at distance d expressed in dBm, C is proportionality constant whose value depends on communication model, P_t is the power used by transmitter to send the message, and α is called distance–power

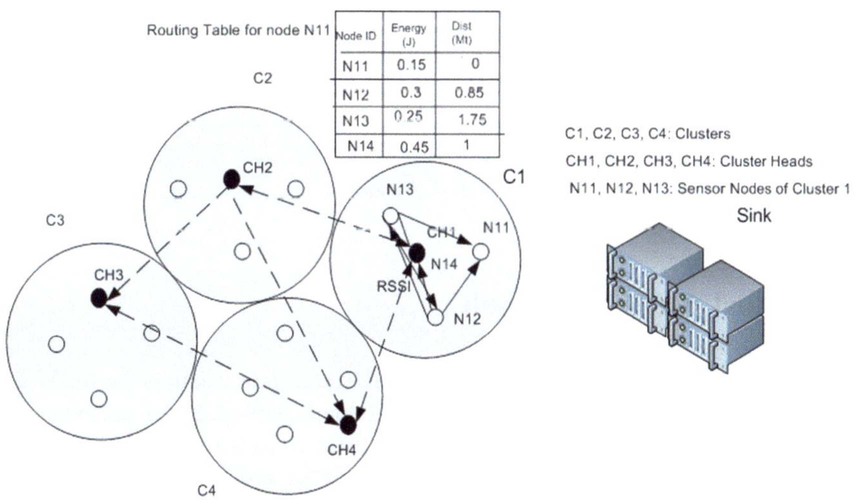

Fig. 2 Cluster setup in MC LEACH protocol

gradient. The broadcast signal also consists of the information of residual energy of sensor node. Hence, a routing table is formed at each node consisting of node id, residual energy, and distance between the nodes and itself. A cluster head is selected for a round on the basis of Eq. 1.

Inter-cluster Setup Phase Each CH transmits their initial route to sink to check whether any other CH is in its path to the sink. If it observes so, it updates its routing table for designing the path to the sink. Once it is done, each sends a message to all remaining CHs of different clusters of the network consisting of node ID and residual energy. Again with RSSI parameter, each CH designs a routing tree keeping itself as a root node and sink as an end node. This sorting is done by binary search algorithm (BSA). This routing tree results in the optimized path between each CH and the sink, and a chain is also formed to transmit the data from different CHs to the sink.

3.4 Steady-State Phase

The proposed algorithm is based on cross-layer optimization. Three layers, physical (PHY), media access control (MAC), and network (routing) layers, work in collaborative manner to fulfill the desired criteria. Once routing tree is formed, communication among the nodes starts.

All sensor nodes of cluster forward their sensed data to CH. CH then processes gathered data, and finally send to the sink. Instead, all CHs of network design the chain according to level of residual energy and distance to sink. The CH, whose residual energy and distance to the sink are least among all CHs available in network, will be the last node of the chain, i.e., nearest node to the sink and vice versa. If the energy level of two nodes is equal, then second parameter, i.e., distance, is the decision parameter. Here, to design an optimized chain route, PHY and MAC layers play their roles. This routing table as well as routing tree will also update in parallel with this calculation of designing the optimized chain (path) to the sink.

$$E_T = k^* E_{\text{cluster}} \tag{3}$$

4 Network Model and Result

Whole experimental setup is done in MATLAB R2017a. Here, a data-gathering WSN is considered, where *N* consistent sensors are randomly deployed in an area with the sink located at the center [19]. The network is experimented with a number of sensor nodes between 200 and 1000 to handle a thin sensor network and a very heavy one. The simulation is done for 3000 rounds. The remaining simulation parameters are summarized in Table 1. The path between the cluster head and sink is shown in Fig. 3.

Table 1 List of parameters considered for MC LEACH protocol

Parameters	Value
Sink location for network	$X_{bs} = Y_{bs} = 50$
Percentage of cluster heads (P) in network	0.1
E_0 (J)	0.5
E_{Tx1}, E_{Rx1} (mJ)	1
E_{Tx2}, E_{Rx2} (mJ)	0.1
E_{DA} (nJ/bit)	5
Maximum number of rounds	3000
E_{elec} (nJ/bit)	50
ε_{amp} (nJ/bit)	50
ε_{fs} (nJ/bit)	10

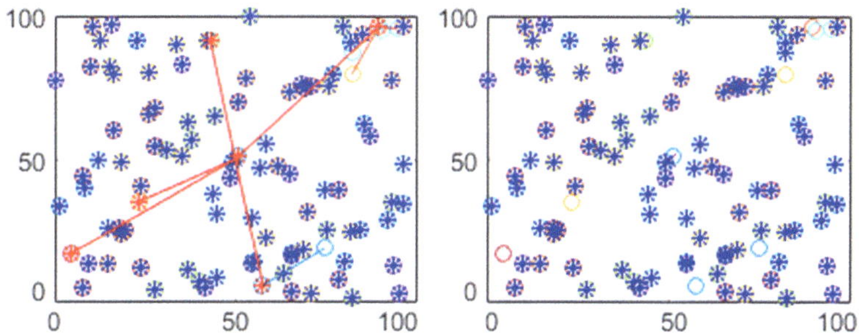

Fig. 3 Path between CH and sink for MC LEACH protocol

The analysis of this protocol is compared with low-energy adaptive clustering hierarchical (LEACH) protocol for residual energy and communication delay parameters (Fig. 4).

5 Conclusion and Future Scope

MC LEACH protocol has been simulated on MATLAB and analyzed. The output results of his protocol show that energy consumption gets affected in a positive manner if the number of hops toward the sink can be changed rather than using a single hop. The introduction of multi-hops communication has substantially improved LEACH by discovering the best path between the source CH and the destination. Simulation results in terms of dead nodes and residual energy. The proposed work can also be analyzed for the delay. Hence, the communication can be faster and can be implemented in delay-conscious applications.

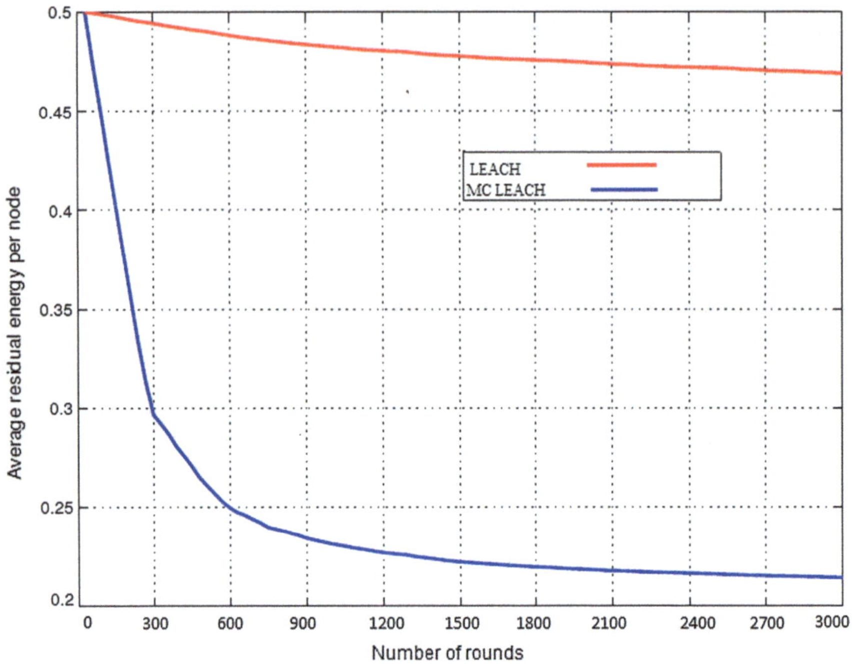

Fig. 4 Comparative analysis of LEACH and MC LEACH protocol for residual energy in the network

References

1. Wu, S., et al.: Delay-aware energy-efficient routing towards a path-fixed mobile sink in industrial wireless sensor networks. Sensors **3**, 899 (2018)
2. Jabbar, S., et al.: Analysis of factors affecting energy aware routing in wireless sensor network. Wirel. Commun. Mob. Comput. **1**, 1–21 (2018)
3. Jin, Z., et al.: A Q-learning-based delay-aware routing algorithm to extend the lifetime of underwater sensor networks. Sensors **17**(1660), 1–15 (2017)
4. Kang, M.W., Chung, Y.W.: A novel energy-aware routing protocol in intermittently connected delay-tolerant wireless sensor networks. Int. J. Distrib. Sens. Netw. **13**(7) (2015)
5. Al-Anbagi, I., et al.: A survey on cross-layer quality of service approaches in WSNs for delay and reliability-aware applications. IEEE Commun. Surv. Tutorials **18**(1), 525–552 (2014)
6. Khiati, M., Djenouri, D.: BOD-LEACH: broadcasting over duty-cycled radio using LEACH clustering for delay/power efficient dissimilation in wireless sensor networks. Int. J. Commun. Syst. **28**, 296–308 (2015)
7. Rao, Y., Deng, C., Zhao, G., Qiao, Y., Fu, L., Shao, X., Wang, R.: Self-adaptive implicit contention window adjustment mechanism for QoS optimization in wireless sensor networks. Elsevier J. Netw. Comput. Appl. **109**, 36–52 (2018)
8. Qaisar, S., Bilal, R.M., Iqbal, W., Naureen, M., Lee, S.: Compressive sensing: from theory to applications, a survey. J. Commun. Netw. **15**(1), 443–456 (2013)
9. Braman, A., Umapathi, G.R.: A comparative study on advances in LEACH routing protocol for wireless sensor networks: a survey. Int. J, Adv. Res. Comput. Commun. Eng. **3**(2), 5683–5690 (2014)

10. Khadivi, A., Shiva, M.: FTPASC: a fault tolerant power aware protocol with static clustering for wireless sensor networks. IEEE International Conference on Wireless and Mobile Computing, Networking and Communications, IEEE, 2006
11. Yarde, P., Srivastava, S., Garg, K.: A modified energy efficient protocol for optimization of dead nodes and energy consumption in wireless sensor networks. In: IEEE, 11th International Conference on Sensing Technology (ICST-2017), pp. 31–36, 2017
12. Akkaya, K., Younis, M.: A survey on routing protocols for wireless sensor networks. Ad hoc Netw. **3**(3), 325–349 (2005)
13. Sohraby, K., et al.: Protocols for self-organization of a wireless sensor network. IEEE Pers. Commun. **7**(5), 16–27 (2000)
14. Younis, M., Youssef, M., Arisha, K.: Energy-aware routing in cluster-based sensor networks. In: Proceedings of 10th IEEE/ACM International Symposium on Modeling, Analysis and Simulation of Computer and Telecommunication Systems (MASCOTS-2002), 2002
15. Yuan, Y., He, Z., Chen, M.: Virtual MIMO-based cross-layer design for wireless sensor networks. IEEE Trans. Veh. Technol. **55**(3), 856–864 (2006)
16. Wang, C., Li, B., Sohraby, K., Daneshmand, M., Hu, Y.: Upstream congestion control in wireless sensor networks through cross-layer optimization. IEEE J. Sel. Areas Commun. **25**(4) (2007)
17. Heinzelman, W., Chandrakasan, A., Balakrishnan, H.: An application-specific protocol architecture for wireless microsensor networks. IEEE Trans. Wirel. Commun. **1**(4), 660–670 (2002)
18. Zheng, J., Liu, Y., Fan, X., Li, F.: The study of RSSI in wireless sensor networks. In: 2nd International Conference on Artificial Intelligence and Industrial Engineering (AIIE2016), Advances in Intelligent Systems Research, vol. 133, pp. 207–209, 2016
19. Neto, J.H.D., Rego, A.D., Cardoso, A.R., Colestino, J.: MH-LEACH: a distributed algorithm for multi-hop communication in wireless sensor networks. In: ICN 2014: The Thirteenth International Conference on Networks, IARIA, pp. 55–61, 2014

PS-2 Controlled Four Wheel Base Drive Badminton Playing Robot

Shivam Mishra and V. K. Sachan

Abstract There are many researchers and engineers who aim to design such robots that can outplay the human players in the field of sports. A design of badminton playing robot is presented by this paper which can be controlled by the PS-2 (play station) controller, and it can also serve shuttlecock in the badminton court similar to a human player. It is done so as to incorporate advancement in the field of sports. It aims at designing a configuration which can assist the players in practicing badminton. A shuttlecock holding system is designed using pneumatics which can be preloaded with three shuttlecocks and can further be increased according to the convenience. A hitting mechanism was designed so that a standard badminton racket can be swung and can be used to hit the shuttlecocks. The problems and all the related solutions that were involved in the designing of this robot are discussed. The practical application of the design indicates that this robot is able to hold three shuttlecocks and can serve with high accuracy and is efficient enough to hit the returning shuttlecocks with high accuracy. This means that a robust robot is designed which can be controlled by any human being.

Keywords Badminton · Motor control · Mobile robot · PS-2 controller
Serving shuttlecock · Microcontroller

1 Introduction

Badminton is a racquet sport played using racquets to hit a shuttlecock across a net. Badminton is one of the most popular sports in the world even it is given priority in almost all the countries. Some of the researchers are currently working on several experiments and theories to make advancements in this field [1–4]. Bringing more and more technology in the field of sports will definitely increase the accuracy and will open up a lot of opportunities for the engineers and the researchers.

S. Mishra (✉) · V. K. Sachan
KIET Group of Institution, Ghaziabad, India
e-mail: mishra.shiva999@gmail.com

L. C. Jain et al. (eds.), *Data and Communication Networks*, Advances in Intelligent Systems and Computing 847, https://doi.org/10.1007/978-981-13-2254-9_9

A design is presented for the PS-2 controlled badminton playing robot. Several subsystems are integrated together to make the complete robot. The robot can play badminton similar to the human player; it can serve the shuttlecock with high accuracy and can also intercept the oncoming shuttlecocks from the opponent. Many robotic competitions like RoboCup is organized so as to motivate the engineers and researchers as well as to acknowledge their contribution and efforts in the field of robotics. Also, the sports robots can also be designed for the learning and teaching purposes [5]. A lot of experiments are going on in developing the robots which can play entertainment games [6].

Badminton is the most dynamic game as the maximum speed of the smash ever recorded is 332 kph in the 2005 Sudirman Cup in Beijing. A player needs very quick and accurate instincts to counter back such high-speed shuttlecocks. The robot should be able to predict the exact trajectory of the shuttlecock and should anticipate according to the prediction accurately so as to achieve the performance equivalent to that of human players.

Modern technologies provide many alternatives in the selection of drive system. It depends on the application whether the fixed speed drives are used or variable speed drive system is used. Here, variable speed drive system has been used to drive the robot efficiently according to the oncoming shuttlecock from the opponent player.

The integration of several subsystems of various engineering sciences to make the robust badminton playing robot is one of the challenges that are prominent. The most important part of the robot is the integration of mechanical design and hardware subsystems with the software subsystems with the help of embedded systems on various computing platforms. These prominent challenges provide opportunities to the researchers to do several experiments to enhance the technology [7].

The paper investigates all the parameters involved in completing this robot. Firstly, the construction of the mechanical part is defined and then the driving electronics circuitry and finally, the implementation of the programming part are defined [8]. As the acceleration and the velocities of the robot and the badminton rackets are very high, so the safety issues are also concerned. So, the preference is given to the safe technology materials as well as parameters. Also, it has future scopes and one can replace the traditional model with some advanced configurations.

2 Methodology

See Fig. 1.

2.1 Mechanical Structure

The structure is made up of aluminum rods with the support of nylon fiber. The structure stands on a four wheel base drive which is mounted with the DC motor.

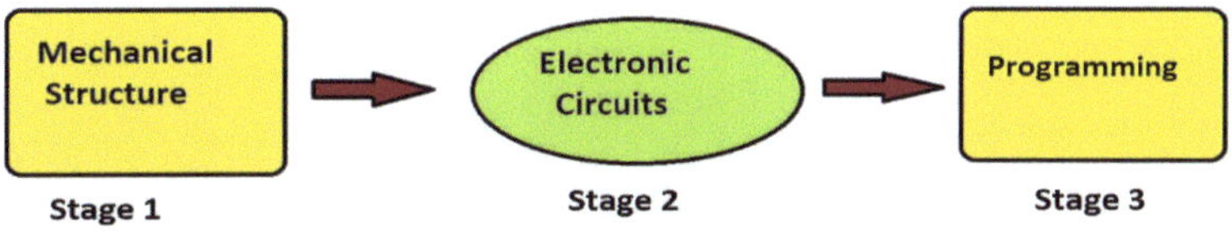

Fig. 1 Block diagram

The four wheel base drive is the combination of Omni wheels and aluminum rods. The effect is that the wheel can be driven with full force, but will also slide laterally with great ease. The linear actuation to the badminton rackets is provided with the help of pneumatic actuators. A PVC (Polyvinyl Chloride) pipe is used to make the shuttlecock holding mechanism.

2.1.1 Serving Mechanism

The serving mechanism consists of the shuttlecock holder and a racket for serving the shuttlecock. The shuttlecock holder is made with the help of PVC pipe. Three slots are made in the PVC pipe and small pneumatic actuators are mounted so as to hold the shuttlecocks in position. When the signal is provided from the transmitter then the lowermost pneumatic actuator releases the Shuttlecock and the serving racket hits the shuttlecock with the required force. The time lapse between releasing the shuttlecock and reaching to the serving racket is calculated and is implemented accordingly so that the racket does not miss the shuttlecock while serving (Fig. 2).

2.1.2 Four Wheel Base Drive

The base is made up of aluminum rods and Omni wheels. The Omni wheels are mounted with DC (Direct Current) motors of 800 rpm (rotation per minute) so as to move the robot throughout the standard badminton court effectively. The structure of the base has the strength to hold 60 kg of weight efficiently. The effect of the Omni wheels is such that the robot can be driven with full force and can also slide laterally with great ease (Fig. 3).

2.1.3 Hitting Mechanism

The hitting mechanism is made up of the pneumatic actuators. There are three pneumatic actuators which are mounted with the badminton rackets on the opposite side of the serving mechanism. These are used so as to provide the linear actuation to the badminton rackets. Using this linear actuation and the force of the pneumatic cylinder the badminton racket hits the shuttlecock. There are two rackets at the lower

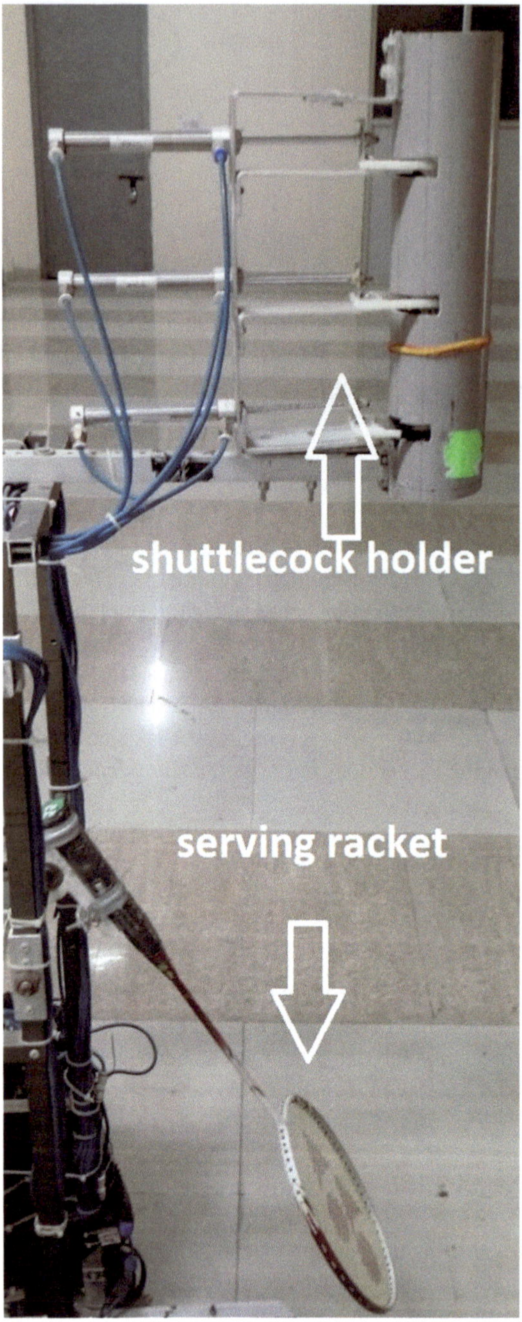

Fig. 2 Serving mechanism

Fig. 3 Four wheel base drive mounted with DC motors

side and one to make the upper hit. The robot has to be adjusted according to the oncoming shuttle by the controller.

The pneumatic actuators are those kinds of actuators which work using air pressure. The air pressure is filled in the black colored bottles placed on both sides of the robot. It is prescribed to fill the bottles up with 8.5 bars of air pressure using the air compressor. The bottles and the pneumatic actuators are connected using 4 mm pipes. Also, a pressure gauge is mounted so as to keep an eye on the air pressure (Fig. 4).

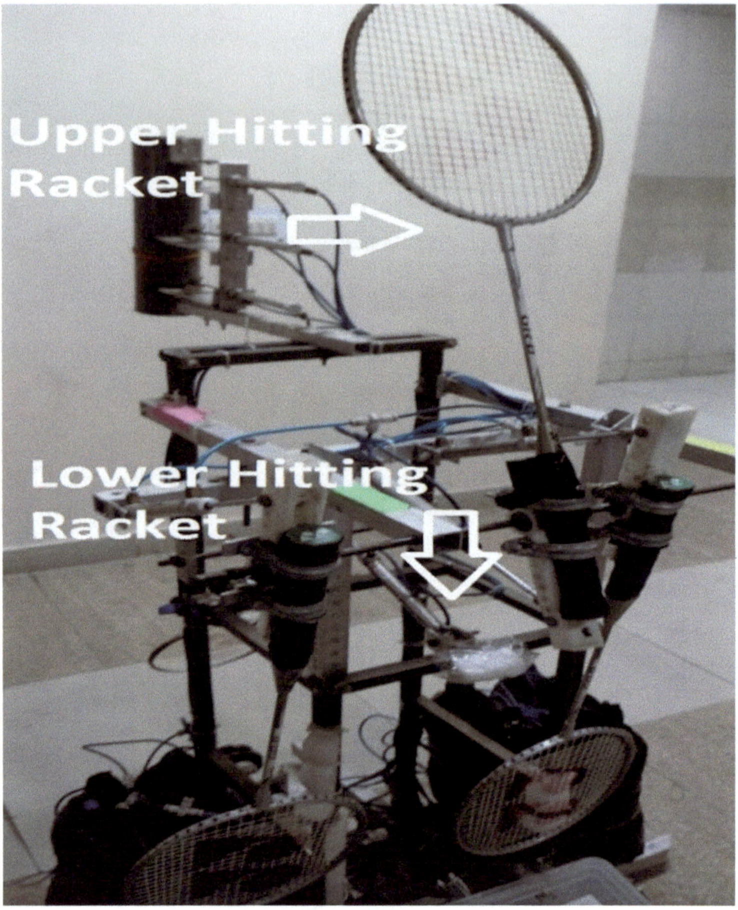

Fig. 4 Hitting mechanism with pneumatic actuators

2.1.4 CAD Design and Actual Robot

See Fig. 5.

2.2 Electronic Circuits

The electronic circuitry is the most important part of the robot which is used to drive the robot. The different parts of the electronics circuitry used in the robot are as follows:

Fig. 5 CAD design and actual robot

2.2.1 Motor Driver

The motor drivers are used to drive the DC motor. These motor drivers are based on the parallel MOSFET H-bridge configurations. It is used in parallel configuration so as to increase the current rating of the motor driver. These motor drivers are designed by us. There are four motor driver circuits used, one for each DC motor. When the signal is provided to the motor driver from the microcontroller, it incorporates motion in the robot and it can easily be controlled by the PS-2 controller (Fig. 6).

2.2.2 Solenoid to Switch Pneumatic Actuators

To direct the airflow into and out of the pneumatic actuators, solenoid valves are used. The pneumatic actuator position is remotely controlled by the solenoids according to the electrical signals provided to them. They have two states energized and de-energized. The position in which the pneumatic actuators will be in, is controlled by the different airflow patterns. Here, the solenoids are switched by providing 24 V dc supply and the switching is done with the help of MOSFET switching circuit and the signal is provided by the microcontroller (Fig. 7).

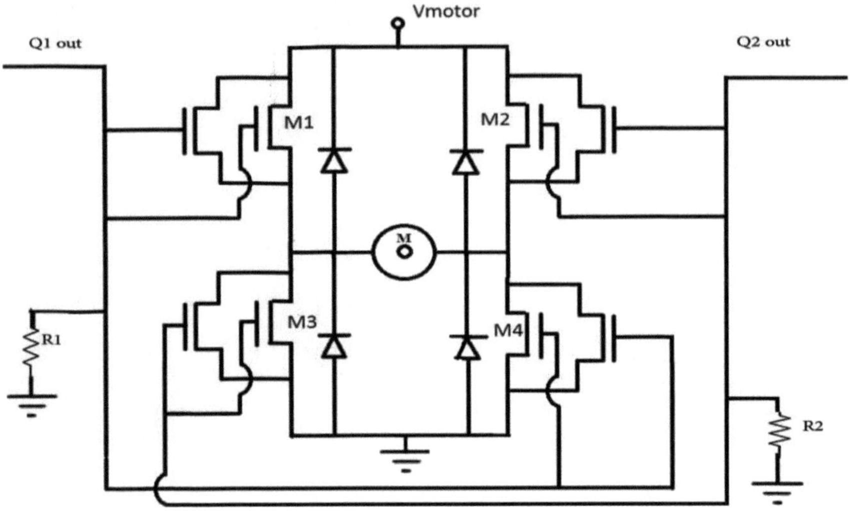

Fig. 6 Motor driver circuit

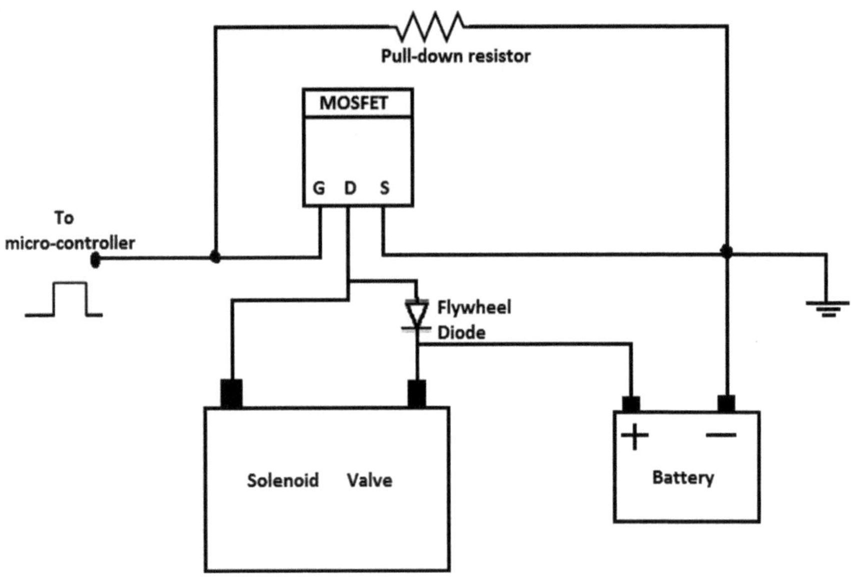

Fig. 7 Solenoid and it's switching circuit

2.3 Programming

Controlling the robot with the help of the PS-2 controller needs software interface. A PS-2 controller is interfaced with the Arduino microcontroller and the signals are transmitted to robot's receiver section wirelessly with the use of Bluetooth module which is also connected to the Arduino microcontroller both in the transmitter section to transmit the signal and to the receiver section so as to receive the signal. The programming section is basically divided into two sections as:

2.3.1 Transmitter Section

The transmitter section basically includes a PS-2 controller which is interfaced with the Arduino microcontroller, and the signal is transmitted wirelessly using the Bluetooth module. In the transmitter section, the Bluetooth module acts as the master and the microcontroller is also the master; i.e., it sends the signal to the robot. There are nine pins in the PS-2 as follows:

1. DATA: This is the data line from controller to PS2.
   This is an open collector output and requires a pull-up resistor (1–10 k). (A pull-up resistor is needed because the controller can only connect this line to ground; it cannot actually put voltage on the line).
2. COMMAND: This is the data line from PS2 to controller.
3. VIBRATION MOTOR POWER.
4. GND: Ground.
5. VCC: VCC can vary from 5 down to 3 V.
6. ATT: To get the attention of the controller, ATT is used. This line must be pulled low before each group of bytes is sent/received and then set high again.
7. CLK: 500 kH/z, normally high on. SPI bus communication appears.
8. Not Connected.
9. ACK: Acknowledge signal from Controller to PS2 [9] (Fig. 8).

2.3.2 Receiver Section

The receiver section includes the Arduino microcontroller which is also connected to the Bluetooth module, and this Bluetooth module acts as the slave. The receiver section receives the signal coming from the transmitter and incorporates the functionality into the robot accordingly (Fig. 9).

3 Block Diagram of Complete Operation

See Fig. 10.

Fig. 8 PS-2 interfaced with Arduino

Fig. 9 Receiver section

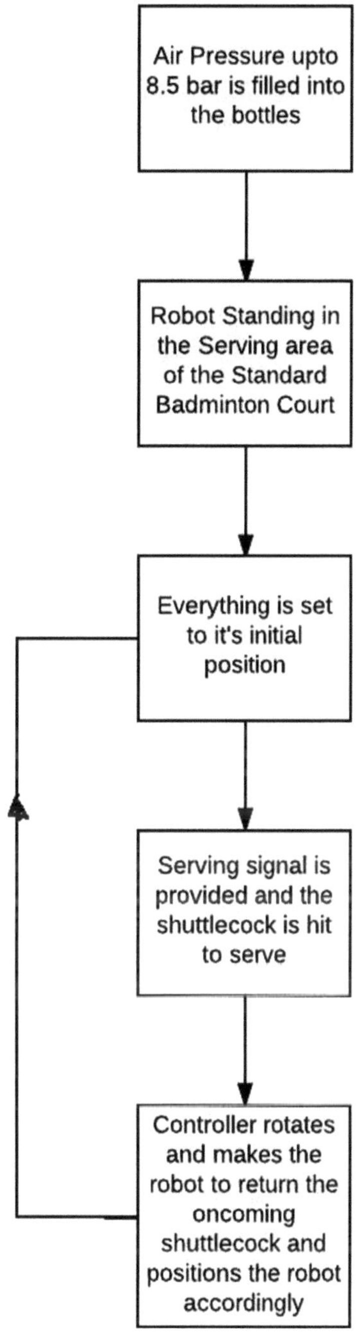

Fig. 10 Operation of the robot

4 Result

The challenge of holding the shuttlecock in the position is done with the help of pneumatic actuators and three pneumatic actuators are used to hold the three shuttlecocks. When the serving signal is provided to the receiver section then the lowermost pneumatic actuator retracts and the lowermost shuttlecock comes down and the badminton racket hits the shuttlecock to serve. Then after a delay of 3 s, the second shuttlecock comes to the lowermost position and the upper shuttlecock comes to the middle position and the topmost position gets empty so that the shuttlecock can be loaded again.

As the air pressure in the bottle keeps on decreasing with the number of hits, so the distance of the shuttlecock is also affected as shown in Table 1 which depicts the accuracy of the serving distance. The accuracy of serving is very high as, till the sixteenth hit the distance of shuttlecock is still more than the distance of the foul line. But after the sixteenth hit, the distance of the serving comes under the foul line and again air pressure is required to be filled into the bottles to again gain the accuracy to that level.

Table 1 Serving distance with the ascending hit number

Ascending hit number	Distance of shuttlecock (cm)	Foul distance (cm)
1	768	400
2	761	400
3	750	400
4	732	400
5	719	400
6	691	400
7	674	400
8	649	400
9	623	400
10	598	400
11	571	400
12	534	400
13	495	400
14	473	400
15	456	400
16	431	400
17	397	400

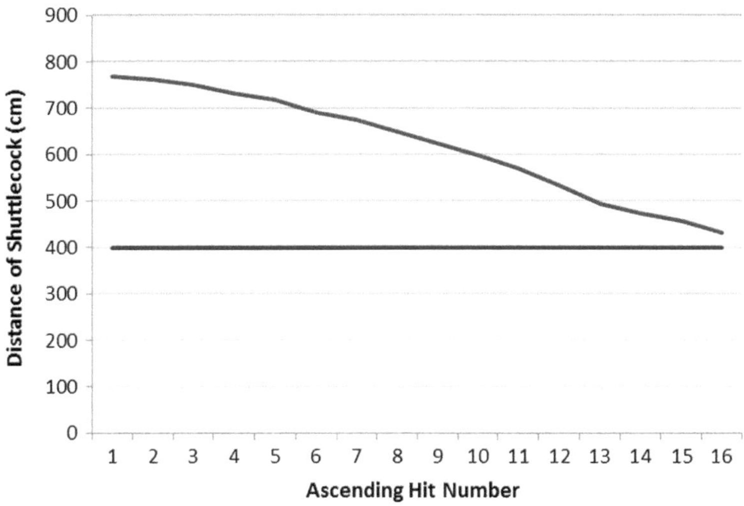

Fig. 11 Success rate of serving and distance versus ascending hit number

4.1 Success Rate of Serving and Hitting

Also, the distance of the shuttlecock varies on hitting it with varying forces which is very important during the game as the controller of the robot can throw shuttlecocks at different positions of the standard badminton court. So Table 2 depicts the distance of the shuttlecock with the varying forces (Figs. 11 and 12).

Table 2 Distance of shuttlecock with the varying force and the hit number

Force (lbs)	Distance of shuttlecock (cm)	Hit number
81.95	768	1
78.33	756	2
77.01	731	3
75.2	710	4
73.78	680	5
71.25	669	6
68.01	651	7
57.26	633	8
50.93	620	9
41.37	573	10
35.89	550	11
26.3	500	12
20.04	460	13
15.73	409	14

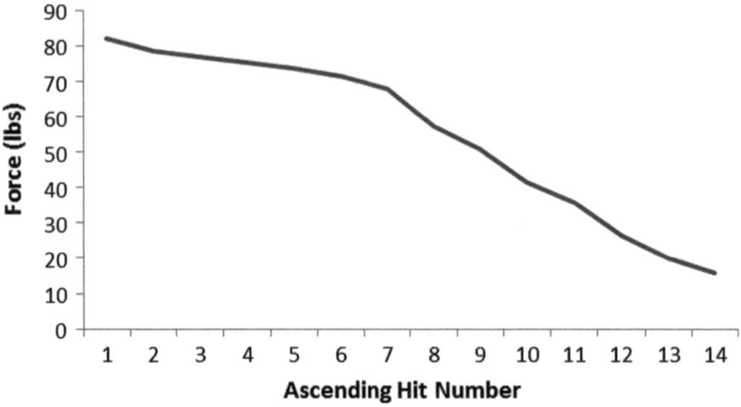

Fig. 12 Hit number versus force of hitting

5 Conclusion

The serving performance and the design of the robot are discussed and analyzed in this paper. The calculation of the time lapse between releasing the shuttlecock from the shuttlecock holder and reaching down to the serving badminton racket is crucial. In this design, the robot is able to serve three shuttlecocks in one loading with a very high accuracy. During service, the distance of the shuttlecock keeps decreasing due to decrement in the air pressure but up to sixteen hits it can easily serve to a distance more than the distance of the foul line, i.e., a valid serve in crosscourt. The maximum and the average distances are 768 and 620.3 cm, respectively. The other three badminton rackets are used to hit the oncoming shuttlecocks from the opponent to the opposite court. It depends on the controller that how effectively the controller can control the robot and position it to hit the oncoming shuttlecock. The DC motors used in the base are of 800 rpm, and it makes the motions of the robot very fast. The structure of the robot is very stable and it will not topple if stopped suddenly.

6 Future Work

More advancement can be done in this robot by incorporating vision system which can trace the trajectory of the shuttlecock as well as the speed of the oncoming shuttlecock. Also, the 3D view of the standard badminton court can be created so as to trace the shuttlecock. For this purpose, we can use LIDAR (Light Detection and Ranging). This can enhance the accuracy of the robot and can also help in making this robot autonomous. Also, the adaptive and fuzzy control system can be incorporated to enhance the performance of the robot.

References

1. Arenas, M., Ruiz-del-Solar, J., Norambuena, S., Cubillos, S.: A robot referee for robot soccer. In: Iocchi, L., Matsubara, H., Weitzenfeld, A., Zhou, C. (eds.) RoboCup 2008: Robot Soccer World Cup XII, vol. 5399, pp. 426–438. Springer, Berlin (2009)
2. Chen, X., Tian, Y., Huang, Q., Zhang, W., Yu, Z.: Dynamic model based ball trajectory prediction for a robot ping-pong player. In: IEEE International Conference on Paper presented at the Robotics and Biomimetics (ROBIO), 2010
3. Yang, H.: Robot football goalkeeper speed control research based on BP neural network algorithm. J. Chem. Pharm. Res. 5(12) (2013)
4. Zitzewitz, J.V., Rauter, G., Steiner, R., Brunschweiler, A., Riener, R.: A versatile wire robot concept as a haptic interface for sport simulation. In: IEEE International Conference Paper presented at the Robotics and Automation ICRA '09, 2009
5. Zhang, J.: Autonomous mobile robot relay race competition. Adv. Sport Sci. Comput. Sci. **57**, 41 (2014)
6. Laue, T., Birbach, O., Hammer, T., Frese, U.: An entertainment robot for playing interactive ball games. In: Behnke, S., Veloso, M., Visser, A., Xiong, R. (eds.) RoboCup 2013: Robot World Cup XVII, vol. 8371, pp. 171–182. Springer, Berlin (2014)
7. Stoev, J., Bartic, A., Gillijns, S., Symens, W.: Badminton playing robot-a multidisciplinary test case in mechatronics. Paper presented at the Mechatronic Systems (2010)
8. Chia, K.S., Yap, X.Y., Low, E.S.: A badminton robot-serving operation design. ARPN J. Eng. Appl. Sci. 11(6), 3968–3974 (2016)
9. Bill Porter: http://www.billporter.info/2010/06/05/playstation-2-controller-arduino-library-v1-0/ (2010)

Detection of RREQ Flooding Attacks in MANETs

B. Nithya, Aishwarya Nair and A. S. Sreelakshmi

Abstract Mobile Ad hoc Networks (MANETs) are prone to vulnerabilities as a result of lack of centralized management, dynamic topologies, and predefined boundary. There are many attacks that affect the performance of MANETs. Flooding attack is a Denial of Service attack, also known as DoS attack that brings down the network by flooding with a huge amount of route request (RREQ) packets, routed to a destination that does not exist. This paper proposes a robust detection system; a Fuzzy-based Flooding Attack Detection System (FFADS) using a first-order Mamdani-type fuzzy inference system to detect RREQ flooding attacks. This proposed system uses network-specific parameters rather than node-specific factors. Comprehensive simulations are carried out to observe the variations of input parameters with respect to the flooding level of the system. The simulation results reveal that any deviation from the normal behavior of nodes is immediately detected by the proposed system.

Keywords MANET · Flooding attacks · Fuzzy logic · RREQ flooding · AODV Network traffic

1 Introduction

Distributed DoS attack is an attack which makes a service unobtainable to users, usually by interrupting or suspending the services temporarily. In an RREQ flooding attack, a malevolent node floods the network by transmitting fake RREQs to non-existent nodes. This leads to congestion, thus resulting in a Denial of Service. It is very

B. Nithya (✉) · A. Nair · A. S. Sreelakshmi
Department of Computer Science and Engineering, National Institute of Technology, Tiruchirappalli, Tiruchirappalli, India
e-mail: nithya@nitt.edu

A. Nair
e-mail: aishwarya05nair@gmail.com

A. S. Sreelakshmi
e-mail: sreelakshmias97@gmail.com

L. C. Jain et al. (eds.), *Data and Communication Networks*, Advances in Intelligent Systems and Computing 847, https://doi.org/10.1007/978-981-13-2254-9_10

important to catch and prevent flooding attacks as these attacks result in congestion of the network. Overflow of routing table in the intermediate nodes is caused by RREQ flooding attacks which disable the nodes from receiving new RREQ packets, resulting in Denial of Service attack. Furthermore, forwarding these bogus RREQ packets consumes valuable node resources like energy and bandwidth.

Many solutions have been put forward to detect and avert flooding attacks in a network including behavioral approach which uses machine learning to classify nodes as malicious, statistical approach to prevent flooding in AODV networks, game theory-based approach, dynamic-based profile approach, and fuzzy logic. The main drawback faced by many of these approaches is that they restrict the number of attacker nodes and detect the maliciousness of the node, fix other parameters in the network, or work effectively only for smaller networks. Many surveys have also been conducted that analyze the effect of this attack on the different network parameters. Most of this research focuses on the nodes of the network and on classifying them as malicious and non-malicious, rather than on how the communication within the network is affected. Also, the dynamic nature of MANETs is often ignored in the development of solutions to detect the attacks on the network.

This paper introduces a system called Fuzzy-based Flooding Attack Detection System (FFADS), to catch RREQ flooding attacks especially in AODV-based networks using the first-order Mamdani-type fuzzy inference system. The solution focuses on the use of parameters that are specific to the network rather than the nodes. To detect the level of flooding attack in the system, the proposed system uses the dynamic input parameters, which are the properties of the network such as throughput, packet loss ratio, and routing overhead. While detecting the level of flooding attack, the proposed FFADS also detects abnormalities in communication within the network which may or may not be due to RREQ flooding. By stopping the malicious node from sending packets, the network can be restored to its previous state with increased throughput and reduced congestion.

This paper is organized into five sections. Section 2 deals with related works besides their detailed comparison. Section 3 deals with our proposed system for detection of flooding attacks. This includes an analysis of the different input parameters, the structure of the fuzzy inference system, and the conversion of the output value of the fuzzy system to the level of flooding in the network. Section 4 deals with performance analysis. It explains the simulation setup, the metrics involved in calculations, and graphs depicting the variations of input parameters with the number of flooding nodes. Section 5 presents the conclusion of this paper.

2 Related Works

There are no fixed topologies in a MANET because the topologies are dynamic in nature. The number of nodes and their relative positions vary greatly, and hence, detection of the number of malicious nodes becomes difficult. Various methods are suggested to detect the presence of flooding nodes in MANET. Some of them are discussed in the following paragraphs.

A Fuzzy-Based Trust Model (FBTM) is proposed [1] to disclose selfish nodes in MANETs. The fuzzy-based analyzer is joined with a trust model to identify the non-cooperative behavior of attacker nodes from the non-attacker nodes. In this scheme, every node in the network continuously checks its nearby nodes for their activities. Every node calculates a value called the trust value of its neighbors, and these values are then passed as input to a fuzzy function. The fuzzy function calculates the net trust value, and the attacker nodes are then detected based on this value.

A Novel Intrusion Detection System (NIDS) for ad hoc network is suggested in [2] using fuzzy logic. The first-order Sugeno-type fuzzy inference system (FIS) is adopted for intrusion detection in a mobile ad hoc network. In this system, different parameters are calculated from the network and fed to the FIS. The FIS returns the verity level for each node. This method is accurate in terms of a high true positive rate and low false positive rate, but requires the number of attacker and non-attacker nodes to remain constant.

The behavioral approach is adopted in [3] to detect malicious attacks. In this system, SVM classifier is used to classify nodes as malicious and non-malicious.

The statistical approach is used in [4] to defend against RREQ flooding attacks in MANETs. The RREQ packets are monitored in real time, and the nodes are compared against their neighbors to check if they exceed default route request limit.

Alleviating route request flooding attack in MANET is suggested in [5] using node reputation scheme to abide by the impact of flooding attack in MANET. This scheme checks the status of a node intermittently and restricts its route request transmitting rate accordingly.

The detection of SYN flooding attack in AODV protocol is performed in [6] to disclose the presence of the SYN flooding attack. The scheme proposed uses game theoretical approach to form a game between the attacker node and the multimedia server node. The performance of the detection algorithm is measured by examining the several qualities of the parameters.

Ad hoc flooding attack is analyzed, and a Flooding Attack Prevention (FAP) mechanism is developed in [7]. The proposed scheme uses a trust function to record the number of route request packets and calculates the trust value. If the calculated trust value exceeds the limit, the network drops these route request packets. In [8], throughput, packet delivery ratio, and round-trip delay are compared with normal network (without attacker nodes) and a network with few attacker nodes. The performance of the network is compared in all the three scenarios.

Dynamic Profile-Based Technique (DPBT) is proposed [9] to detect the flooding attacks in MANET. The proposed scheme defines a profile value based on the performance of MANET. It recognizes the attack and tries to stop it every time the node attempts to exceed the threshold value. This value changes with respect to the request placed by the network.

A flow-based discovery mechanism against flooding attacks is developed in [10]. Two flow-based detection features are designed, and the algorithm used on them precisely detects flooding attacks.

The above-discussed methods use a number of methods to detect the presence of flooding nodes in the system. Most researches concentrate exclusively on either detecting the presence of the malicious nodes or on being aware of the presence of these nodes and mitigating the effects of the attack. The main difficulties faced are due to the unpredictable nature of a MANET, in terms of number of harmless nodes, mischievous nodes, network topology, and mobility pattern. Most studies tend to restrict one or more of the above parameters to compute the result. Motivated by these factors, our paper tries to vary as many parameters as possible, while maintaining a maximum level of accuracy. Our paper uses fuzzy-based approach to detect the level of severity of flooding attack in a network independent of the nodes in the network, topology of the network, or the percentage of flooding nodes.

3 Proposed Fuzzy-Based Flooding Attack Detection System (FFADS)

The proposed Fuzzy-based Flooding Attack Detection System (FFADS) uses fuzzy logic to detect the number of flooding nodes in a MANET. Fuzzy logic is beneficial in this situation as it can handle uncertainties and make decisions in a given range. This paper uses three-input, single-output-based first-order fuzzy Mamdani inference system for composing the decision. The fuzzy parameters are elicited from the network traffic and then passed on to the fuzzy interface. In the fuzzy interface, the fuzzy rules are applied and the number of flooding nodes is estimated.

3.1 Fuzzy Controller in the Proposed System

A MANET network is simulated with an arbitrary number of nodes and various parameters of the network—routing overhead, throughput, and packet loss ratio are extracted from the traffic. These parameters are then fed to fuzzy inference system as shown in Fig. 1, which uses a set of rules to define the output. This can be reported by the system in terms of the extent of flooding attack in the system, which is classified as low, medium, and high.

From the simulated MANET environment, the parameters—throughput, packet loss ratio, and routing overhead—to be used in the intrusion detection system are extracted. The inputs are then fed to the fuzzy controller. The fuzzification module converts input data to values of the membership functions and matches data with conditional rules in the rule base. The Mamdani-type fuzzy inference engine applies the rules and returns a fuzzy set for the defuzzification block. The output values are converted to crisp values through the defuzzification module. The output values are used to decide the extent of flooding attack in the system. The fuzzy output is

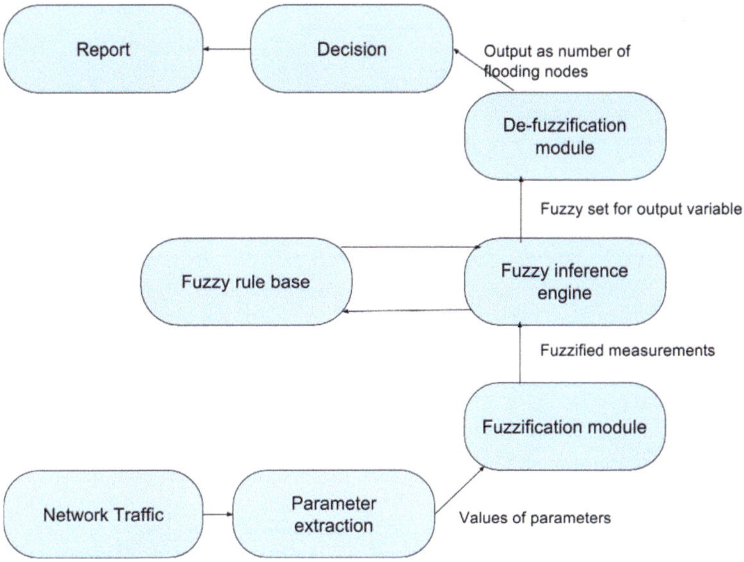

Fig. 1 Fuzzy inference system

generated within the range of [0, 1] which indicates the flooding level where 0–0.3 is classified as low, 0.3–0.7 is classified as medium, and 0.7–1 is classified as high.

3.2 *Fuzzy Inputs*

Routing Overhead: Routing overhead is determined as the ratio of route request packets (RREQ packets) among the total number of packets sent. It is observed that, as the number of flooding nodes increases, the number of route request packets relative to the total number of packets sent increases. Since routing overhead depends on the number of RREQ packets, the value will dynamically change depending on whether the bandwidth is utilized by RREQ packets. The membership function of this fuzzy input variable is depicted in Fig. 2.

$$\text{Routing overhead} = \frac{\text{Number of Route request packets sent}}{\text{Total number of packets sent}}$$

Packet Loss Ratio: Packet loss ratio is the ratio of the number of route request packets dropped relative to the total number of RREQ packets sent. It is observed that, as the number of flooding nodes increases, the packet loss ratio increases as the packets have to share bandwidth with the increasing number of route request packets. In a flooding attack, fake RREQs are sent to a destination that does not exist. Therefore, the fake RREQs are dropped since the node is unable to forward the

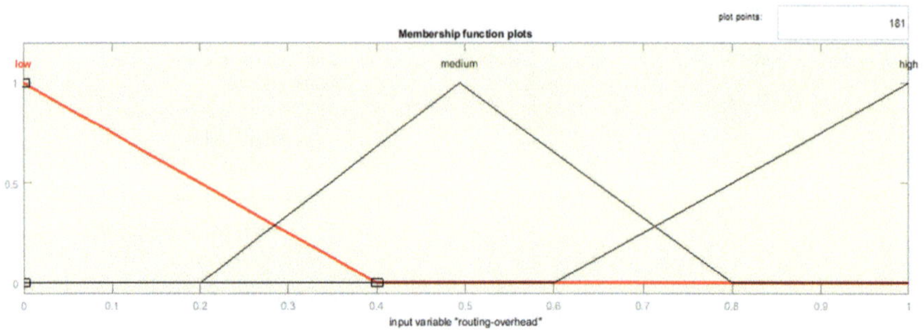

Fig. 2 Routing overhead membership function

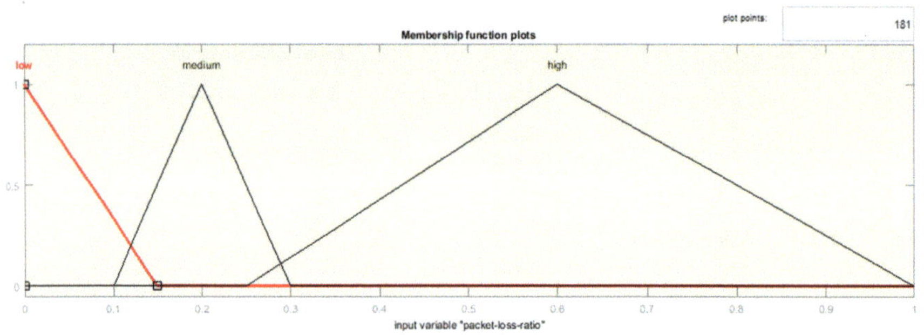

Fig. 3 Packet loss ratio membership function

RREQ packet. This increases packet loss ratio. The membership function is depicted in Fig. 3.

$$\text{Packet loss ratio} = \frac{\text{Number of RREQ packets dropped}}{\text{Number of RREQ packets sent}}$$

Throughput: In this simulation, throughput is determined as the fraction of TCP/UDP packets received at the destination to the sent packets. It is observed that, as the number of flooding nodes increases, the throughput decreases as the TCP/UDP packets are forced to share bandwidth with the increasing route request packets. As the number of fake RREQ increases, the channel experiences congestion and throughput gradually reduces. The membership function of this fuzzy input variable is depicted in Fig. 4.

$$\text{Throughput} = \frac{\text{Number of data packets}}{\text{Time}}$$

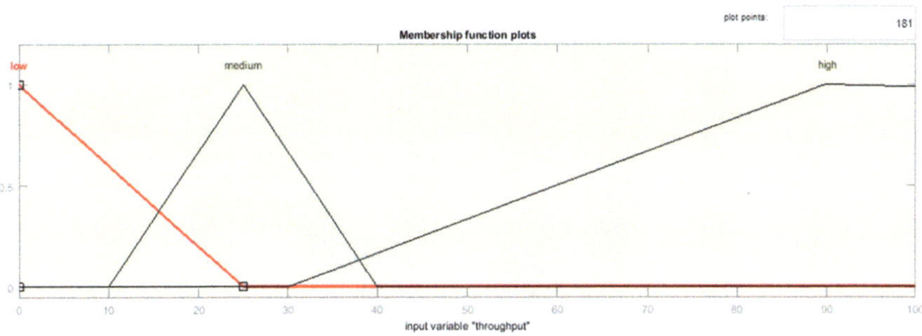

Fig. 4 Throughput membership function

Table 1 Fuzzy inference rules

Rule number	Routing overhead	Packet loss ratio	Throughput	Flooding level
1	Low	Low	High	Low
2	High	High	Low	High
3	Medium	Medium	Medium	Medium
4	High	High	Medium	High
5	Low	Low	Medium	Low
6	Low	Medium	High	Low
7	High	Medium	Medium	High
8	Medium	Medium	High	Medium
9	High	Medium	Low	High

3.3 Fuzzy Inference Rules

This detection system uses Mamdani-type inference system and takes in three parameters as input values—routing overhead, packet loss ratio, and throughput. The rule base considered in the proposed FFADS is given in Table 1.

3.4 Fuzzy Surface View

The surface view between two input parameters and the output parameter (flooding level) is shown in Figs. 5, 6, and 7. Different colors represent the severity of flooding. Figure 5 shows the output surface (flooding level) versus the two parameters—routing overhead and packet loss ratio. Figure 6 depicts the output surface (flooding level) versus the two parameters—routing overhead and throughput. The output surface (flooding level) versus the two parameters—throughput and packet loss ratio—is plotted in Fig. 7. The different colors of the surface graph—blue, green

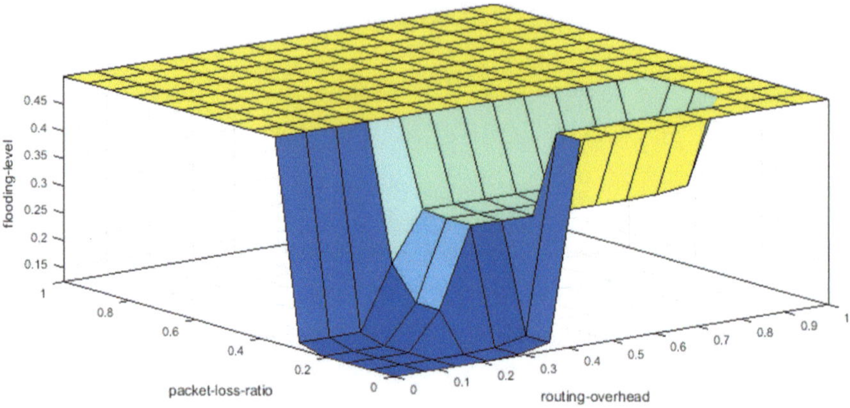

Fig. 5 Routing overhead, packet loss ratio

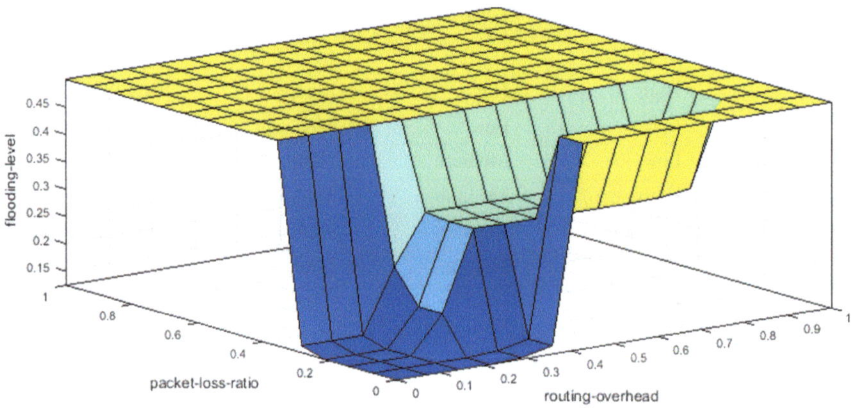

Fig. 6 Packet loss ratio, routing overhead

and yellow—represent the different levels of the output for the two given inputs, that is, the different intensities of flooding attack.

3.5 *Fuzzy Output*

The controller takes in the above-mentioned fuzzy inputs and maps them to their membership functions. This is then fed to the fuzzy rules. The obtained truth values are then defuzzified. The output values represent the extent of flooding attack in the system as shown in Fig. 8.

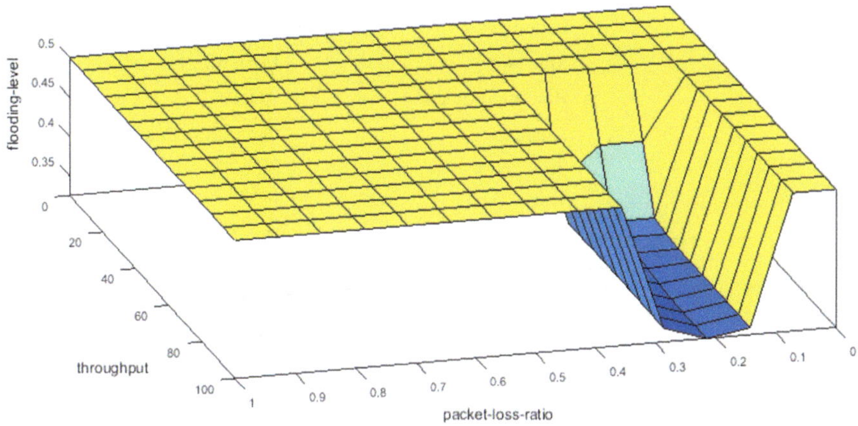

Fig. 7 Throughput, packet loss ratio

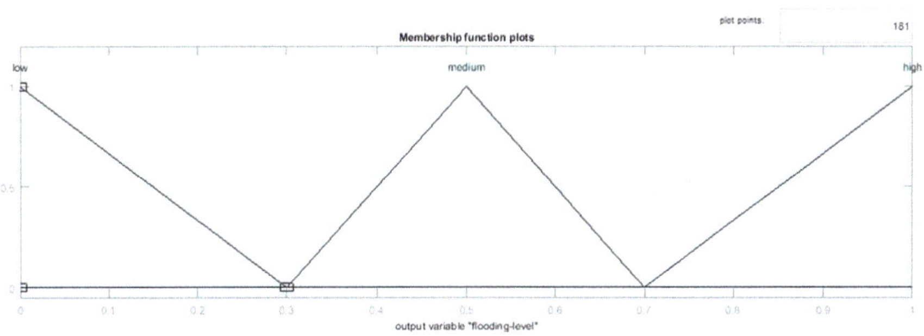

Fig. 8 Fuzzy output

4 Performance Analysis

This section briefs about the simulation setup and the performance analysis depending on the routing overhead, packet loss ratio, and throughput.

4.1 Simulation Setup

For this system, Ns2 simulator is used to simulate MANET. The simulation parameters are shown in Table 2. The parameters are extracted from the trace files and are fed to the Fuzzy Toolbox in MATLAB.

Table 2 Simulation parameters

Parameter	Value
Mac type	IEEE 802.11
Routing protocol	AODV
Antenna	Omni-directional
Simulation time	20 s
Traffic type	FTP
Topology	Random
Simulation area	500 m × 500 m
Number of nodes	Random

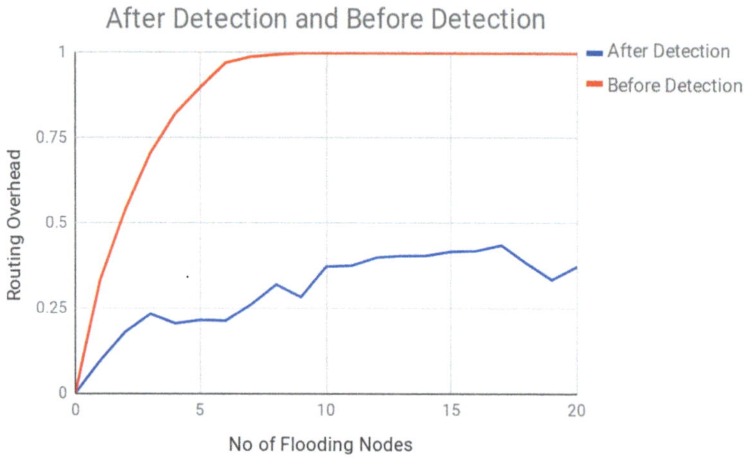

Fig. 9 Routing overhead versus number of flooding nodes

4.2 Metrics

In the proposed system, the following three parameters are considered to affect the level of flooding attack in the system.

Routing Overhead: Routing overhead is observed to increase with the extent of flooding attack in the system. The graph plotted in Fig. 9 shows a setup consisting of 25 nodes and different number of flooding nodes ranging from 0 to 20. It is observed that routing overhead increases linearly till a certain point beyond which it is observed to be constant. After detection of flooding attack, if the flooding node is correctly identified and the RREQ packets sent by the flooding node are ignored by all the other nodes, the effect of flooding attack is greatly reduced.

Packet Loss Ratio: It is observed from Fig. 10 that there is gradual increase in packet loss ratio with the increase in number of flooding nodes. Detection of flooding attacks is simulated by dropping any RREQ packets with unknown destination. The number of RREQ packets dropped is the same in both cases. Since packet loss ratio

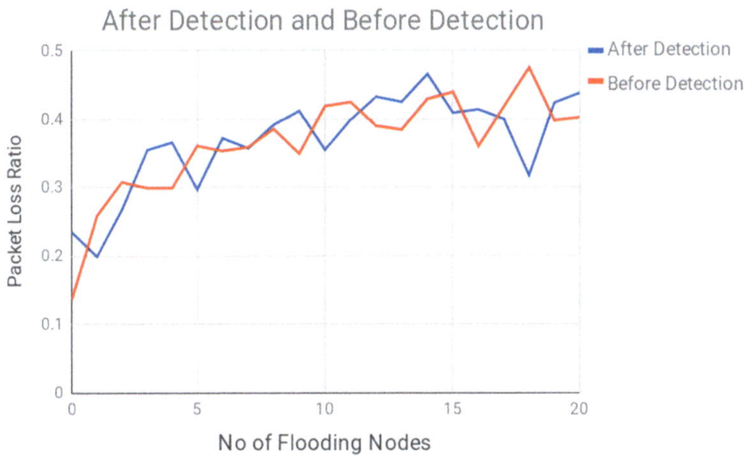

Fig. 10 Packet loss ratio versus number of flooding nodes

is the number of RREQ packets dropped, the packet loss ratio remains almost the same. Figure 10 shows that the RREQ packet loss ratio before and after detection.

Throughput: It is inferred from Fig. 11 that the throughput decreases as the number of flooding nodes increases. Since the fake RREQ packets utilize the bandwidth, the overall throughput decreases. It is observed that throughput increases significantly after detection of flooding attack as the entire bandwidth can now be used by the data communication taking place. The RREQ packets sent by the flooding node are dropped by all other nodes as long as it continues flooding the medium. Figure 11 depicts the variation of throughput with number of flooding nodes before and after detection of the flooding attack.

Predicted Flooding Level: The graph plotted in Fig. 12 depicts the flooding level of the network which consists of nodes ranging from 0 to 30. We have taken different setups with different number of nodes and measured the input parameters. It is observed that the predicted flooding level increases with the increase in number of flooding nodes. It is also observed that the predicted flooding level is independent of number of nodes, simulation time, and number of connections.

5 Conclusion

In this paper, a detection system—Fuzzy-based Flooding Attack Detection System (FFADS)—is proposed for detecting flooding attacks using fuzzy logic. The proposed system predicts the flooding level using three input parameters—routing overhead, throughput, and packet loss ratio. The flooding level is independent of the number of nodes, simulation time, and number of connections and depends only on the externally measured parameters. The system has an edge over existing systems that

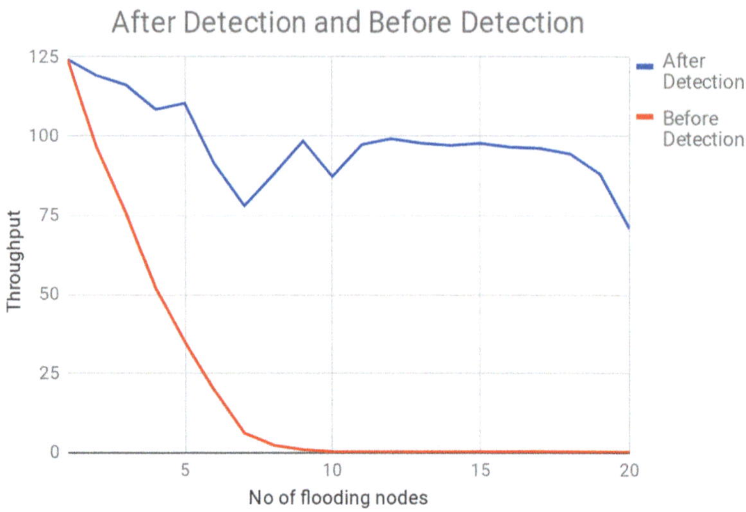

Fig. 11 Throughput versus number of flooding nodes

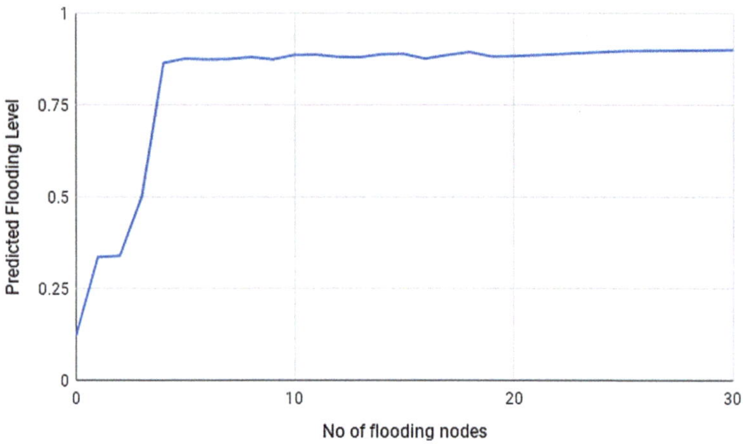

Fig. 12 Predicted flooding level versus number of flooding nodes

detect intrusions as it is dynamic in nature and the predicted flooding level changes with the performance of the network. Also since the system does not restrict the size or topology of the network, it can be applied on real-world networks to detect flooding attacks.

References

1. Ullah, Z., Khan, M.S., Ahmed, I., Javaid, N., Khan, M.I.: Fuzzy-based trust model for detection of selfish nodes in MANETs. In: 2016 IEEE 30th International Conference on Advanced Information Networking and Applications (AINA), Crans-Montana, pp. 965–972 (2016)
2. Chaudhary, A., Tiwari, V., Kumar, A.: A novel intrusion detection system for ad hoc flooding attack using fuzzy logic in mobile ad hoc networks. In: International Conference on Recent Advances and Innovations in Engineering (ICRAIE-2014), Jaipur, pp. 1–4 (2014)
3. Patel, M., Sharma, S.: Detection of malicious attack in MANET a behavioral approach. In: 2013 3rd IEEE International Advance Computing Conference (IACC), Ghaziabad, pp. 388–393 (2013)
4. Rmayti, M., Begriche, Y., Khatoun, R., Khoukhi, L., Gaiti, D.: Flooding attacks detection in MANETs. In: 2015 International Conference on Cyber Security of Smart Cities, Industrial Control System and Communications (SSIC), Shanghai, pp. 1–6 (2015)
5. Choudhury, P., Nandi, S., Pal, A., Debnath, N.C.: Mitigating route request flooding attack in MANET using node reputation. In: IEEE 10th International Conference on Industrial Informatics, Beijing, pp. 1010–1015 (2012)
6. Geetha, K., Sreenath, N.: Detection of SYN flooding attack in mobile ad hoc networks with AODV protocol. Arab. J. Sci. Eng. **41**, 1161 (2016)
7. Chouhan, N.S., Yadav, S.: Flooding attacks prevention in MANET. Int. J. Comput. Technol. Electron. Eng. (IJCTEE) **1**(3) (2013)
8. Bhalodiya, S., Balakrishnan, V., Chauhan, R., Chouhan, N.S., Fitri, M.H., Ghonge, M.M., Hong, F., Jawandhiya, P.M., Joshi, K., Lomte, V., Murthy, S.R., Pushpalatha, M., Rai, M.K., Song, J., Tupakula, U., Vaghela, K., Varadharajan, V., Venkataraman, R.: Enhanced detection and recovery from flooding attack in MANETs using AODV routing protocol (2015)
9. Sathish, T., Sasikala, E.: Dynamic profile based technique to detect flooding attack in MANET. Int. J. Innov. Res. Comput. Commun. Eng. **2**(1) (2014)
10. Guo, Y., Gordon, S., Perreau, S.: A flow based detection mechanism against flooding attacks in mobile ad hoc networks. In: 2007 IEEE Wireless Communications and Networking Conference, Kowloon, pp. 3105–3110 (2007)

Selection of Software Development Model Using TOPSIS Methodology

Dayanand Gaur and Sakshi Aggarwal

Abstract In software industry, the software project failure is a serious concern for stakeholders. The company suffers a huge financial loss in a year due to the team's negligence and irresponsibility. It can happen because of mismanagement in decision making at any stage of project development. But the study conducted so far cites several causes such as unarticulated project goals, mishandling of requirements, poor estimation of resources, sloppy software development life-cycle model. The paper tries to reduce the effort of decision-makers and project team by outlining the significance of TOPSIS model. The statistical and quantitative analysis is the main feature of TOPSIS. It accomplishes the experts' job by validating their opinions. It prioritizes the defined options after evaluating them against confliction and multiple attributes. The research proposes a decision-making framework for the selection of software development model using one of the widely accepted multicriteria decision-making tools, i.e., TOPSIS.

Keywords Software engineering · MCDM · Decision-making methods · TOPSIS

1 Introduction

1.1 Software Engineering

The traditional concept of software engineering, one of the prominent disciplines of computer science, merely defines the detailed and systematic study to the development of software [1, 2]. But in the past 15 years, the emerging trends and technologies, diversities in customer requirements, market demand, and much more complied the

D. Gaur
Computer Science, Galgotias University, Greater Noida 201310, India
e-mail: er.daya2014@gmail.com

S. Aggarwal (✉)
Software Engineering, Galgotias University, Greater Noida 201310, India
e-mail: agg.sakshi77@gmail.com

L. C. Jain et al. (eds.), *Data and Communication Networks*, Advances in Intelligent Systems and Computing 847, https://doi.org/10.1007/978-981-13-2254-9_11

engineers to change their perception for software engineering. The term does not remain only for the software development practices like elicitation, designing, modeling, codification, debugging, maintenance, and documentation but data collection, decision support system, knowledge discovery are some of the aspects also incorporated with the software engineering. Now, data accumulation, gaining experiences, knowledge extraction, building statistical systems go hand-in-hand with software development activities.

Domain experts, researchers, engineers, client, trade union involvement, and those who directly or indirectly influence the development process have become a part of the software engineering. All of them constitute as *stakeholders* [1]. They take every promising step in order to make a successful project as the success or failure of software is the responsibility of whole team. But apart from these entire endeavors, the studies conducted by the researchers over the past decade state that the failure rate of software has crossed the threshold. They identified certain causes of failure [3] which are enumerated as (1) unarticulated project goals, (2) unclear requirements, (3) sloppy software development models, (4) inaccurate estimation of resources, and (5) poor communication among customers and developers.

One of the reasons they cited is the wrong selection of software development model (SDM). The paper underlines this problem as a subject of study and proposes a solution using TOPSIS methodology—an operative approach of multicriteria decision making (MCDM) [4].

1.2 Basics of Multicriteria Decision Making (MCDM)

MCDM is a qualitative and quantitative technique for making effective decisions when multiple and conflicting parameters are taken into account [5]. It offers experts an opportunity to utilize its significance in the areas where decision making is considered as tedious job. The wide variety of MCDM methods provides framework which selects the best suited alternative after evaluating different criteria, defined under a particular scenario [6, 7]. The paper focuses on the application of MCDM approach in software engineering field.

The whole research is divided into several sections. The first section broadly describes the MCDM and its methods followed by the literature review in software engineering scenario. The flow of TOPSIS is described in the third section. The following section is the main concern of this paper. It focuses on the proposed model of TOPSIS for SDM selection. Results and discussion are briefly presented in the next section. Finally, the above content is supported by the conclusion of this study.

2 MCDM

MCDM has gained the popularity because of its intuitive nature. It begins with the decomposition of the complex problem into hierarchical structure which is easy for humans to understand. It is classified into various methods which provide optimal solution for selecting the best among alternatives [5]. The following section highlights the MCDM supporting methods and how they have been turned into a reliable and robust decision-making tool in software engineering applications.

2.1 *MCDM Methods*

Figure 1 represents the classification tree of MCDM methods. The successive heading gives a glance over the working of some methods extensively used in corporate sector for making judgments.

2.1.1 Outranking-Based Methods

The outranking-based methods, such as ELECTRE [7, 8] and PROMETHEE, are based on concordance analysis. It establishes the "outranking relationship" between concordance and discordance matrices and uses their indices to choose the best alternative.

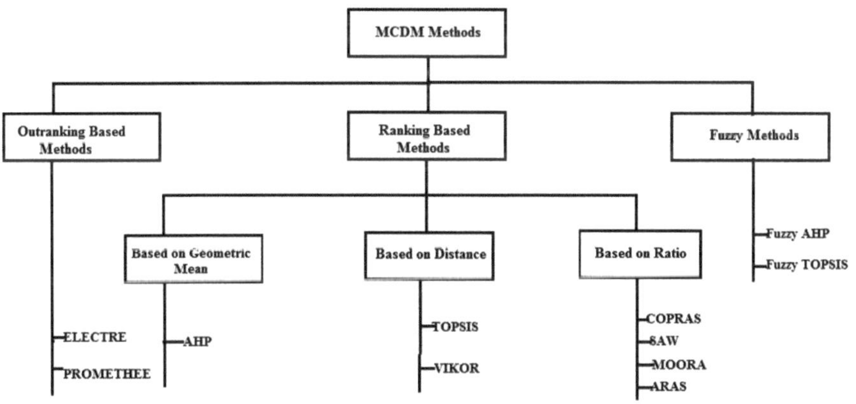

Fig. 1 Hierarchical structure of MCDM methods [6]

2.1.2 Ranking-Based Methods

The ranking model prioritizes the alternatives after rigorous analysis based on either geometrical mean, Euclidean distance or ratio. It performs pairwise comparison among criteria as well as for each alternative. Two approaches, AHP and TOPSIS [9], are most popular among decision-makers and experts. The reason behind is ease in the steps of algorithm. In AHP, the Saaty Table 1 [10] plays a crucial role in constructing pairwise comparison matrices. The weights are computed with the help of geometric mean. The different alternatives are ranked after analyzing eigenvalues and eigenvectors.

On the contrary, TOPSIS model uses Euclidean distance for weight computation. Instead of using the Saaty scale, the preference is given to the view of experience holders, decision-makers, and experts. Its plus factor is found in the consistency of result even when the small degree of uncertainty cannot be neglected.

2.1.3 Fuzzy Methods

What happens when decision-makers stuck in certain circumstances where the uncertainties, complexities, vagueness, incompleteness are high? In that case, the classic approach does not have suitable answers. The introduction of fuzzy set theory offers stakeholders a novel concept to not only take the discrete values, rather consider the membership of defined variables. It aims to find a solution in between *True and False, Yes or No, 0 and 1, High or Low*. The researchers embedded the fuzzy application into traditional methods of decision making. Some of the fuzzy approaches are fuzzy AHP and fuzzy TOPSIS.

Fuzzy AHP, more or less, is similar to AHP. However, the Saaty scale [10] had been modified by researcher Chang [11]. The simple scale is converted into triangular fuzzy number (TFN) for pairwise comparison matrices. TFN translates linguistic term into a tuple containing *optimistic, moderate, and pessimistic* values [12]. The entire algorithm runs alike AHP over TFNs.

As TOPSIS is another version of AHP, similarly fuzzy TOPSIS is close to fuzzy AHP.

Table 1 Defined scale and their corresponding definition

Saaty scale	Description
1	Equally important
3	Weakly important
5	Fairly important
7	Strongly important
9	Absolutely important
2, 4, 6, 8	The intermediate values between two adjacent scales

2.2 MCDM in Software Engineering Scenario

Recently in March 2018, TOPSIS approach is used as a tool in selection of location for property development. The author [13] prescribed 32 criteria for analysis including soil texture, drainage system, distance from city. Alternatives for locations are sites A, B, and C which are part of this study through decision support system (DSS). The steps of TOPSIS process are done as follows: (1) Collect relevant data about locations, (2) location selection, (3) survey, (4) data processing using TOPSIS, (5) discussion of results, and (6) recommend the location.

In August 2017, the commendable job is done by the team of research scholar for the selection of best software engineering practices [4] using PAPRIKA, a decision-making approach. The study covers 11 criteria as software development practices for designing a medical application. The proposed PAPRIKA framework is used for the project development which does not conform to a definite software methodology.

In March 2017, another popular technique fuzzy AHP (or FAHP) has been applied in the software effort estimation in scarcity of data [14]. The results were validated with IVR dataset of a software industry. The relative ranking of 20 projects from the dataset has been achieved after evaluating their weights. The equation for estimating effort is based on the variables including estimated effort, known effort, and their respective weights.

In 2016, the application of integrated techniques comes into existence. AHP and TOPSIS methodologies are integrated for ETL software selection [15]. ETL, stands for Extract, Transfer and Load, is a powerful action for data preprocessing before it loaded into data warehouse. The framework is composed of three steps: (1) AHP technique, responsible for making pairwise comparison among criteria and determining their weights; (2) TOPSIS technique, responsible for constructing decision matrices, determining ideal and negative ideal solution and prioritizing alternatives; (3) final selection of the best alternative, i.e., decision making.

3 TOPSIS Methodology

The paper revolves around the implementation of TOPSIS in software project planning to answer "*Which SDM is best fitted into the scenario?*" The successive heading describes the TOPSIS work flow.

TOPSIS, acronym for Technique for Order Preference by Similarity to Ideal Solution, follows two rules [13, 15]. One, selected alternative is closest to ideal solution, and second, it should have the distance from the negative ideal solution. Ideal solution involves best criteria (*maximum profit, minimum time, high value,* etc.), whereas negative ideal solution contains worst criteria (*minimum profit, maximum time, low value*, etc.).

3.1 Major Steps

Suppose there are '*k*' numbers of decision-makers, '*m*' number of alternatives, and '*n*' number of criteria, then the steps of algorithm are as follows:

Step 1 Generation of decision matrices

For every alternative *m*, the decision matrix *D* is given by Eq. (1),

$$D^p = \left[d_{ij}\right] \quad 1 \le p \le m, \ 1 \le i \le n, \ 1 \le j \le k \tag{1}$$

where d_{ij} is the rating scale from 1 to 10 or 1 to 100, where 1 stands for *low* and 10 for *very good*.

Similarly, the decision matrix is constructed for rating the criteria and represented by the form given below.

$$W = \left[w_{ij}\right] \quad 1 \le i \le n, \ 1 \le j \le k \tag{2}$$

where w_{ij} is the rating scale from 1 to 10 or 1 to 100, where 1 stands for *low* and 10 for *very good*.

The mean along each row of *W* is calculated and assign them to criteria as their weights.

$$\overline{W} = [\bar{w}_x] \quad 1 \le x \le n \tag{3}$$

where $\overline{w_x}$ is the computed mean along each row.

Step 2 Reduction of m decision matrices into single matrix

Taking mean along each row in *m* matrices, these are reduced into one decision matrix of $n \times m$ dimension.

Step 3 Standardize the decision matrix

To standardize the matrices, the columns are divided with root of sum of the squares computed along each row.

$$D = \left[d_{ij}\right] \quad 1 \le i \le n, \ 1 \le j \le m \tag{4}$$

Step 4 Weighted decision matrix

To construct the weighted matrix, ratings achieved in step 3 are multiplied with the respective criteria weights from Eq. (3). The modified decision matrix D from step 3 is multiplied with the criteria weights $\overline{w_x}$ as,

$$D = \left[d_{ij} \times \overline{w_x}\right] \quad 1 \le i \le n,\ 1 \le j \le m,\ 1 \le x \le n \tag{5}$$

Step 5 Determining ideal and negative ideal solution

For ideal solution, compute the maximum for each criteria as,

$$I = \left\{\max\left[d_{ij}\right]\right\} \quad 1 \le i \le n,\ 1 \le j \le m \tag{6}$$

For negative ideal solution, compute the minimum for each criteria as,

$$NI = \left\{\min\left[d_{ij}\right]\right\} \quad 1 \le i \le n,\ 1 \le j \le m \tag{7}$$

Step 6 Determining distance from ideal and negative ideal solutions

Firstly, each weight is subtracted from concerned value in I and squared. Then, root mean distance is evaluated along columns for determining closeness to the ideal solution.

$$D = \left[\left(d_{ij} - y\right)^2\right] \quad 1 \le i \le n, 1 \le j \le m,\ y \in I \text{ for respective } i \tag{8}$$

For particular j, root mean distance for ideal solution S_i is

$$S_i = \left\{\left(\sqrt{d_{1j} + d_{2j} + d_{3j} + \ldots\ldots + d_{nj}}\right)\right\} \tag{9}$$

Similarly, the above equation is used for determining separation from negative ideal solution denoted by S_{ni}

Step 7 Relative closeness to ideal solution

$$\mathrm{RC} = \left\{\frac{S_{nl}}{S_i + S_{ni}}\right\} \tag{10}$$

The RC denotes the relative closeness of each alternative to ideal solution.

Table 2 TOPSIS for SDM selection [13]

Step 1	Determine number of alternatives, criteria, and decision-makers' involvement
Step 2	Construct decision matrices
Step 3	Standardize the decision matrix
Step 4	Compute weighted standardize decision matrix
Step 5	Determine ideal solution and negative ideal solution
Step 6	Find relative closeness to ideal solution
Step 7	Order the alternatives in descending

Table 3 Decision matrix for criteria

	E1	E2	E3	Weights
Flexibility of requirements	10	9	9	$\frac{10+9+9}{3} = 9.33$
Risk analysis	9	9	8	$\frac{9+9+8}{3} = 8.67$
Customer feedback	9	8	7	$\frac{9+8+7}{3} = 8.00$

Table 4 Decision matrix for alternative Waterfall Model

	E1	E2	E3	Weights
Flexibility of requirements	3	4	3	$\frac{3+4+3}{3} = 3.33$
Risk analysis	3	3	2	$\frac{3+3+2}{3} = 2.67$
Customer feedback	4	3	3	$\frac{4+3+3}{3} = 3.33$

4 Proposed Work—TOPSIS for SDM Selection

The study traces the steps of TOPSIS to propose a model for choosing appropriate SDM. The flow of our TOPSIS model is depicted in Table 2.

Consider

Alternatives: (1) Waterfall Model, (2) Spiral Model, (3) Agile Model

Criteria: (1) Flexibility of requirements, (2) Risk analysis, (3) Customer feedback

Experts: (1) E1, (2) E2, (3) E3 (Tables 3, 4, 5, 6, 7, and 8).

Table 5 Decision matrix for alternative Spiral Model

	E1	E2	E3	Weights
Flexibility of requirements	6	7	8	$\frac{6+7+8}{3} = 7.00$
Risk analysis	9	8	9	$\frac{9+8+9}{3} = 8.67$
Customer feedback	7	7	8	$\frac{7+7+8}{3} = 7.33$

Table 6 Decision matrix for alternative Agile Model

	E1	E2	E3	Weights
Flexibility of requirements	10	10	9	$\frac{10+10+9}{3} = 9.67$
Risk analysis	9	9	9	$\frac{9+9+9}{3} = 9.00$
Customer feedback	8	8	9	$\frac{8+8+9}{3} = 8.33$

Table 7 Standard decision matrix

	Waterfall Model	Spiral Model	Agile Model	
Flexibility of requirements	3.33	7	9.67	$\sqrt{3.33^2 + 7^2 + 9.67^2} = 12.39$
Risk analysis	2.67	8.67	9	$\sqrt{2.67^2 + 8.67^2 + 9^2} = 12.77$
Customer feedback	3.33	7.33	8.33	$\sqrt{3.33^2 + 7.33^2 + 8.33^2} = 11.58$

Table 8 Standard decision matrix (continued)

	Waterfall Model	Spiral Model	Agile Model
Flexibility of requirements	3.33/12.39 = 0.26	7/12.39 = 0.56	9.67/12.39 = 0.78
Risk analysis	2.67/12.77 = 0.20	8.67/12.77 = 0.67	9/12.77 = 0.70
Customer feedback	3.33/11.58 = 0.28	7.33/11.58 = 0.63	8.33/11.58 = 0.71

5 Results and Discussion

Weight matrix is achieved by multiplying the weights of respective criteria. It is represented as,

Table 9 Relative closeness of alternatives

Alternative	Waterfall Model	Spiral Model	Agile Model
Relative closeness	0	0.72	1

$$\begin{bmatrix} 0.26 \times 9.33 = 2.42 \\ 0.20 \times 8.67 = 1.73 \\ 0.28 \times 8.0 = 2.24 \end{bmatrix}, \begin{bmatrix} 5.22 \\ 5.80 \\ 5.04 \end{bmatrix}, \begin{bmatrix} 7.27 \\ 6.06 \\ 5.68 \end{bmatrix}$$

Ideal solution, {7.27, 6.06, 5.68}, and negative ideal solution, {2.42, 1.73, 2.24}

Distance from ideal solution $S_i = \{7.35, 2.15, 0\}$ and negative ideal solution $S_{ni} = \{0, 5.67, 7.35\}$

Finally, relative closeness is calculated using Eq. (10), and we get (Table 9).

Therefore, the order of ranking is defined as below:

Agile Model > Spiral Model > Waterfall Model

So, the TOPSIS model recommends *Agile* Model because its closeness is maximum to best criteria.

6 Conclusion

Multicriteria decision making (MCDM) has been becoming indispensable in the software engineering discipline for more than a decade [16]. Its presence is vital when making decisions is uncompromising and stern for company. It does not limit to software industry but also in ecological assessment [17]. The methods of MCDM involve quantitative and qualitative analyses for linguistic variables in contrast to discrete variables.

As we prove, the TOPSIS provides robust and simple framework for prioritizing the alternatives and present the best suitable option before stakeholders. It does not much require the domain knowledge but the statistical analysis. The results are acceptable and clear to experts. It “best discriminates” the superior and inferior alternative on the basis of closeness to attributes.

TOPSIS has wide potential to mark its presence in another area. Its scope is high in the applications of software engineering and other subfields of computer science which are yet to be explored.

References

1. Pressman, R.S.: Software Engineering: A Practitioner’s Approach, 7th edn. McGraw Hill, New York (2010)
2. IEEE Standard Glossary of Software Engineering Terminology: Computer Society of IEEE, New York (1990)

3. Verner, J., Sampson, J., Cerpa, N.: What factors lead to software project failure? In: 2nd IEEE International Conference on Research Challenges in Information Science, Morocco, pp. 85–93 (2008)
4. Hernández-Ledesma, G., Ramos, E.G., Fernández-y-Fernández, C.A., Aguilar-Cisneros, J.R., Rosas-Sumano, J.J., Morales-Ignacio, L.A.: Selection of best software engineering practices: a multi-criteria decision making approach. Res. Comput. Sci. 47–60 (2017)
5. Velasquez, M., Hester, P.T.: An analysis of multi-criteria decision making methods. Int. J. Oper. Res. **10**, 56–66 (2013)
6. Aruldoss, M., Lakshmi, T.M., Venkatesan, V.P.: A survey on multi criteria decision making methods and its applications. Am. J. Inform. Syst. **1**, 31–43 (2013)
7. Triantaphyllou, E., Shu, B., Sanchez, S.N., Ray, T.: Multi-criteria decision making: an operations research approach. Encycl. Electr. Electron. Eng. **15**, 175–186 (1998)
8. Ermatita, Hartati, S., Wardoyo, R., Harjoko, A.: Electre methods in solving group decision support system bioinformatics on gene mutation detection simulation. IJCSIT **3**, 40–52 (2011)
9. Anitha, M., Sathiya, M.: Improved technique by integrating AHP-TOPSIS for prioritization. IJETST **2**, 1911–1915 (2015)
10. Saaty, T.L.: How to make a decision: analytic hierarchy process. Eur. J. Oper. Res. **48**, 9–26 (1990)
11. Chang, D.-Y.: Applications of the extent analysis method on fuzzy AHP. Eur. J. Oper. Res. **95**, 649–655 (1996)
12. Ayhan, M.B.: A fuzzy AHP approach for supplier selection problem: a case study in a gear motor company. IJMVSC **4**, 11–23 (2013)
13. Jollyta, D.: TOPSIS technique for selecting of property development location. Softw. Eng. **6**, 20–26 (2018)
14. Sehra, S.K., Brar, Y.S., Kaur, N.: Applying fuzzy-AHP for software effort estimation in data scarcity. IJETT **45**, 4–9 (2017)
15. Hanine, M., Boutkhoum, O., Tikniouine, A., Agouti, T.: Application of an integrated multi-criteria decision making AHP-TOPSIS methodology for ETL software selection. In: SpringerPlus, pp. 1–17 (2016)
16. Dipti, S.: Maintainability estimation of component based software development using fuzzy AHP. IJETST **1**, 280–285 (2014)
17. Ersayin, K., Tagil, S.: Ecological sensitivity and risk assessment in the Kizilirmak Delta. Fresenius Environ. Bull. **26**, 6508–6516 (2017)

Situation-Aware Conditional Sensing in Disaster-Prone Areas Using Unmanned Aerial Vehicles in IoT Environment

J. Sathish Kumar, Mukesh A. Zaveri, Saurabh Kumar and Meghavi Choksi

Abstract Environmental sensing is the most crucial task that needs to be performed in order to analyze the situation of a region during a disaster. The devices deployed in such regions are responsible for sensing and communication effectively. During a disaster, the operation of these devices may be affected by the environmental conditions and their respective power constraints. Moreover, the mobility of these devices in the network leads to a challenging task to perform sensing and communication in such an environment. The disaster recovery may need different sensor data at various points of time. In such cases, the selectivity of data from different sensors and its dissemination in real time are the most important tasks. In this paper, the proposed algorithm is based on the situation-aware conditional sensing for disaster-prone areas using unmanned aerial vehicles. The technique presented in this paper focuses on the control of way points of the aerial vehicles based on the events detected in the Internet of Things environment.

Keywords Disaster management · Unmanned aerial vehicles · Internet of things

1 Introduction

The disaster management is a crucial research problem that needs attention to address various critical problems such as environmental monitoring, real-time data collection,

J. Sathish Kumar (✉) · M. A. Zaveri · S. Kumar · M. Choksi
Computer Engineering Department, Sardar Vallabhbhai National Institute of Technology, Surat 395007, India
e-mail: ds14co001@coed.svnit.ac.in

M. A. Zaveri
e-mail: mazaveri@coed.svnit.ac.in

S. Kumar
e-mail: ds14co006@coed.svnit.ac.in

M. Choksi
e-mail: meghavichoksi@gmail.com

L. C. Jain et al. (eds.), *Data and Communication Networks*, Advances in Intelligent Systems and Computing 847, https://doi.org/10.1007/978-981-13-2254-9_12

rescue recovery operations. It is reported that in the last two decades, approximately 400,000 lives were lost all over the globe in total [1]. However, due to the technological advancements, it is possible to handle the effects of the disaster to a certain extent. The most important challenge in such situations is the real-time surveillance [2] and communication [3]. In addition, the adversity faced in the case of a disaster demands that all the above-mentioned challenges must be solved intelligently, preferably using automation. The real-time support must be provided irrespective of the disaster type, i.e., natural or man made [4]. A glimpse of various disaster types such as fire and smoke in a building due to terrorist attack, extreme fire in the forest, road damages due to landslide, heavy flood, etc., is shown in Fig. 1 [5]. These situations need to be sensed both in pre-disaster and post-disaster cases. The sensed data in the disaster environment must be communicated to the rescue team for preparing the solution strategy, by bridging the gap between distance and time. The Internet of things (IoT) environment provides a mechanism, with which the real-time sensing can be communicated to any place at anytime from anywhere [6]. However, the IoT environment must be configured and organized in such a way that an optimized communication should result into automation of tasks, reducing the faults and failures in the system. During the disaster, the real-time data collection needs sensor devices, which are responsible to sense the situations under surveillance. However, these sensors are very low-powered and energy-constrained devices. Due to this limitation of the sensing devices, there is a need of IoT device to be computationally efficient ultimately making it energy efficient using Internet. The IoT environment provides collaborative processing among sensing devices and IoT devices that best suits these requirements [7]. The rescue operations and the response in the case of a disaster must be automated using these IoT devices, so that human lives are not lost in this process. Due to the adversity and danger in the case of a disaster, these IoT devices must be protected from the physical damage which is a precautionary requirement.

The unmanned aerial vehicles (UAVs) [8, 9] may serve a great purpose in such scenarios as discussed above. These vehicles may be operated from anywhere, at anytime, and to the distances that depend on the requirements of the situation under observation. UAV is a mobile device and can be controlled by an IoT device for fulfilling the requirements of real-time communication over the Internet from anywhere to the desired place [10]. Figure 1a depicts the fire and smoke situation in a building. In such a case, the sensor devices are deployed in and around this building. These devices sense different parameters and report the abnormality conditions to the ground control station. In this regard, the dynamic network is to be set up for the ad hoc data acquisition and on the spot monitoring from all possible directions. The UAV is responsible for monitoring and data collection using these deployed devices because of the hostile and restricted entry situations. It is mandatory to have knowledge of the safe distance of operation for the UAV as incidences of fire and smoke may affect the operations of the sensing and communication. Further, in the case of any abnormality observed by the UAV, it must be able to respond and rescue itself. Therefore, the two basic requirements that may be outlined in a disaster scenario are (1) for real-time response, the effective sensing, and communication operation and (2) the protection of UAV devices involved in the operation.

(a) 26/11 Terrorist attack

(b) Forest fire disaster

(c) Road damage due to landslide disaster

(d) Heavy flood disaster

Fig. 1 A glimpse of various disaster types [5]

This paper focuses on the use of UAV with the help of an IoT device to collect the sensed data from the sensor devices in case of a disaster. The situation-aware conditional sensing mechanism is considered to address the problems as specified above. There is a necessity to communicate the information in real time, even in adverse situations instantly and protect themselves so as not to burden the operation in terms of cost. The validation of various situations under observation is performed by providing different conditions to the UAV device for enhancing its automation in total. The simulation is performed in IoT environment using DroneKit-Python simulation tool [11], and the experiments were performed on an actual drone.

The rest of the paper is organized as follows: Section 2 discusses the proposed algorithm that is used to validate the effectiveness of the UAV operation. Section 3 presents the implementation of the proposed algorithm and explains the situational awareness of the UAV in different scenarios. Finally, Sect. 4 concludes the work.

2 Proposed Algorithm

In this section, the proposed algorithm performs the sensing and communication using aerial vehicles by including various conditions. In this context, the problem is formulated using different notations. Let us assume a geographical region, R. This region may be a city, a forest, a rural area, a coastal area. Let us assume that the region R is subdivided into n smaller subregions such that $R=(R_1, R_2, \ldots R_i \ldots, R_n)$. Such divisions are needed to perform the area monitoring effectively and with ease. The division of the region depends on different factors related to the type of region or the type of disaster. In this context, the disaster may be a fire in a building [12], a flood [13], or a damage to structures such as flyovers, heavy machineries, highways [14] or rescue of a group of people lost in the forest [15]. The glimpse of various disasters is shown in Fig. 1. It is further assumed that the sensors deployed in the region R are for sensing different environmental parameters, such as temperature, pressure, humidity, stress, strain, atmospheric pressure, luminosity.

Under such circumstances, there is a need to identify the subregion R_i, where a critical situation has occurred. This is done based on the abnormality observed in the sensing parameters of the sensors operating in R_i. Next, for monitoring the situation specifically in the region R_i, the mission planning for UAV navigation at different locations within R_i is performed at the ground control station. A ground control station is a place where the trajectory planning and tracking of the UAVs are performed. For the real-time analysis, communication between UAV and ground station is performed using the Internet. In this context, the proposed algorithm works in two steps. In the first step, the UAV senses the situation of environment on the fly using various sensor devices. In the second step, the algorithm helps the UAV to make the decision dynamically based on the environmental conditions. After the surveillance and real-time communication of various parameters that are required, the relief response team is dispatched to the most required place in the disaster site.

In the second step, for the dynamic navigation of the UAV, various conditions are considered and incorporated in the UAV. To perform this dynamic navigation operation of the UAV, let us assume that there are m different parameters sensed in the region such that $P=(P_1, P_2, \ldots P_j \ldots, P_m)$. These values are collected by the aerial vehicle, and the collected values are sent to the ground control station. For instance, it may be possible that the temperature sensor shows more abnormality in its reading as compared to the other parameters. Let this be the jth parameter in the region where the disaster has occurred. Based on this gathering of the sensed data, two types of threshold can be determined viz maximum and minimum values of the parameter P_j. There are four different conditions that may be used to navigate the UAV successfully. If the value of P_j is greater than the threshold maximum, observed at the region, then an alert is generated. Similarly, if the sensed value of P_j is lesser than the threshold minimum, then there may exist two different scenarios. Scenario 1 may suggest that the region is not under disaster, whereas on the contrary scenario 2 suggests that there is a false alarm. In both the cases, the alert is generated. Once the alert is triggered, then the UAV may skip the remaining locations given in the mission plan and make an immediate decision to return back to its launch location. If the above-mentioned cases do not occur, then the UAV needs to continue the navigation through the planned path as per the mission to reach the next location for surveillance. The collected sensed information is communicated to the ground station in real time. Following any sensed anomaly, the relief operation at the site is executed. The relief operation can be initiated in two ways: either after complete monitoring of all the locations in the disaster region or at each and every location in the planned mission. In the second case, the UAV sends the recorded parameters continuously to the ground station in real time to prepare for the further solution strategy relevant to the problems faced at the disaster site. The proposed algorithm is explained in brief in Algorithm 1.

During a disaster, there are three most crucial operations that need to be performed. First, the real-time data must be sensed and communicated to the ground control station. This helps in formulating the strategy for the disaster relief. Second, the disaster site R_i must be isolated from the other regions of R, such that the operation surveillance is focused on only R_i. Third, the relief response must be provided in real time, with no loss of life. The proposed algorithm uses UAV as a part of the disaster response team to collect real-time data from the sensors deployed at the disaster site and communicate this data to the ground station using IP-based communication. Finally, the relied response may be provided in real time using UAV without causing a threat to human life. This process of situation-aware conditional sensing of the environment using IoT network may prove to be an effective mechanism in disaster situations. The implementation of the proposed methodology is discussed with different scenarios in the following section.

Algorithm 1: Condition-based Sensing Algorithm
Require: Sensing device, UAV, and m reading parameters
Ensure: Relief response at disaster site, protection of UAV during relief operation

```
Define a region R. Subdivide R such that R = (R1, R2, …, Rn).
for disaster observed in Ri region do
/*Define Thresholds */
Estimate max and min values of parameter Pm.
if val(Pm) ≥ max(Pm) then
/*Alert Is generated */
Navigate UAV to launch location.
else if val(Pm) ≤ min(Pm) then
/*Alert Is generated */
Navigate UAV to launch location.
else
Continue the navigation through the mission path.
end if
Communicate the data collected to ground station.
end for
Perform the relief operation.
```

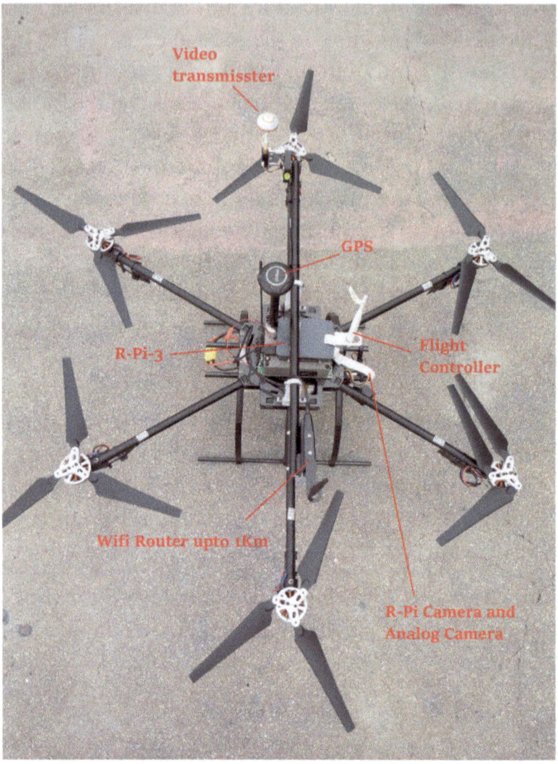

Fig. 2 UAV with its specifications

3 Scenarios and Results

In this section, for demonstrating the proposed algorithm using UAV, DroneKit-Python [11] is used and simulation is performed. DroneKit-Python is a simulation tool that allows the developer to control the navigation of the UAVs. Additionally, the mission planner tool [16] is used to track the navigation of the UAV in the region. After successful simulations, the scenarios are emulated using the actual UAV. As shown in Fig. 2, the Hexa Drone with high stability incorporated with GPS module, Raspberry Pi 3 onboard microcontroller with pi camera, high accurate flight controller, high-resolution video transmitter, analog video capture device, and lithium-powered battery of 20 min fly time is designed for the purpose of post-disaster management and surveillance. The glimpse of the drone on the fly for surveillance is shown in Fig. 3.

For the purpose of demonstrating the different scenarios, initially, a region of interest is defined as shown in Fig. 4. In the case presented in this paper, the region of interest, R is outlined using Google maps on mission planner, which consists of three department buildings, residential hostel, playground, health center, and few forest areas. Further, the whole region is partitioned into nine different subregions given

Fig. 3 Snapshot of on-the-fly UAV

by $R = (R_1, R_2, \ldots, R_9)$. These subregions are as shown in Fig. 5 with boundaries. Let us assume that a disaster occurs in region R_1 and the unnatural fire condition is considered as the disaster for experimental purpose.

In such a scenario, the unmanned aerial vehicle is dispatched to perform the surveillance operation, and at the same time, real-time communication of the sensed information is sent to the ground control station. The UAV is instructed to navigate through the region of disaster, which is shown in red color and performs the sensing and real-time communication operations at eight different locations, $L = (L_1, L_2, \ldots, L_8)$.

The planned mission with the waypoints of the UAV is shown in Fig. 6, and their respective locations are summarized in Table 1. These eight locations are the critical points from where the required parameters are needed to be sensed. In our case, the abnormality in temperature value needs to be sensed by the onboard system in the UAV. The basic mission plan instructs the UAV to start its flight from the home

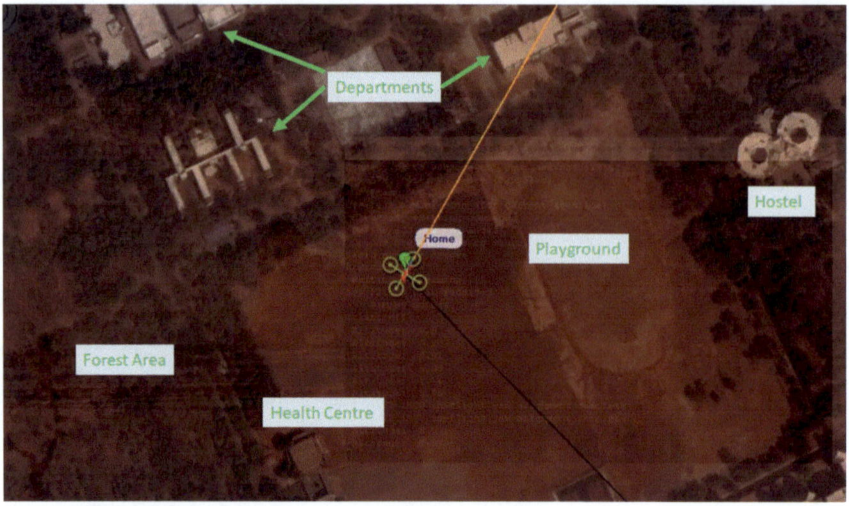

Fig. 4 A demonstration of region of interest in our campus

Fig. 5 Occurrence of disaster in region R_1

location and sense the parameters at eight locations before returning back to its home location again.

The ground station provides the UAV with an approximate maximum and minimum values of temperature in the disaster region. The threshold of the minimum and the maximum values is assigned during the experiment with 25 and 35 °C, respectively. During its flight from one location to another, the UAV senses the temperature value and communicates this information to the ground station. For the verification

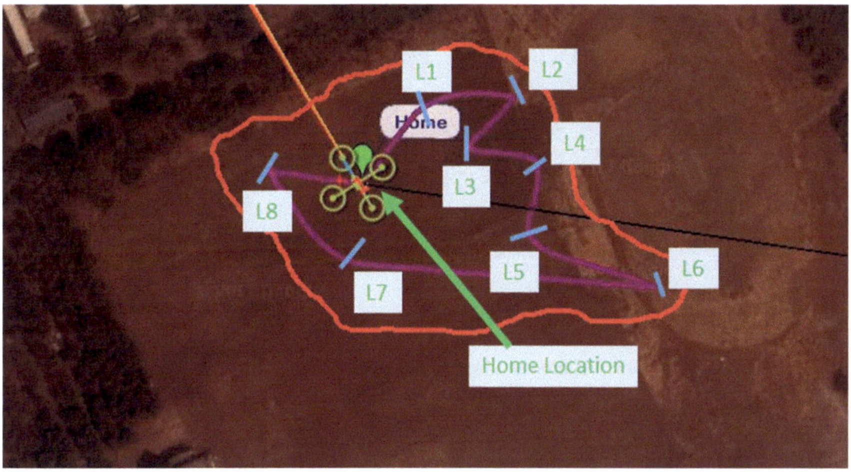

Fig. 6 Basic mission plan of UAV to sense the disaster region

Table 1 A summary of locations for navigation of the aerial vehicle

Location ID	Latitude	Longitude
Home location	21.161580	72.785524
L_1	21.161925	72.785813
L_2	21.161915	72.786162
L_3	21.161720	72.785974
L_4	21.161680	72.786280
L_5	21.161370	72.786210
L_6	21.161180	72.786776
L_7	21.161275	72.785417
L_8	21.161625	72.785132

of our proposed algorithm, three scenarios are demonstrated. In the first scenario, a false alarm is created by altering the temperature value in the code, i.e., by giving 23 °C, which is lesser than the minimum temperature threshold provided to the UAV. In such a case, the UAV generates the alert and returns to the home location at L_3, as shown in Fig. 7, where the UAV path tracking is depicted. Similarly, the second scenario is demonstrated as shown in Fig. 8, where the sensed temperature by the UAV is more than the maximum threshold. In this case, the UAV senses 39 °C at a location L_6 that is more than the maximum temperature range provided to the UAV. In such a scenario, again the UAV generates the alert and returns to the home location, as the UAV is vulnerable to get affected by the ambient condition and there is a possibility of crash. The UAV tracking in mission planner is shown in Fig. 8. In the third scenario, if the UAV senses the temperature within the specified range, then the UAV should navigate through each location of the disaster region. While navigation, real-time communication of the sensed information to the ground control

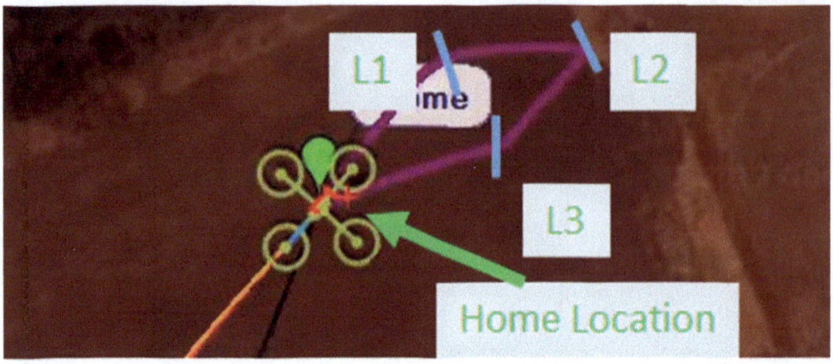

Fig. 7 UAV returns to home location in the case false alarm at L_3

station is performed. In this way, the control station knows the exact condition of the site through parametric sensing report received from the UAV. The full mission traversed by the UAV is as depicted in Fig. 9. Once the navigation area of the disaster site is decided successfully, there is a requirement to carry out disaster relief operation. However, the UAV performs the situation analysis by considering the various parameters and continuously evaluating the given conditions during the disaster. The disaster control and corresponding recovery are very crucial operations and need to be carried out with minimum risk of human lives. In this paper, the real communication using Internet enables the UAV to send the sensed data to the ground control station or to the cloud. Further, the real-time streaming of video using Raspberry Pi camera and analog camera is sent to the ground control station for accurate analysis. Therefore, the application of the IoT in such a scenario is efficient by using the automated devices as an integral part of the rescue team. This paper addresses these concerns by the use of the UAVs for relief operation and providing a mechanism for dynamic navigation, which is an essential requirement for such applications in order to save human lives.

4 Conclusion

The disaster management and the response are very critical that need to be performed in an effective way so that many lives can be saved. There is a need of real-time response in the disaster scenarios. Moreover, the sensing and communication tasks must be performed in such a way that the rescue can be carried out speedily. In this paper, the use of the UAV is demonstrated for strategy formulation of rescue team during disaster to sense the critical parameters at disaster site and accomplish the real-time communication to the ground station. In this context, the use of the UAV for real-time sensing and communication and an algorithm to protect these vehicles during extreme conditions is addressed. The implementation of the proposed work is performed using DroneKit-Python simulation tool and emulated the scenarios using

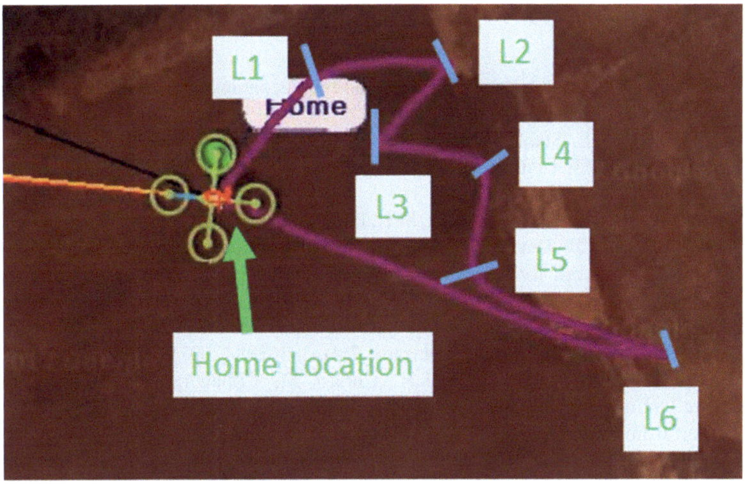

Fig. 8 UAV returns to home location while exceeding the threshold range at L_6

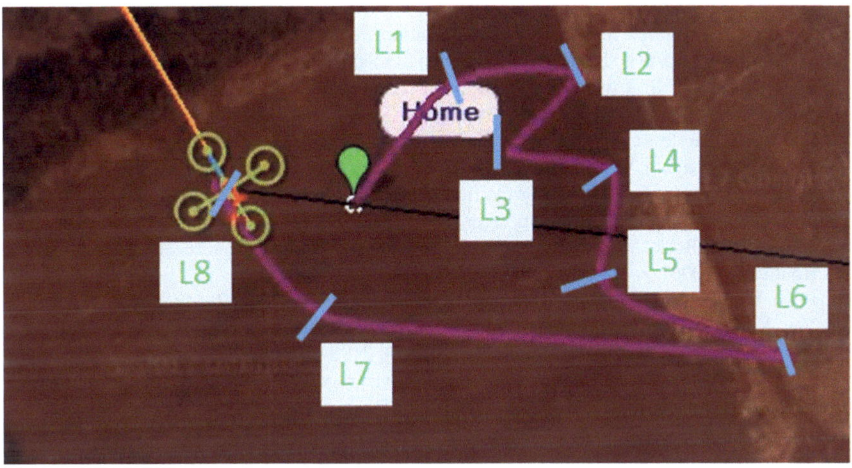

Fig. 9 UAV sensing values at eight different locations

actual drone developed for post-disaster management and aerial surveillance with the aid of mission planner tool. In the future, the proposed work can be tested for its efficiency with several parameters in extreme conditions.

Acknowledgements This work is supported by the Ministry of Electronics and Information Technology (MeitY), funded by Government of India (Grant no. 13(4)/2016-CC&BT).

References

1. Cavallo, E., Noy, I., et al.: Natural disasters and the economy—a survey. Int. Rev. Environ. Resour. Econ. **5**(1), 63–102 (2011)
2. Benfold, B., Reid, I.: Stable multi-target tracking in real-time surveillance video. In: Proceedings of the IEEE Conference on Computer Vision and Pattern Recognition (CVPR), pp. 3457–3464 (2011)
3. Houston, J.B., Hawthorne, J., Perreault, M.F., Park, E.H., Goldstein Hode, M., Halliwell, M.R., Turner McGowen, S.E., Davis, R., Vaid, S., McElderry, J.A., et al.: Social media and disasters: a functional framework for social media use in disaster planning, response, and research. Disasters **39**(1), 1–22 (2015)
4. Rose, A.: Economic resilience to natural and man-made disasters: multidisciplinary origins and contextual dimensions. Environ. Hazards **7**(4), 383–398 (2007)
5. Zaveri, M.A., Kumar, J.S., Pandey, S.K., Choksi, M.: Collaborative Data Processing and Resource Optimization for Post Disaster Management and Surveillance using IoT. Technical Report from (MeitY), India (2016)
6. Bandyopadhyay, D., Sen, J.: Internet of Things: applications and challenges in technology and standardization. Wireless Pers. Commun. **58**(1), 49–69 (2011)
7. Kumar, J.S., Zaveri, M.A.: Graph based clustering for two-tier architecture in Internet of Things. In: Proceedings of 9th IEEE International Conference on Internet of Things (iThings), pp. 229–233 (2016)
8. Pajares, G.: Overview and current status of remote sensing applications based on unmanned aerial vehicles (UAVs). Photogram. Eng. Remote Sens. **81**(4), 281–329 (2015)
9. Tuna, G., Nefzi, B., Conte, G.: Unmanned aerial vehicle-aided communications system for disaster recovery. J. Netw. Comput. Appl. **41**, 27–36 (2014)
10. Reiter, G.: Wireless connectivity for the Internet of Things. Europe 433, 868 MHz (2014)
11. Mason, I.A., Nigam, V., Talcott, C., Brito, A.: A framework for analyzing adaptive autonomous aerial vehicles. In: Proceedings of the International Conference on Software Engineering and Formal Methods, pp. 406–422. Springer (2017)
12. Hasofer, A., Beck, V.R., Bennetts, I.: Risk Analysis in Building Fire Safety Engineering. Routledge (2006)
13. Plate, E.J.: Flood risk and flood management. J. Hydrol. **267**(1–2), 2–11 (2002)
14. Yan, Y., Cheng, L., Wu, Z., Yam, L.: Development in vibration-based structural damage detection technique. Mech. Syst. Signal Process. **21**(5), 2198–2211 (2007)
15. Davids, A.: Urban search and rescue robots: from tragedy to technology. IEEE Intell. Syst. **17**(2), 81–83 (2002)
16. Hadi, G.S., Varianto, R., Trilaksono, B., Budiyono, A.: Autonomous UAV system development for payload dropping mission. J. Instrum. Autom. Syst. **1**(2), 72–77 (2014)

An Empirical Analysis of Collaborative Filtering Algorithms for Building a Food Recommender System

Ashique Mohaimin Ornab, Sakia Chowdhury and Seevieta Biswas Toa

Abstract Recommender system has been playing a great role in almost every sectors starting from online shopping Web sites to online movie sites and social networking sites. However, the use of recommendation engine has been very little in the food sector. Sometimes people become tired of having the same kind of meals everyday because of several reasons. Some people need to consume fixed food due to their illness; others consume same meals everyday to stay healthy despite having any diseases. In this paper, we have first discussed two collaborative filtering algorithms that can be used to build a food recommender system for the people who have been leading a monotonous food consumption lifestyle and are bored of having the same kind of meals every day. After that, we have analyzed the two approaches of building a food recommender system and finally concluded that the model-based approach is more reliable than the memory-based approach.

Keywords Recommender system · Collaborative filtering · Cosine similarities Alternating least squares (ALS)

1 Introduction

Recommender system is basically a system that enables items to be suggested to similar users based on their personal common preferences and choices. Recommender system searches for the similarities between items or users and suggests the most similar items to the users. In a world of full of information and variety of choices, it becomes difficult for people to choose the best out of the bests which match their

A. M. Ornab · S. Chowdhury (✉) · S. B. Toa
Department of Computer Science and Engineering, BRAC University, Dhaka, Bangladesh
e-mail: sakiacm@gmail.com

A. M. Ornab
e-mail: amornab@gmail.com

S. B. Toa
e-mail: seevieta@gmail.com

L. C. Jain et al. (eds.), *Data and Communication Networks*, Advances in Intelligent Systems and Computing 847, https://doi.org/10.1007/978-981-13-2254-9_13

interests. This is where recommender system comes into play; it helps people find out the best for them.

Generally, recommender system is an information filtering method finding out related information. Recommender system can be divided mainly into two types—content-based and collaborative filtering. There is another type called hybrid which is usually done by the combination of content-based and collaborative filtering. In our work, we have used collaborative filtering to build the food recommender system.

1.1 Collaborative Filtering

People have been sharing ideas and opinions for centuries and that is the origin of what we call collaborative filtering. It is a process that evaluates and filters people's likes and dislikes for a particular product or item. Collaborative filtering is a common and popular recommendation algorithm that predicts and recommends based on past behavior of the system's users or ratings given by the users [1, 2].

Collaborative filtering can be divided into two types: memory-based and model-based. In memory-based approach, the system makes up a relationship between users and items in order to suggest users an item. The whole dataset is being used to find out the similar values between users and/or items. Methods such as Euclidean distance, Pearson correlation [3], Jaccard similarity, cosine similarity, can be used to implement memory-based system. For model-based system, linear regression, matrix factorization, Bayesian clustering, associative classification can be used.

2 Related Works

The online food menu recommender system proposed by Bundasak et al. [4] has been built using collaborative filtering and slope one predictor. It is a memory-based recommender system that works with users' ratings and their preferences to suggest menus. There are mainly three steps in this system—menu items rating, collaborative filtering, and slope one predictor. And at the end, the system recommends top N menu items to the user.

Li et al. [5] proposed a recommender system which uses web-based data mining and the system recommends healthy meals to the users. This is a recommender system especially for health-related issues that analyze health conditions and food intake of a user. They proposed that there is a Web site where users had been ordering different food items of different recipes and from there they collected all the users' information using data acquisition. Data mining algorithms like clustering, classification, association rules have been used to extract the necessary user data.

A personalized suggestion system proposed by Agapito et al. [6] monitors a user's health profile and recommends him or her healthy food meal. The profile of the user has been built based on real-time questionnaires by medical doctors, and the answers are provided by the users. The system has been built up by their self made algorithm which focuses on users' profile and their health status and suggest meals based on that. The result of this system is typical Calabrian foods are suggested by DIETOS to the users in three ways: (i) according to the user's health profile, foods are suggested spontaneously; (ii) showing the place on a map where the Calabrian foods are made; and (iii) displaying the nutritional properties of each product stored in the database, benefits, and side effects of pathologies and specific health conditions.

Gaur et al. [7] majorly focused on two initial dimensions of food recommendations: calories intake by the person and the calories which are left unused for that person. The author proposed a model that used information taken from the university's students belonging to the age group 18–24. This information includes the basic details of the users like age, weight, height, lifestyle, disease (if any), and food taken by that user till evening. This information is given as input to calorie and BMI calculators for calculating the BMI and calorie consumed by that person, and it is based on the information of a person defined above which will result in the body mass index calculator (BMI calculator) is used to calculate the total fat of the body on the basis of weight and height. After calculating these and taking into accounts, the user's BMI, DNA, and genetic disease—a suitable recommendation of food is given to the user that will be beneficial for his or her health.

3 Methodology

Our food recommender system has been built in two ways—memory-based and model-based approaches of collaborative filtering [8]. The memory-based approach is used cosine similarity [9, 10], and model-based approach is used matrix factorization [10, 11] which is a machine learning technique.

3.1 Memory-Based Recommender System

Memory-based collaborative filtering can be divided into two types: user–user collaborative filtering and item–item collaborative filtering which are basically neighborhood methods.

For our memory-based recommender system, we have used item–item collaborative filtering. In item–item collaborative filtering, the similarity values between items are found and the most similar item to an item is suggested to the user.

The most significant part of memory-based recommender system is creating the user–item matrix and then finding the similarity values between the items.

In order to build a food recommender system, we have accumulated an approximate of 100 users' information that contains the users' personal details such as age, height, weight. We gave the users some selected 100 meal sets with respective calories for each meal set, in context of Bangladesh, for which they had to rate the meal sets on a rating scale of 1–5. Any meal set rated a value of 5 means that meal set is most healthy for the user and any meal set rated a value of 1 means that meal set is least healthy for the user.

Our food recommender system will contain user profiles who have already rated our given meal sets in the system by creating a profile of their own. They can also add new meal sets and rate them.

The following example, with ten users and five meal sets, shown below describes how our memory-based food recommender system works

I. A meal set (X) is given as user input. The meal set must be present in our database in order to get a recommendation as replacement for the input meal set.
II. Creation of user–item matrix from the dataset. This small portion of dataset (Fig. 1) has been used for better illustration and explanation of how our food recommender system works.

In Fig. 1, ROWS are the users and COLUMNS are the meal sets (items). The user–item matrix illustrates all the ratings given by each user to the meal sets they consume. The higher the rating, the healthier is the food for them. For example, User 1 gave Item 1101 (bread, omelet) a 4 out of 5 and gave Item 1102 (omelet, boneless chicken) a 5 out of 5. All the ratings are in the range of 1–5.

The meal sets (items) which are not given any rating are assumed to be rated a zero by default. For example, User 2 has not given any rating to Item 1101 so the rating for Item 1101 by User 2 is a zero.

	Meal set (Item)				
User	bread, omelette	omelette, boneless chicken	bread, fruit jam, green tea	bread, low fat milk, poached egg	cereal, low fat milk
	1101	1102	1103	1104	1105
1	4	5			
2				5	
3	5				2
4	1				4
5					
6			5		
7			1		
8			2		
9					
10				2	

Fig. 1 user–item matrix of ten users and five meal sets

In case of cosine similarity, a similarity value of 0 means least similar and a value of 1 means most similar. In our user–item matrix, all our ratings are positive. So all the item–item cosine similarity values will be in the range between 0 and 1.

III. Finding similarity between meal sets (items)

The similarity values between meal sets have been found using cosine similarity. Cosine similarity finds similarity by calculating the angle between two vector lines. Two items are considered as two vector lines in an n-dimensional space, and the angle between them is the cosine angle. The smaller the angle, the more similar the two items are. If the number of users is 'n', then the vector lines of the two items will be in n-dimensional space. Here we have considered ten users so the similarity values between any two items will be found in ten-dimensional space.

If we want recommendation for the meal set Item 1101, then we need to find out the similarity values between Item 1101 and all other remaining items in the dataset.

The cosine similarity value between two items i_m and i_b is calculated by

$$s_u^{\cos}(i_m, i_b) = \frac{i_m . i_b}{||i_m||||i_b||} = \frac{\sum x_{a,m} x_{a,b}}{\sqrt{\sum x_{a,m}^2 x_{a,b}^2}} \tag{1}$$

where

$i_{m=}$ Item m
$i_{b=}$ Item b
$x_{a,m=}$ rating of 'x' of user 'a' on item 'm'
$x_{a,b=}$ rating of 'x' of user 'a' on item 'b'

Now let us apply this formula on our user–item matrix (Fig. 1).

Item 1101 = (4,0,5,1,0,0,0,0,0,0)
Item 1102 = (5,0,0,0,0,0,0,0,0,0)
Item 1103 = (0,0,0,0,0,5,1,2,0,0)
Item 1104 = (0,5,0,0,0,0,0,0,0,2)
Item 1105 = (0,0,2,4,0,0,0,0,0,0)

$$s_u^{\cos}(\text{Item } 1101,\ \text{Item } 1102) = \frac{(4\times 5)+(0\times 0)+(5\times 0)+(1\times 0)+(0\times 0)+(0\times 0)+(0\times 0)+(0\times 0)+(0\times 0)+(0\times 0)}{\sqrt{(4^2+0^2+5^2+1^2+0^2+0^2+0^2+0^2+0^2+0^2)\times(5^2+0^2+0^2+0^2+0^2+0^2+0^2+0^2+0^2+0^2)}} = 0.6172133998$$

$$s_u^{\cos}(\text{Item } 1101,\ \text{Item } 1103) = 0$$
$$s_u^{\cos}(\text{Item } 1101,\ \text{Item } 1104) = 0$$
$$s_u^{\cos}(\text{Item } 1101,\ \text{Item } 1105) = 0.4830458915$$

From the above calculation, we can conclude that Item 1101 and Item 1102 are the most similar meal sets in this small portion of dataset. So when the user will give

Item 1101 as an input, our food recommender system will suggest him or her to have Item 1102 among the other four meal sets.

3.2 Model-Based Recommender System

For building the model-based food recommender system, we have used the latent factor model of collaborative filtering. Latent factor model characterizes both the user and the item to explain ratings. The rudimentary idea is that vectors of latent factors are inferred from ratings' pattern and matrix factorization characterizes users and items by this. The higher the correspondence between the users and items, the greater the chance of recommendation.

The most significant part of model-based recommender system is filling up the sparse matrix by alternating least square algorithm minimizing the error.

We have used the same dataset that has been used before to implement the memory-based food recommender system.

Our model-based food recommender system will contain user profiles who have already rated our given meal sets in the system by creating a profile of their own. They can also add new meal sets and rate them.

The following example, with ten users and five meal sets, shown below describes how our model-based food recommender system works

I. Give the user id as input. The user must be present in our database in order to get a recommendation for meal sets that he or she has not eaten yet.
II. Creation of user–item matrix from the dataset (same user–item matrix as shown in Fig. 1).
III. Factorize the user–item matrix into two different matrices.

In this step, the user–item matrix (Fig. 1) is factorized into individual user matrix and item matrix by alternating least square algorithm. And if the user matrix and item matrix are multiplied back together, it produces an approximation of the original user–item matrix.

For illustration, we have used Fig. 2.

After formation of individual matrices U and P (Fig. 2), alternating least square (ALS) puts random values in the cells of the matrices, calculating an error term and then continues to alternate the values back and forth between matrix U and matrix P until the error is minimized. Once this process is completed, the matrices U and P are multiplied back together and the blank cells in the user–item matrix (Fig. 2) are filled up with predicted ratings as shown in Fig. 3.

In Fig. 3, the predicted ratings N are in the range of 1–5 which is calculated randomly by the algorithm. For instance, if the latent factors of Item 1101 and Item 1104 are similar and the predicted rating of Item 1104 is one of the highest values among all other items for User 3, then User 3 will get Item 1104 as one of the top meal sets. Here latent factors can be the type of meal set; that is, both the items are for breakfast; the calories of the meal sets are similar.

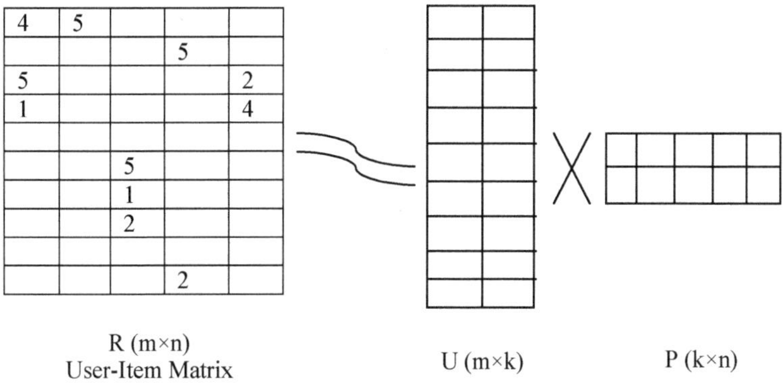

Fig. 2 Matrix factorization of user–item matrix

	Meal set (Item)				
User	bread, omlette	omlette, boneless chicken	bread, fruit jam, green tea	bread, low fat milk, poached egg	cereal, low fat milk
	1101	1102	1103	1104	1105
1	**3.99**	**4.95**	N	N	N
2	N	N	N	**4.99**	N
3	**4.78**	N	N	N	**1.93**
4	**0.98**	N	N	N	**3.98**
5	N	N	N	N	N
6	N	N	**4.90**	N	N
7	N	N	**0.88**	N	N
8	N	N	**1.85**	N	N
9	N	N	N	N	N
10	N	N	N	**1.99**	N

Fig. 3 Filled up user–item matrix

3.3 Apache Spark

Apache Spark is a big data processing framework which enables us to use some of the most popular tools to implement big data-related tasks. We used Apache Spark to utilize Spark MLlib which is a machine learning library that consists of different algorithms and utilities. We have implemented alternating least square (ALS) algorithm for our recommendation system using MLlib. Moreover, we used Spark so that in the future this system can take mass volume of data as input and for making it efficient, scalability, and fast processing are needed.

4 Result and Analysis

For the memory-based food recommender system, if a user gives the unique meal set id of any existing meal set of the system as input, then he or she will get a list of top ten meal sets as recommendation that match the user's preferences. You can see the user has given 2201 as input for the meal set rice, mixed vegetables as shown in Fig. 4.

For the model-based food recommender system, the user has his or her personal unique user id using which he or she can rate any meal set in the system and can get recommendation for new meal sets which the user has not tasted yet. For that only the user requires to give his or her user id as input and then he or she will get top ten meal sets' recommendation that has been found by comparing the user's previous preferences. Figure 5 shows that for the user who has id 8 has got a list of top ten meal sets which he or she might like to consume according to his or her preferences.

The memory-based food recommender system mainly focuses on how much an item (meal set) is liked by the users. It figures out the similar ratings of an item (meal set) pair and checks whether they are similar meals or not. This means if there is

```
c:\spark\bin>spark-submit food-similarities.py 2201

Loading food names...
Top 10 similar foods for rice|mixed vegetables
flour bread|mixed vegetables    score: 0.9561828874675149
corn soup|stir fry vegetables   score: 0.8944271909999159
boiled vegetables|omlette       score: 0.7324670207647144
chicken corn soup       score: 0.7254762501100117
bread|boiled egg|orange juice   score: 0.5595028849441883
flour bread|mixed vegetables|salad      score: 0.558156305651438
rice|beef|fish  score: 0.529150262212918
rice|stir fry spinach|lentils   score: 0.3162277660168379
chicken salad|green tea score: 0.29361010975735174
rice|chicken curry      score: 0.24806946917841693
```

Fig. 4 Top ten meal sets recommendation of memory-based food recommender system

```
c:\spark\bin>spark-submit food-als-input-user-id.py 8
Loading food names...

Top 10 recommendations for user ID 8:
oats|milk  score 4.97781590676389
cereal|milk|banana  score 4.957016308295558
vegetable oats  score 3.9904579362164734
oats|apple|milk  score 3.9891894117898232
cereal|milk|almonds  score 2.971070602355025
rice|chicken|lentils  score 2.0076163269128826
bread|fruit jam|green tea  score 1.9913363771694286
chicken salad|green tea  score 1.9897783116558676
chicken corn soup  score 1.9817899511876291
rice|boiled vegetables|fish curry  score 1.9771210298904378
```

Fig. 5 Top ten meal sets recommendation of model-based food recommender system no similarity of ratings between any pairs of meal sets, then the system will give an average recommendation for the meal set alternative to what the user is looking for. So there must be such dataset that has at least one similar meal set for a given meal set that is both the meal sets must be given similar values by at least one user. Here no matter how poor or good ratings are given by the users to the meal sets it is not taken into account. This means if two meal sets are given a rating of 1 by the same user, then these two meal sets are very similar. But if one meal set is rated a 1 and another meal set is rated a 5, then these are very dissimilar meal sets thus in such case we will get average recommendation for a meal set. Even if a meal set is rated very high by the users if that meal set does not have any similar rating with other meal sets, then the recommendation of alternative meal sets will be very poor. Moreover, if a meal set is rated by only one user and the user has not rated any other meal sets, then the recommendation system will have no similar meal sets for that meal set and therefore no recommendation for that meal set. The memory-based recommender system using cosine similarity also assumes that the rating of a meal set is zero if it has not been rated by any user. So it finds very dissimilarity among items which are rated and which are not rated. Thus, again makes poor recommendation for a given meal set.

On the other hand, model-based food recommender system not only focuses on the values of the ratings given by the users. The machine learning technique matrix factorization uses the existing ratings of meal sets to predict the ratings of other meal sets that have not been rated by the users taking into account of other parameters of the meal sets such as the type of meal set, the calories of the meal set. These predicted ratings are fully done by the system, and there is no scope of any error if the ratings on the meal sets given by the users are absolutely meaningful and correct. But in a case where an item X is rated a 5.0 by user A, a 4.0 by user B, a 1.0 by user C, and D has not rated item X. Now when the values of the ratings are computed for training and getting the predicted ratings; sometimes the lowest rating value compared with other ratings of an item is not considered. Here the rating value 1.0 might not be computed for training the existing values and computing the predicted ratings. This can make the recommendation very inefficient sometimes, and thus, scaling is really necessary in such cases.

Another big factor that works for both the approach is how the users rate the meal sets. If there is any wrong or misleading rating on the meal sets despite the meal sets being very good for health, then also there is a possibility of getting wrong or poor recommendation for meal sets. So the dataset must be arranged in such a way that there is no scope of incorrect or false ratings on the meal sets.

Here we cannot exactly compare between these two types of recommender system. But we can conclude that the model-based recommender system works best since there is no scope of assuming an unrated meal set a zero that minimizes the dissimilarity among different meal sets which happen in case of memory-based recommender system. Moreover, in the model-based food recommender system the latent factors like calories, type of meal sets are also taken into account during recommendation which is not done by the memory-based food recommender system. These kinds of meal information play an important role in recommending people with healthy food. It narrows down the food recommendation by suggesting users only with the meal sets that are within the calorie range of their usual food intake.

5 Conclusion

To conclude, we tried to build a food recommender system by using two different collaborative filtering algorithms. We have seen that the model-based recommender system is more effective than that of the memory-based system which suggests items to users based on only the similar ratings no matter how poor the similar ratings are. In other words, even if a meal set is not healthy for person in terms of calories, it might be recommended to him or her. But the model-based system takes into account the other parameters such as calories and suggests users with meal sets that are within the users' preferred calorie intake. This model-based recommendation engine will hopefully help those people who have a boring, fixed eating pattern and make their food consumption an interesting one by helping them to choose variety of meal sets.

References

1. Leskovec, J., Rajaraman, A., Ullman, J.D.: Mining of Massive Datasets, 2nd edn. Cambridge University Press, New York, NY, USA (2014)
2. Schafer, J.B., Frankowski, D., Herlocker, J., Sen, S.: Collaborative filtering recommender systems. In: Brusilovsky, P., Kobsa, A., Nejdl, W. (eds.) The Adaptive Web. Lecture Notes in Computer Science, vol. 4321. Springer, Berlin, Heidelberg (2007)
3. Arsan, T., Köksal, E., Bozkuş, Z.: Comparison of collaborative filtering algorithms with various similarity measures for movie recommendation. Int. J. Comput. Sci. Eng. Appl. (IJCSEA) **6**(3), 1–20 (2016). https://doi.org/10.5121/ijcsea.2016.6301
4. Bundasak, S., Chinnasarn, K.: eMenu recommender system using collaborative filtering and slope one predictor. In: The 2013 10th International Joint Conference on Computer Science and Software Engineering (JCSSE), pp. 37–42. IEEE, Maha Sarakham, Thailand (2013). https://doi.org/10.1109/jcsse.2013.6567316
5. Li, X., Liu, X., Zhang, Z., Xia, Y., Qian, S.: Design of healthy eating system based on web data mining. In: 2010 WASE International Conference on Information Engineering, pp. 346–349. IEEE, Beidaihe, Hebei, China (2010). https://doi.org/10.1109/icie.2010.89
6. Agapito, G., Calabrese, B., Guzzi, P.H., Cannataro, M., Simeoni, M., Car´e, I., Lamprinoudi, T., Fuiano, G., Pujia, A.: DIETOS: a recommender system for adaptive diet monitoring and personalized food suggestion. In: 2016 IEEE 12th International Conference on Wireless and Mobile Computing, Networking and Communications (WiMob), pp. 1–8. IEEE, New York, NY, USA (2016). https://doi.org/10.1109/wimob.2016.7763190
7. Gaur, N., Singh, A.: Recommender system—making lifestyle healthy using information retrieval. In: 2016 2nd International Conference on Next Generation Computing Technologies (NGCT-2016), pp. 479–484. IEEE, Dehradun, India (2016). https://doi.org/10.1109/ngct.2016.7877463
8. Sarwar, B., Karypis, G., Konstan, J., Riedl, J.: Item-based collaborative filtering recommendation algorithms. In: WWW '01 Proceedings of the 10th international conference on World Wide Web, pp. 285–295. ACM, New York, NY, USA (2001). https://doi.org/10.1145/371920.372071

9. Nandi, M.: Recommender systems through collaborative filtering. (2017). https://blog.dominodatalab.com/recommender-systems-collaborative-filtering/
10. Levinas, C.A.: An analysis of memory based collaborative filtering recommender systems with improvement proposals. UPCommons (2014)
11. Koren, Y., Bell, R., Volinsky, C.: Matrix factorization techniques for recommender systems. Computer **42**(8), 30–37 (2009). IEEE. https://doi.org/10.1109/mc.2009.263

Performance Analysis of Iris Recognition System

Ruqaiya Khanam, Zohreen Haseen, Nighat Rahman and Jugendra Singh

Abstract A biometric system offers automatic identification of an individual based on a unique feature or characteristic possessed by the individual. Iris recognition is regarded as the most reliable and accurate biometric identification system available. Although iris identification system is based on pattern recognition technique but due to poor iris boundary detection and high computational time in previous work, we used neural network and discriminant machine learning technique to obtained high accuracy. In this work, we implement neural network and discriminant analysis of machine learning method for iris recognition in iris images to implement in day-to-day life, using MATLAB 2016a. The emphasis will be only on the software for performing recognition and not hardware for capturing an eye image. The proposed method gives better recognition rate than SVM technique with less computational complexity. Neural network and discriminant methods are used for matching and finding recognition accuracy. Thus, the accuracy obtained from neural network is 94.44%, whereas from discriminant analysis the accuracy obtained is 99.99%.

Keywords Iris recognition · Neural network · Machine learning

1 Introduction

The advancement of technology and expand importance of security have gained more awareness toward biometric identification system. Biometric systems are used based

R. Khanam (✉) · Z. Haseen · N. Rahman · J. Singh
Galgotias University, Greater Noida 201310, India
e-mail: dr.ruqaiya@gmail.com

Z. Haseen
e-mail: zovahaseen93@gmail.com

N. Rahman
e-mail: nighatrehman08@gmail.com

J. Singh
e-mail: jugendra12031993@gmail.com

L. C. Jain et al. (eds.), *Data and Communication Networks*, Advances in Intelligent Systems and Computing 847, https://doi.org/10.1007/978-981-13-2254-9_14

on face, iris, fingerprint, gait parameter, etc. Application of biometric systems is used for unique identity, border security, airport security, etc. Biometric identification is the process of an automatic identification based on characteristic or unique feature of an individual or a person. In 1980s two ophthalmologists, Leonard Flom and Aran Safir proposed that even in twins no two irises are alike thus making them a good biometric. By 1994, the algorithms have been developed and patented and further now used as the basis for iris recognition systems [1, 2]. The iris starts to develop in third month of gestation and complete its structure by the eighth month, however, pigment accretion continues to the first postnatal years. Each individual has a unique iris even the left and right eye of a particular individual differs in iris. Through the strategies of image processing, uniqueness of an iris pattern can be extracted from digitized image of the eye and thus encode it into a biometric template, which saved in database. This biometric template contains a mathematical representation of the unique data stored in the iris and enables comparisons to be made between templates.

2 Related Work

The first scientist who developed the algorithm for iris recognition system was Daugman integrodifferential operator was used for iris localization and for iris normalization rubber sheet model of Daugman was used. The matching of two iris codes was done by Hamming distance. Lim proposed an efficient method of personal identification having high level of stability and distinctiveness. In this paper, Haar wavelet is used to extract the features from iris image. Navjot provides the review of existing methods as proposed by various researchers for iris recognition. Mohd. Tariq proposed an algorithm by using 1D Gabor filter for the extraction of feature, normalizing, and segmenting the iris and pupil boundaries of eye from database images. Further, Proenca proposed a method which encloses three main parts that are eye detection, iris segmentation, and discrimination of noise-free iris texture. An algorithm implemented by Mayank was used to enhance both accuracy and speed of iris recognition. Samir relates a GACs (Geodesic Active Contours) with an iris segmentation scheme which extract the iris from nearby structures. The scheme invokes iris texture further assisted by local and global properties of the image. Zhaofeng presented an algorithm for both fast and accurate iris segmentation. FAWAZ is known for proposing multi-algorithmic approach to enhancing the security of iris recognition system and can be achieved by fusing the data acquired at the feature level and applying the K-nearest neighbor classifier (K-NN).

3 Background Theory of Proposed Work

3.1 Biometric Technology

Biometric generally describes a characteristic or a process. For automatic recognition, it is used. Its framework provides a programmed recognition of a person taking same kind of special elements or trademark governed by the person. A biometric system generally involves three modules as: recognition, verification, identification (Fig. 1).

Sensor Module: It defines the connection of the human with the system and thus vital to the execution of the biometric system.

Feature Extraction and Quality Assessment Module: The quality of the biometric data attained by the sensor is assessed for feature extraction. During enrollment, the extracted feature is stored in the database known as "template", representing the identity of an individual.

Matching Module: The identity of an individual is verified by comparing the template to the input (query) biometric data of an individual.

Decision-Making Module: This is to verify an exact identity or to provide ranks to the enrolled identities. Usually, the match score is compared against the "threshold" to determine the authenticity of an individual.

3.2 The Human Iris

To control the amount of light entering through the pupil is the function of iris. The average diameter of the iris is 12 mm, and the pupil size can vary from 10 to 80% of the iris diameter [3]. The pigmentation of iris depends on genetics which determines its color.

Fig. 1 Working of biometric system

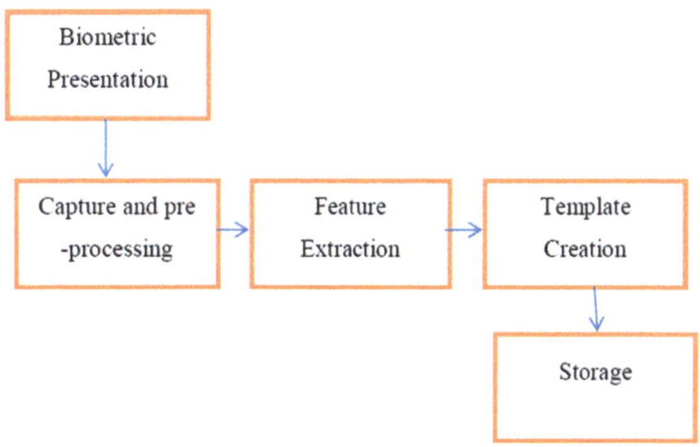

3.3 Different Recognition System

Biometric recognition, or biometrics, refers to the automatic identification based on anatomical or behavioral characteristic or unique feature of an individual. Some of the recognition systems are fingerprint recognition, speech recognition, iris recognition.

3.4 Work Methodology

Steps in Iris Recognition System

Image Acquisition: It is the first step of the image processing and can be defined as recovering an image from some source. High-quality image acquisition techniques are used for iris recognition to make accurate models of different surfaces.

Image Preprocessing: It aimed to enhance ability to interpret quantitatively image components. It removes low-frequency noise, normalizes the intensity of individual images, and removes reflections.

Segmentation: The accomplishment relies on upon the imaging nature of images of eye. Because of the exertion of close infrared light for enlightenment, images in the CASIA iris database [4] do not contain specular reflections.

Normalization: The next stage after iris region segmentation is to change the iris region, so it has fixed measurements to allow comparisons. There are dimensional irregularities between eye images due to extending of iris created by pupil expansion from shifting levels of brightening. The normalization methodology will create iris regions, which have the same measurements, so two photos of the identical iris under exclusive conditions will have characteristic features at the same spatial area.

Feature Encoding and Matching: The template that is created in the feature encoding procedure will require a relating matching metric, which allow comparison between two iris templates (Fig. 2).

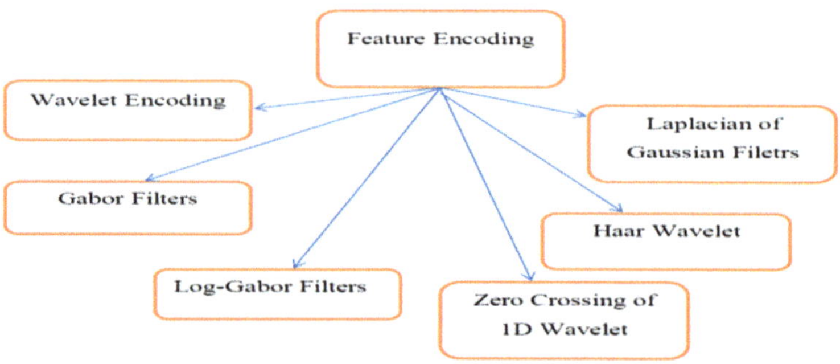

Fig. 2 Technological perspectives in feature encoding module

- **Feature Encoding Algorithms**: Wavelets can be deployed to decompose the information in iris region into constituent that show up at different resolutions.
- **Gabor Filters**: Daugman encodes iris pattern data by using 2D variant of Gabor filters. A 2D Gabor filter over the image space (x, y) is represented as:

$$G(x, y) = \mathrm{e}^{-\pi\left[(x-x_o)^2/\alpha^2+(y-y_o)^2/\beta^2\right]}\mathrm{e}^{-2\pi i[u_o(x-x_o)+v_o(y-y_o)]}$$

where (x_o, y_o) is position in the image, (α, β) is the effective width and length, and (u_o, v_o) determine modulation, which has spatial frequency. $\omega_o = \sqrt{\left(u_o^2 + v_o^2\right)}$.

- **Haar Wavelet**: Oppenheim and Lim [5] use the wavelet change to extract features from the iris region. From multi-dimensionally filtering, a feature vector with 87 estimations is featured. Since every estimation has a real value expanding from −1.0 to +1.0, the feature vector is sign quantized so that any positive value is represented by 1 and negative by 0.

3.5 *Matching Algorithms*

Matching Algorithms are used for verification as well as identification functions. Three algorithms which are repeatedly used in iris recognition technology discussed below are:

- **Hamming Distance**: The measure of same number of bits between two bit patterns is given by hamming distance. The hamming distance for two bit patterns X and Y is defined as the sum of discarding bits (sum of the exclusive-OR between X and Y) over N, the total number of bits in the bit pattern.

$$\mathrm{HD} = \frac{1}{N}\sum_{i-1}^{N} X_i(\mathrm{XOR})Y_i$$

- **Weighted Euclidean Distance**: In order to compare two templates composed of integer values WED is used. The metric is employed by Zhu et al. [6] and is specified as

$$\mathrm{WED} = \frac{1}{N}\sum_{i=1}^{N}(f_i - f_i^{(k)})^2/(\delta_i^{(k)})^2$$

where f_i is the ith feature of the unknown iris, and $f_i^{(k)}$ is the ith feature of iris template, k and $\delta_i^{(k)}$ is the standard deviation of the feature in iris template k.
- **Normalized Correlation**: It is represented as

$$\sum_{i=1}^{n}\sum_{j=1}^{m}(p_1[i,j]-\mu_1)(p_2[i,j]-\mu_2)/nm\sigma_1\sigma_2$$

where p_1 and p_2 are two images of size $n \times m$, μ_1 and σ_1 are the mean and standard deviation of p_1, and μ_2 and σ_2 are the mean and standard deviation of p_2.

4 Proposed Work and Techniques Used

The implementation of proposed work is shown in Fig. 3.

4.1 Feature Extraction Technique

The feature map of iris image is extracted with the help of Daugman rubber sheet model which is used for the normalization of image and then Haar wavelet is used for the feature extraction.

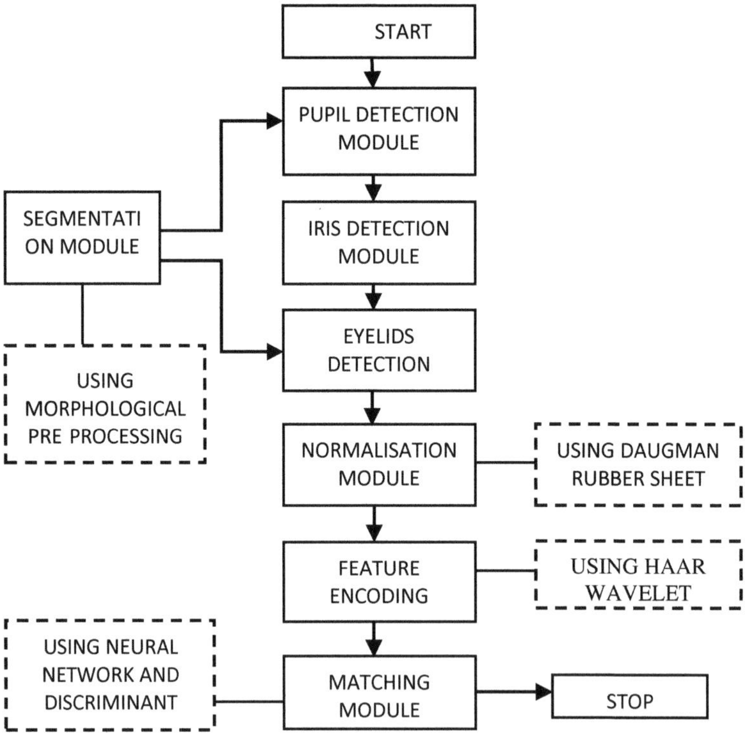

Fig. 3 Flow chart of proposed algorithm

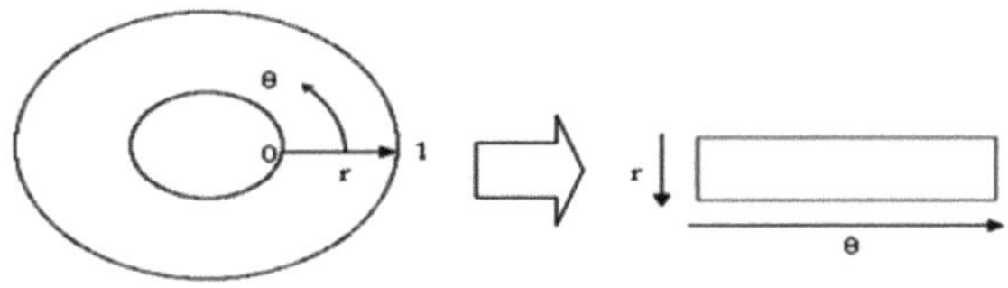

Fig. 4 Daugman's rubber sheet model

Daugman's Rubber Sheet Model: The homogenous rubber sheet model elaborate by Daugman relocates every point inside the iris area to a couple of polar directions (r, θ) where r is on the interval [0, 1] and θ is angel $[0, 2\pi]$ (Fig. 4).

The relocating of iris region from Cartesian coordinates (x, y) to the normalized non-concentric polar representation is modeled as

$$I(x(r, \theta), y(r, \theta)) \rightarrow I(r, \theta)$$

With, $x(r, \theta) = (1 - r)x_p(\theta) + rx_1(\theta)$

$$\mathrm{y}(r, \theta) = (1 - r)y_p(\theta) + ry_1(\theta)$$

where, $I(x, y)$ is iris region picture, (x, y) are the Cartesian coordinates, (r, θ) are the relating standardized polar directions, and are the directions of the pupil and iris limits along the θ direction.

4.2 Matching and Accuracy Determination

In this proposed work, the accuracy of the feature extraction is determined with the help of discriminant analysis and neural network.

Machine Learning Algorithm: A type of artificial intelligence learning algorithm which helps the computers the ability to learn without being clear-cut programmed. There are various types of machine learning algorithm: supervised, unsupervised, semi-supervised, reinforce.

5 Experimental Result

Every first image in result modules represent our best performance as when we compare characteristically features of our first image with others we find considerable difference in image quality but the catch is that for most of images taken we have got such technical results which are relevant to our above assumptions. Outputs or results are shown in sequential manner (Fig. 5).

The image is shown in Fig. 6 is binarized behavior of original iris grayscale images. The image of this type is used for further segmentation of iris and then its localization.

In Fig. 7, the image quality indicates a good level of clarity in the image representation. The background is noise free and representations of pixels are 0. This is a standard image which will be used for masking and making the image noise free.

The current image has been used to show the red circles that are marked in the image. Here when the image is seen highlighted with circles, the radii so evaluated are marked only for a certain region or at an angular area shown in Figs. 8 and 9.

Fig. 5 Image after binarization of iris image

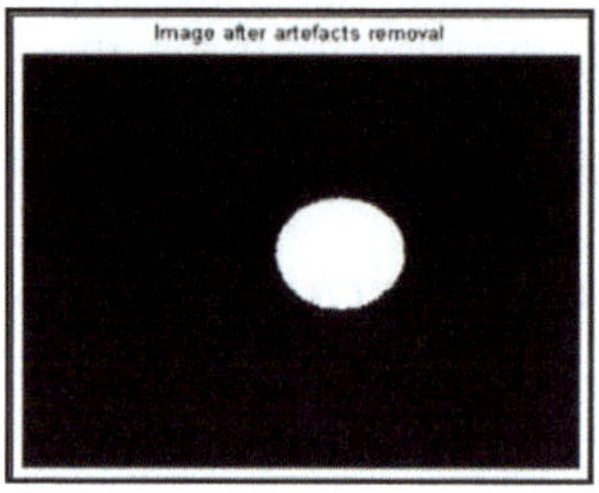

Fig. 6 Image after artefacts removal

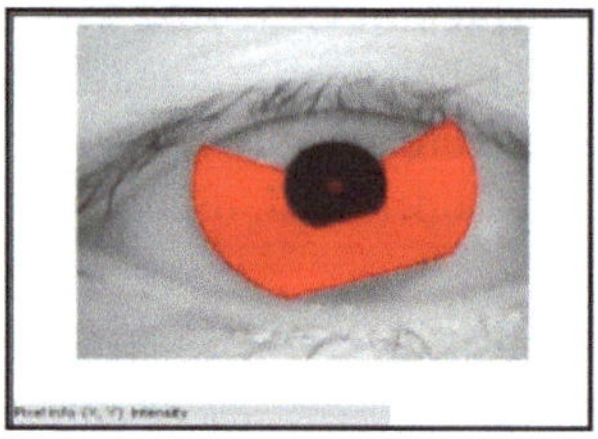

Fig. 7 Original iris image holding the red circles with a certain angle

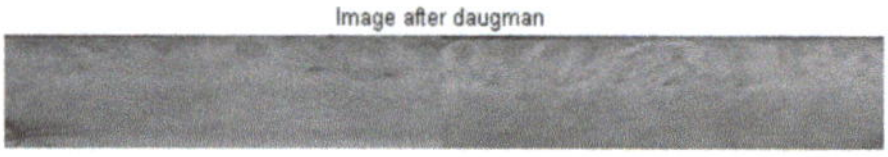

Fig. 8 Image after Daugman filtering

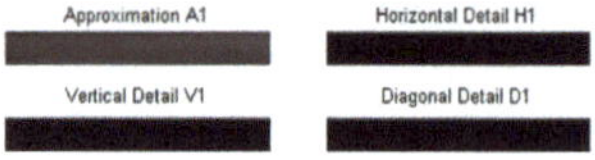

Fig. 9 Detailed, approximated, horizontal, and diagonal components of wavelets

Fig. 10 Image shows neural network training

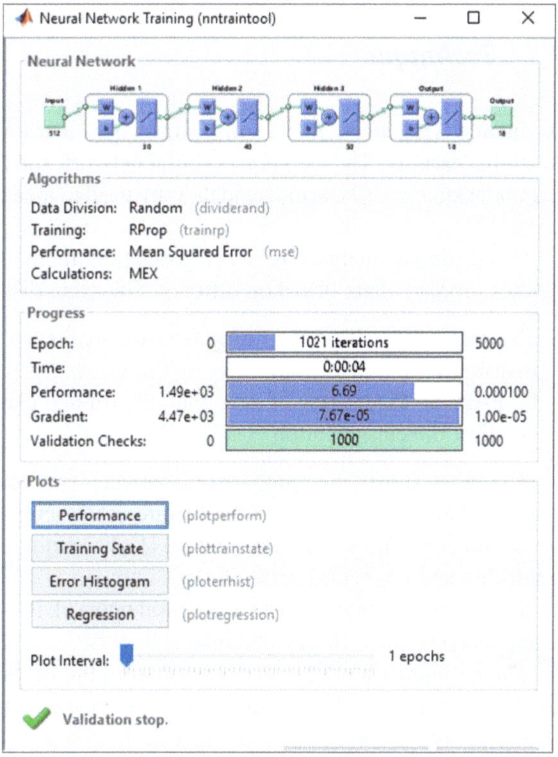

To enable the transform of circular region into a horizontal region the image so calculated is divided in two regions. Now if Fig. 10 is compared to Fig. 9, the circular iris region will be considered as polar coordinates and then the red marked semicircles are reduced to Daugman Cartesian coordinates. The image so seen here is a combination of left and right region of iris.

This image is imperative reason being that image here is a segmented image or partitioned image which will reduce the image size to either half dimension or to a lower dimension. If this wavelet analysis is done then ultimate aim of the image is to calculate the results as features to identify the prime features. In this research the approximate image seen at the upper left corner will be reduced and used, the total pixels will be approximately 512.

The feature of four images of subject 1 selected shown by number of columns, here the number of images will be four and for each image 100 prime features are shown out of 512. If calculations are traced, the features are evaluated using Haar wavelet.

5.1 Accuracy Measurement and Comparison with Other Technique

Accuracy Measurement: In this supervised learning is used which is a part of machine learning. The techniques neural network and discriminant analysis are coming under supervised learning and they are used to measure the accuracy in the present work.

Discriminant analysis is a categorization method. It concludes that different classes produce data, based on different Gaussian distributions:

- To create a classifier, the fitting function evaluates the parameters of a Gaussian distribution for each class.
- To estimate the classes of new data, the trained classifier finds the class with the smallest misclassification cost.

Figure 11 shows the comparisons between two techniques used in the work for the calculation of accuracy of the features abstracted from the images of iris of the given subjects. The techniques used are (1) discriminant analysis of machine learning algorithm and (2) neural network.

The accuracy obtained from the neural network is 94.44%, whereas from discriminant analysis the accuracy obtained is 99.99%.

The comparison of results obtained from the research work is well described in Table 1. Accuracy of the features of human iris images has been derived in this

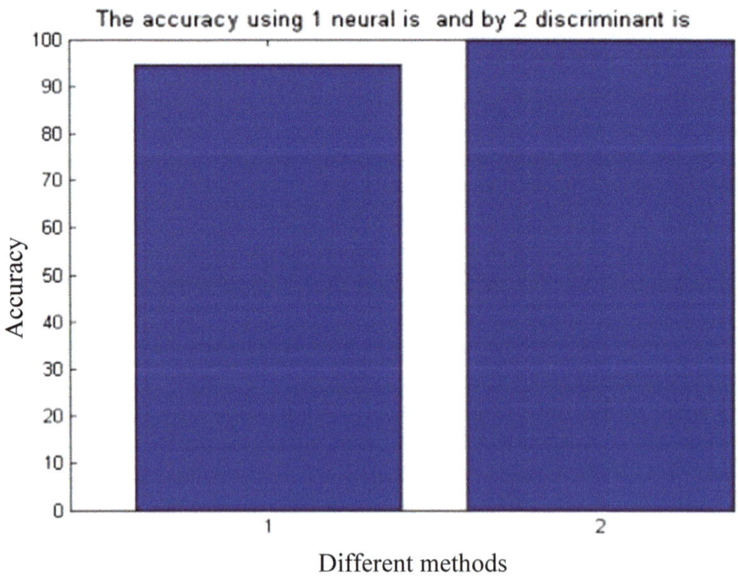

Fig. 11 Matching accuracy of two mentioned methods

Table 1 Recognition accuracy with different feature vector on CASIA image database

Methodology	Accuracy (%)	False acceptance rate (%)
SVM with Haar	91.33	8.66
Hamming distance	99.60	0.33
SVM with 1D log Gabor	99.65	0.33
Proposed work 1 (Haar + neural network)	94.76	5.21
Proposed work 2 (Db1 + neural network)	94.76	5.22
Proposed work 3 (Haar + discriminant analysis)	99.99	0.01
Proposed work 4 (Db1 + discriminant analysis)	99.99	0.01

proposed work after enhancing the acquired images of different subjects obtained from the CASIA data source [7].

6 Conclusion

In this proposed work, an efficient technique for recognition and feature extraction was explained. The crisscross collarette region of iris was picked because it is the most significant region of iris complex layout due to which high acknowledgment rate has been taken out. Haar wavelet and Daubechies wavelet were used for evicting out the features; these extracted features were utilized in the iris recognition which was using the feedforward neural network technique for recognition. The proposed system also used the discriminant analysis approach for the matching stage with the use of same extracted feature. This also gives better recognition rate than SVM technique with less computational complexity. The performance accuracy of present work is best in the favor of CASIA as well as check image database. So for both identification and verification, the proposed work is efficient.

7 Experimental Result

Every first image in result modules represent our best performance as when we compare characteristically features of our first image with others we find considerable difference in image quality but the catch is that for most of images taken we have got such technical results which are relevant to our above assumptions.

References

1. Daugman, J.: High confidence visual recognition of persons by a test of statistical independence. IEEE Trans. Pattern Anal. Mach. Intell. **15**(11), 1148–1161 (1993)
2. Daugman, J.: Biometric personal identification system based on iris analysis. U.S. Patent No. 5,291,560, 1 Mar 1994
3. Wolff, E.: Anatomy of the Eye and Orbit, 7th edn. H. K. Lewis & Co. LTD
4. Chinese Academy of Sciences—Institute of Automation: Database of 756 grayscale eye images. http://www.sinobiometrics.com Version 1.0 (2003)
5. Oppenheim, A., Lim, J.: The importance of phase in signals. Proc. IEEE **69**, 529–541 (1999)
6. Zhu, Y., Tan, T., Wang, Y.: Biometric personal identification based on iris pattern. In: ICPR2000: The 15th International Conference on Pattern Recognition, pp. 805–808, Barcelona, Spain (2002)
7. CASIA iris image database. http://www.sinobiometrics.com
8. Lim, S., Lee, K., Byeon, O., Kim, T.: Efficient iris recognition through improvement of feature vector and classifier. ETRI J. **2**, 61–70 (2001)
9. Daugman, J.: How iris recognition works. In: Proceedings of 2002 International Conference on Image Processing, vol. 1 (2002)
10. Kaur, N., Juneja, M.: A review on iris recognition. In: Proceedings of 2014 RAECS, UIET Panjab University, Chandigarh, 6–8 Mar 2014
11. Khan, M.T., Arora, D.: Feature extraction through iris images using 1-D Gabor filter on different iris datasets. IEEE (2013). ISBN 978-1-4799-0192-0/13
12. Proenca, H.: Iris recognition: on the segmentation of degraded images acquired in the visible wavelength. IEEE Trans. Pattern Anal. Mach. Intell. **32**(8) (2010)
13. Vatsa, M., Singh, R., Noore, A.: Improving iris recognition performance using segmentation, quality enhancement, match score fusion, and indexing. IEEE Trans. Syst. Man Cybern. (2008). IEEE
14. Shah, S., Ross, A.: Iris segmentation using geodesic active contours. IEEE Trans. Inf. Forensic Secur. **4**(4) (2009)
15. He, Z., Tan, T., Sun, Z., Qiu, X.: Towards accurate and fast iris segmentation for iris biometrics. IEEE Trans. Pattern Anal. Mach. Intell. (2008)
16. Ashwini, M.B., Imran, M., Alsaade, F.: Evaluation of iris recognition system on multiple feature extraction algorithms and its combinations. Int. J. Comput. Appl. Technol. Res. **4**(8), 592–598 (2015). ISSN 2319–8656
17. http://biometrics.cse.msu.edu/info.html
18. Wildes, R., Asmuth, J., Green, G., Hsu, S., Kolczynski, R., Matey, J., McBride, S.: A system for automated iris recognition. In: Proceedings IEEE Workshop on Applications of Computer Vision, pp. 121–128, Sarasota, FL (1994)
19. Kong, W., Zhang, D.: Accurate iris segmentation based on novel reflection and eyelash detection model. In: Proceedings of 2001 International Symposium on Intelligent Multimedia, Video and Speech Processing, Hong Kong (2001)
20. Tisse, C., Martin, L., Torres, L., Robert, M.: Person identification technique using human iris recognition. In: International Conference on Vision Interface, Canada (2002)
21. Ma, L., Wang, Y., Tan, T.: Iris Recognition Using Circular Symmetric Filters. National Laboratory of Pattern Recognition, Institute of Automation, Chinese Academy of Sciences, Beijing (2002)
22. Boles, W., Boashash, B.: A human identification technique using images of the iris and wavelet transform. IEEE Trans. Signal Process. **46**(4) (1998)
23. Lee, T.: Image representation using 2D Gabor wavelets. IEEE Trans. Pattern Anal. Mach. Intell. **18**(10) (1996)
24. Jain, A.K., Ross, A.: Introduction to biometrics. In: Jain, A.K., Flynn, P., Ross, A. (eds.) Handbook of Biometrics, pp. 1–22. Springer (2008). ISBN 978-0-387-71040-2

25. Sreekala, P., Jose, V., Joseph, J., Joseph, S.: The human iris structure and its application in security system of car. In: IEEE International Conference on Engineering Education: Innovative Practices and Future Trends (AICERA) (2012)
26. Shah, N., Shrinath, P.: Iris recognition system—a review. Int. J. Comput. Inform. Technol. **3**(2) (2014)
27. Wibowo, E.P., Maulana, W.S.: Real-time iris recognition system using a proposed method. In: International Conference on Signal Processing Systems, pp. 98–102, 15–17 May 2009
28. Zhiping, Z., Maomao, H., Ziwen, S.: An iris recognition method based on 2DWPCA and neural network. In: Chinese Control and Decision Conference (CCDC '09), pp. 2357–2360, June 2009
29. Sciencedirect.com: A new texture analysis approach for iris recognition. Online Available: http://en.wikipedia.org/wiki/Iris_recognition (2015)

Challenges in Mining Big Data Streams

Veena Tayal and Ritesh Srivastava

Abstract Big data deals with data of very large data size, heterogeneous data types and from different sources. The data is very complex in nature and having growing data. Dealing with big data is one of the emerging areas of research which is expanding at a rapid rate in all domains of engineering and medical sciences. A major challenge imposes on the analysis of big data is originated from big data generation source, which generate data with very fast speed with varying data distribution due to which the classical methods are unable to process big data. This paper discusses the characteristics, challenges, and issues with big data mining. It also illustrates the examples taken from various fields like medical, finance, social networking sites, stock exchange, etc. to realize the application and importance of big data mining. This paper explains about the use of parallel computing in data mining security issues and how to deal with them. Furthermore, this paper also discusses challenges associated with big streaming data with concept drifts.

Keywords Big data · Data mining · Machine learning · Online learning

1 Introduction

Data processing is a challenging task in big data mining because data is sourced from multiple sources; moreover, they have very complex and evolutionary relationship that is increasing at a very rapid rate. Today is an era of big data [1–3]. A very large amount of data is produced every day. It is about thousands of billion bytes and that data is produced, recently, within 5 years ago [4]. The capability of data generation has enhanced so much due to advent of information technology. The example of big data is presidential debate between President Barack Obama and Governor Mitt

V. Tayal (✉) · R. Srivastava
CSE Department, FET, MRIIRS, Faridabad, India
e-mail: veena.mittal06@gmail.com

R. Srivastava
e-mail: ritesh21july@gmail.com

L. C. Jain et al. (eds.), *Data and Communication Networks*, Advances in Intelligent Systems and Computing 847, https://doi.org/10.1007/978-981-13-2254-9_15

Romedy triggered more than 10 billion tweets in just two hours [5]. It is how fast the rate on which data is expanding. There are many more examples available in these tweets, the moments that generated discussions of public interests related to Medicare and vouchers. It provides new means to judge the public opinion and generates feedback in real time which is more valuable. It gives better response as compared to radio or TV broadcasting.

It is one of the old saying "A picture depicts thousand words." The pictures play a very important role in exploring human society, public affairs, etc. The Flickr is an example of public sharing sites [6]; it requires storage in various terabytes every day. Due to large amount of data, the commonly used tools are not sufficient to capture and manage the data, within allowed time limit. It is very challenging to extract useful information out of such a large size data [7]. As the storage is required in large capacity, which is almost infeasible. For example—Square Kilometer Array [8] is used for large storage. The data generated from it is very large. It requires a data analysis and prediction that is effective in order to achieve timely response. The data can be sourced from multiple sources and the improvement in the efficiency of single-source knowledge discovery methods are discussed in [9]. The dynamic data mining methods along with its analyzes are discussed in [10]. The multisource perspective of data mining methods as the large amount of data is sourced from multiple locations, therefore multisource data mining as discussed in [11, 12]. The theory proposed in these papers provides the solution for the problem of full search but also helps in finding of global model of which traditional data mining approach are not capable. The data classification and clustering gain a lot of importance in data mining. Furthermore, clustering is one of the analytical methods in data mining and k-means are one of the popular techniques of clustering. There are various improvements suggested in the base algorithm of k-means in the past few years and one of such improvement is discussed in the [13]. There are many machine learning algorithms that are used for classification and prediction in various applications like text classification, spam detection, etc. The Naïve Bayesian is an example of machine learning approach, and the survey of this approach for the classification of text document is discussed in [14].

2 Challenges with Big Data Mining

The challenges of big data are manifold. It focuses on how to access data and how to apply computational operations on it related with arithmetic computing. The big data storage location is not single but sparse at different places and the characteristics of big data is that it is continuously growing in nature; therefore, there is a requirement of distributed data storage at large scale. It is also expensive to move data across different locations. It is required in case of data mining algorithms, the data, to be loaded in main memory for computing purposes. The challenges are also related with semantics and domain knowledge of big data. It will provide benefits to data mining algorithm but disadvantage is that it increased complexity. It is very important to understand the semantics at both low level and high level for example—in social networking— users are linked and shared dependency structures but in some other applications it can be represented using highly correlated items [15]. The challenges in the analysis of big data mining can be discussed by considering factors [16] like complexity, tools, data integration, cost of solution, availability of skills, etc. as shown in Fig. 1.

These challenges also focus on designing of algorithms in handling the difficulties of complex and dynamic data. The sharing of information is required for better development and processing at each stage. The challenges of big data can also be studied by considering the four V's of big data, i.e., volume, velocity, variety, veracity as shown in Fig. 2. The volume is increasing continuously in the form of clickstream, log, event, speech, social media, etc. The velocity is speed of generation of data; it can also be called as rate of analysis. The variety deals with different types of data like structured, unstructured, and semi-structured. The veracity is data in doubt, distrusted, and unclean. The uncertainty is due to data inconsistency and incompleteness of data.

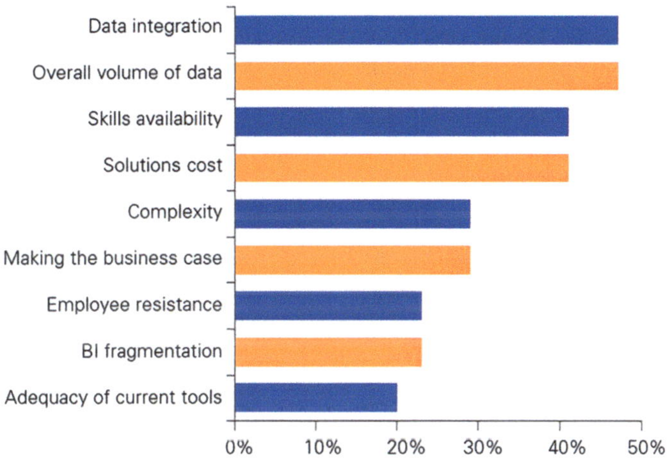

Fig. 1 Biggest challenges for success in big data and analytics

Fig. 2 Challenges of big data

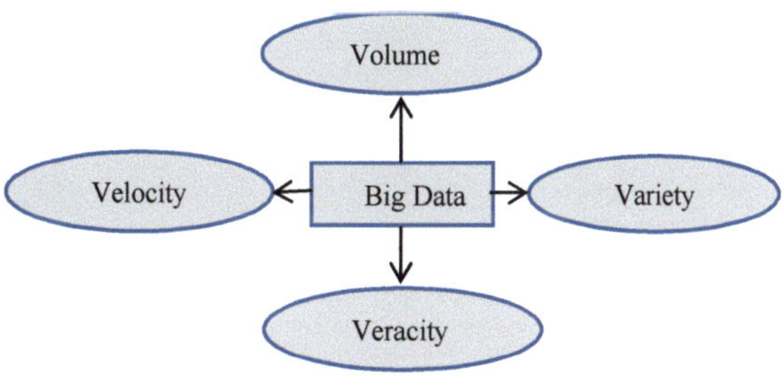

There are many application areas of big data mining like smart health care, finance, log analysis, traffic control, telecom, manufacturing, trading analytics, retails, improving search quality, sentiment analysis [17–20], etc. There are various crowdsourcing models like Google Maps, Wikipedia, etc. The information is shown in Table 1, considering volume, velocity, variety, veracity of various crowdsourcing models.

The framework of processing big data can be studied in various tiers as shown in Fig. 3. Tier 1 discusses low-level detail that deals with accessing of data and how to perform computations on them. Tier 2 concentrates on high level in the form of domain knowledge, semantics, sharing of data, and privacy. Tier 3 deals with the algorithms of data mining having complex and dynamic data.

2.1 Tier 1: Platform of Big Data

The mining requires intensive computing units and there for data and computing processors are required. There is a requirement of large-scale, medium-scale, and small-scale resources for data mining tasks. In case of small scale, a personal computer is sufficient but in medium scale, data is large and cannot fit into memory.

Table 1 Crowdsourcing models

Crowdsourcing models	Volume	Velocity	Variety	Veracity
Google maps	Terabytes	High	Low	Medium
Duollingo	Terabytes	Medium	High	High
reCAPTCHA	Petabytes	Very high	Very high	Very high
Amazon Turk	Petabytes	Medium	Very high	High
Wikipedia	Petabytes	Medium	High	Very high

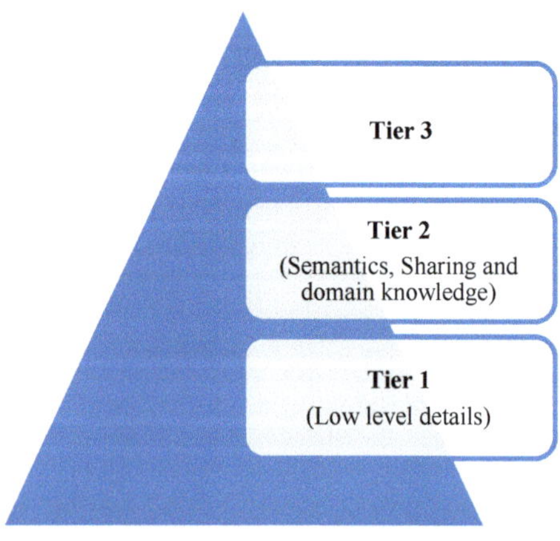

Fig. 3 Framework of big data

It emerges the need of parallel computing [21, 22] and collective mining [23]. For large scale, programming tools such as MapReduce is required. There is a need of cluster computers which are having high speed. There is a need of parallel computing infrastructure for the computing of large size data. For example—data size in Exabyte, petabyte and furthermore they have complex computing process. It changes the quantity to quality. The parallel processing models which used to process such a large data size includes MapReduce, cloud computing, etc. The efficiency of MapReduce is further enhanced to deal with large-scale data along with its real-time nature. There are many algorithms available for data mining. For example—k-means, Naive Bayes, linear support vector machines, regression, etc. The data is divided into small subsets which are independent of each other and then combine their results to produce the desired result and intermediate results can be taken using summation on mapper nodes. There are various tools available for data mining like R, Weka, Hadoop, etc. The integration of R and Hadoop [24] improves the scalability of traditional analysis and also the poor analysis capabilities. The integration of Weka and MapReduce [25] also improves the capability of standard Weka tool. The capability of Weka tool is to run only on single machine and that too with limited memory of 1 Gb. By its integration with MapReduce, its memory can be extended to 100 Gb, and it is achieved by exploring the advantage of parallel computing.

2.2 Tier 2: Semantics of Big Data

Semantics of big data is related with sharing of data, privacy, domain knowledge, and knowledge of application. The parallel computers require sharing of information, but it can be difficult in case of sensitive information, for example, medical records, banking transactions credit cards, etc. The privacy issues involved in such type of information and there can be some measures to resolve such issues, for example, by adding information such as certificate authority that limits the access of data. The role of domain knowledge [26] plays a very important role in big data mining algorithms. For example, hormones test gives better and clear indication about the detection of deceases such as NCAH, adult acne rather than just external view. Also in stock exchange, there is great contribution to the stakeholders if the domain knowledge related with upward/downward is there, it would be much easier to predict the market trend rather than random guess [27]. The privacy of data is ensured by introducing randomness in the data so that actual information can be kept hidden. There are many methods for protecting the data and also presented in [28] such as:

- **Anonymity**: The suppression and generalization are two techniques used for imposing anonymity in the data.
- **Deleting sensitive values**: The data values having sensitive information can be deleted so that it can be protected. Adding random noise: The addition of some random information that is not required is also effective method in protection of information.

- **Swapping of values**: The matching of values can be avoided if they are swapped. It is very useful in the protection of sensitive information.
- **Replacing of values**: The original values can be replaced with some synthetic values that are generated from simulations.

The distribution of data is at different locations; therefore, users do not have physical access to such data. The third-party data miners require efficient data access mechanism. There should no downloading allowed of any local data copies. There should not be having direct access to original data.

2.3 Tier 3: Complexity of Data: Relationships and Associations

The activities related to distributed sites are also an issue to produce unbiased view from them. It is done at various levels; data level, model level, and the knowledge level. The data level is related with local sites, remote site and to aggregate the information at different sites. The data is multisource, dynamic and massive in nature. There is emerging requirement for the expansion of data mining methods. The gradual improvement in the hardware allows improving the knowledge discovery methods, so that they can work well for massive data. There are a lot of differences between single source and multisource data mining due to its real time, heterogeneous and complex in nature like characteristics. The application of data mining such as financial analysis, medical field, online trading diagnostic, etc. are inherently very dynamic in nature where static knowledge is not capable of adapting the change of dynamism. Evolving knowledge is a continuous process in real time and dynamic applications, for example, in a class of average students any student among them can perform extraordinary well, may be because of various external factors like affording of various expensive coaching, guidelines receive from expert, etc. and various other factors with the passage of time. It gives birth to knowledge discovery in case of concept drift. There can be various types of drifts possible in it. It gives rise to new area of research which is called as concept drifting data stream mining [29–31], the detail discussion on concept drifting data stream is present in Sect. 3. There are various types of complex data in big data [15].

- **Complex and heterogeneous data**: It is having structured data and unstructured data. It is having data in the form of table, image, audio, video, hypertext, etc. The data models are incapable of handling such a complex nature of data.
- **Complex relationship networks in data**: There is existence of social relationship between individual and that comes under the category of complex social network. The examples include Facebook, twitter, LinkedIn, and other social networks.
- **Complex intrinsic semantics associations in data**: The "text–image–video" is having strong semantics associations. It helps to improve the performance of applications. It is very challenging to describe the semantic features.

3 Big Data Streams and Concept Drifts

3.1 Data Stream and Types of Concept Drifts

Big data stream is time-ordered sequence of instances generated by a data generating source at very high speed. It is often in the case of data stream that its underlying statistical distribution also changes very frequently with time. Such underlying distribution data stream is also known as concept drifting data stream. The concept drifts are formally defined as the change in underlying statistics of data stream. The classical methods of classification are not suitable for concept drifting data streams. The concept drifts can be broadly categorized under the following five categories as shown in Fig. 4.

3.2 Methods for Dealing with Concept Drifting Data Stream

The approaches for concept drifting data stream mining can be broadly classified into three categories as shown in Fig. 5 namely window-based methods, drift detection-based methods and ensemble-based methods.

- **Window-Based Methods**: The window-based method is a kind of incremental learning in which the classifier is trained by incoming instances for a fixed length of window. Mostly, sliding window is used to track the time-varying data instances. The forgetting strategy is the key practice of such method. By forgetting the old learned concept is dropped and the new concept is learned.
- **Drift detection-based methods**: In drift detection-based methods, the rate of change in the accuracy of the classifier is monitored continuously and any significant drop in the accuracy above a threshold value is taken as a drift in the concept. If drift is detected, the classifier is rebuilt to accommodate the change in concept.
- **Ensemble-based methods**: In ensemble-based methods, a set of diverse component classifiers are pooled together. While performing the classification, all component classifiers participate in decision making according to their weights and the decision about the any income instances is taken by using decision integration rule like weighted majority.

4 Conclusion

The big data mining is very complicated task and challenging too. It is because of real time, complex, multisource, and hetero generous nature of data. The mining of such a big data and with such type of characteristics requires big mind to fully exploit the benefit out of it. We have discussed the challenges faced in big mining at data

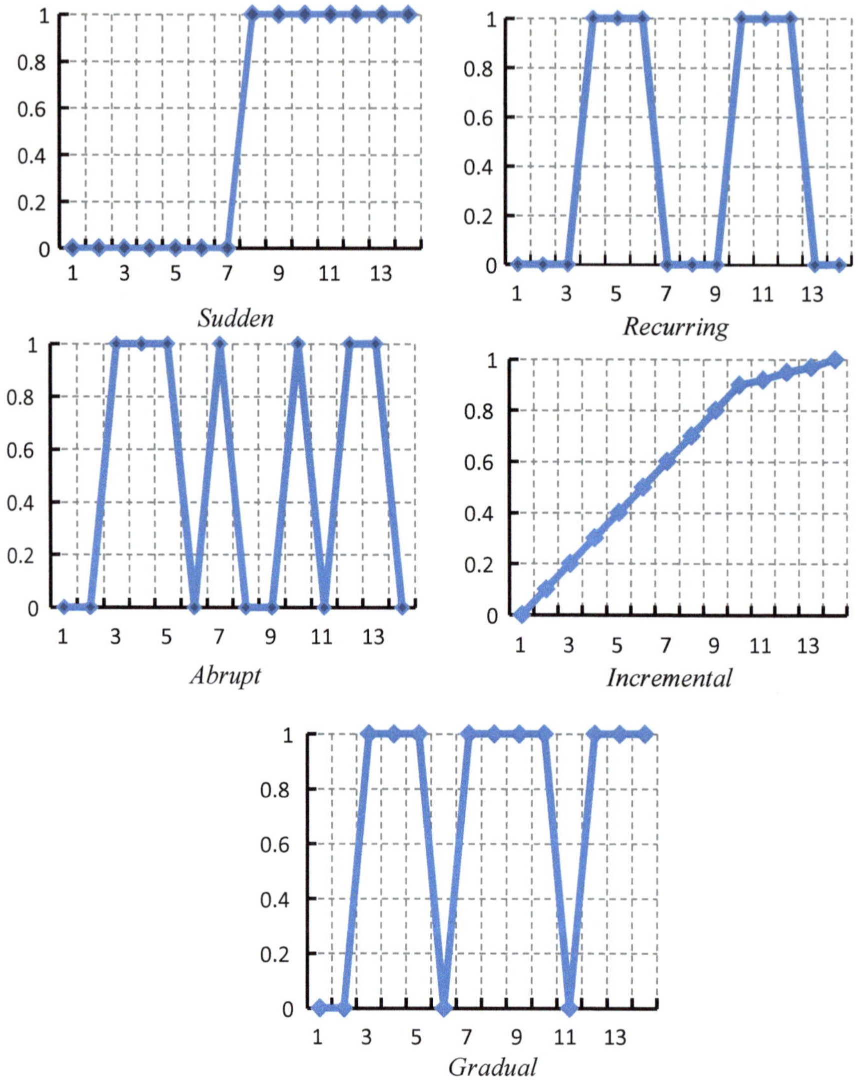

Fig. 4 Types of concept drifts in data stream

level, models level, and system level. It requires devices with high computing power which is both effective and efficient. The challenging situation in big data mining is issue of privacy, random noise and errors, missing data, etc. which are the problems faced at data level. The challenge faced at model level is how to integrate the subsets locally to form a large unified view. The problem of communication at distributed sites which are multisource arises. The problem at system level is to consider the relationship between complex data, semantics, along with their evolving changes.

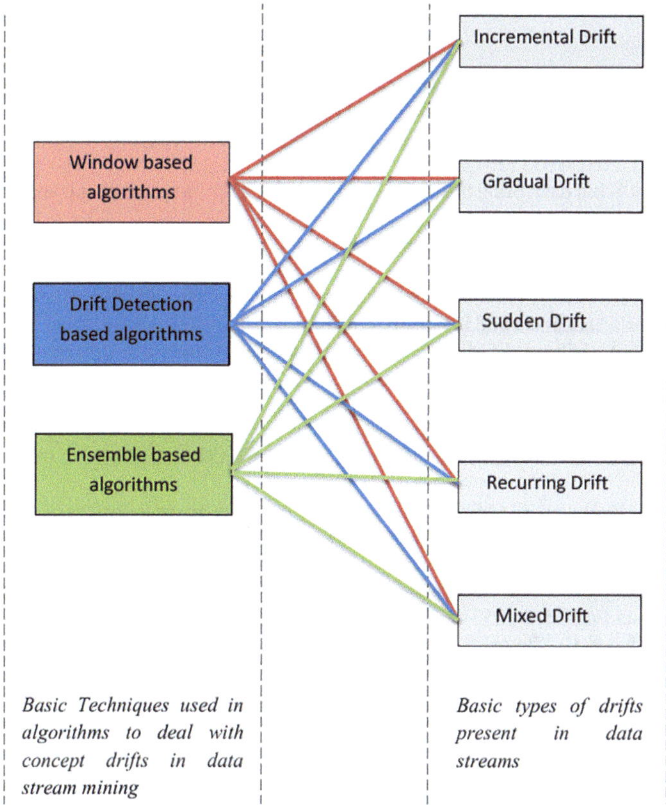

Fig. 5 Methods for dealing with concept drifting data stream

The system should be designed in such a manner that it can link unstructured data to extract useful information from it considering their complex relationships. It should from the basis of prediction of future trend. Lastly, big data is an emerging field and its need is arising in all the fields of engineering and science domains. With the advent of big data technique, we can better understand our society by analyzing the big data available on various social networking sites. It increases the participation of public audiences to receive response in the various social-economic events. It proves today is an era of big data. This paper also explains the challenges associated with concept drifting big data streams, types of concept drifts, and the methods used to deal with concept drifting data streams.

References

1. Nature Editorial: Community cleverness required. Nature **455**(7209), 1 (2008)
2. Howe, D., et al.: Big data: the future of biocuration. Nature **455**, 47–50 (2008)
3. Labrinidis, A., Jagadish, H.: Challenges and opportunities with big data. Proc. VLDB Endowment **5**(12), 2032–2033 (2012)
4. IBM: What is big data: bring big data to the enterprise. http://www.01.ibm.com/software/data/bigdata/ (2012)
5. Blog, T.: Dispatch from the Denver debate. http://blog.twitter.com/2012/10/dispatch-from-denver-debate.html (2012)
6. Michel, F.: How many photos are uploaded to flickr every day and month? http://www.flickr.com/photos/franckmichel/6855169886/ (2012)
7. Rajaraman, A., Ullman, J.: Mining of Massive Data Sets. Cambridge University Press (2011)
8. Dewdney, P., Hall, P., Schilizzi, R., Lazio, J.: The square kilometre array. Proc. IEEE **97**(8), 1482–1496 (2009)
9. Chang, E.Y., Bai, H., Zhu, K.: Parallel algorithms for mining large-scale rich-media data. In: Proceedings of the 17th ACM International Conference on Multimedia (MM'09), pp. 917–918 (2009)
10. Domingos, P., Hulten, G.: Mining high-speed data streams. In: Proceedings of the Sixth ACM SIGKDD International Conference on Knowledge Discovery and Data Mining (KDD'00), pp. 71–80 (2000)
11. Wu, X., Zhang, S.: Synthesizing high-frequency rules from different data sources. IEEE Trans. Knowl. Data Eng. **15**(2), 353–367 (2003)
12. Wu, X., Zhang, C., Zhang, S.: Database classification for multi-database mining. Inf. Syst. **30**(1), 71–88 (2005)
13. Na, S., Xumin, L., Yong, G.: Research on k-means clustering algorithm: an improved k-means clustering. In: Third International Symposium on Intelligent Information Technology and Security Informatics, pp. 63, IEEE (2010)
14. Vidhya, K.A., Aghila, G.: A survey of Naïve Bayes machine learning approach in text document classification (IJCSIS). Int. J. Comput. Sci. Inf. Secur. **7**(2) (2010)
15. Wu, X., Zhu, X., Wu, G., Ding, W.: Data mining with big data. IEEE Trans. Knowl. Data Eng. **26**(1), 97–106 (2014)
16. TM Forum: Challenges of big data (2012)
17. Srivastava, R., Bhatia, M.: Offline vs. online sentiment analysis: issues with sentiment analysis of online micro-texts. Int. J. Inf. Retrieval Res. (IJIRR) **7**(4), 1–18 (2017)
18. Srivastava, R., Bhatia, M.: Real-time unspecified major sub-events detection in the twitter data stream that cause the change in the sentiment score of the targeted event. Int. J. Inf. Technol. Web Eng. (IJITWE) **12**(4), 1–21 (2017)
19. Srivastava, R., Bhatia, M.: Challenges with sentiment analysis of on-line micro-texts. Int. J. Intell. Syst. Appl. **9**(7), 31 (2017)
20. Srivastava, R., Bhatia, M.: Ensemble methods for sentiment analysis of on-line micro-texts. Presented at the International Conference on Recent Advances and Innovations in Engineering (ICRAIE) (2016)
21. Shafer, J., Agrawal, R., Mehta, M.: SPRINT: a scalable parallel classifier for data mining. In: Proceedings of the 22nd Conference on VLDB (1996)
22. Luo, D., Ding, C., Huang, H.: Parallelization with multiplicative algorithms for big data mining. In: Proceedings of the IEEE 12th International Conference on Data Mining, pp. 489–498 (2012)
23. Chen, R., Sivakumar, K., Kargupta, H.: Collective mining of Bayesian networks from distributed heterogeneous data. Knowl. Inf. Syst. **6**(2), 164–187 (2004)
24. Das, S., Sismanis, Y., Beyer, K.S., Gemulla, R., Haas, P.J., McPherson, J.: Ricardo: integrating R and Hadoop. In: Proceedings of the ACMSIGMOD International Conference on Management Data (SIGMOD'10), pp. 987–998 (2010)

25. Wegener, D., Mock, M., Adranale, D., Wrobel, S.: Toolkit-based high-performance data mining of large data on MapReduce clusters. In: Proceedings of the International Conference on Data Mining Workshops (ICDMW'09), pp. 296–301 (2009)
26. Kopanas, I., Avouris, N., Daskalaki, S., The role of domain knowledge in a large scale data mining project. In: Vlahavas, I.P., Spyropoulos, C.D. (eds.) Proceedings of the Second Hellenic Conference on AI: Methods and Applications of Artificial Intelligence, pp. 288–299 (2002)
27. Bollen, J., Mao, H., Zeng, X.: Twitter mood predicts the stock market. J. Comput. Sci. **2**(1), 1–8 (2011)
28. Machanavajjhala, A., Reiter, J.P.: Big privacy: protecting confidentiality in big data. ACM Crossroads **19**(1), 20–23 (2012)
29. Mittal, V., kashyap, I.: Online methods of learning in occurrence of concept drift. Int. J. Comput. Appl. **117**(13), 18–22 (2015)
30. Mittal, V., Kashyap, I.: Empirical study of impact of various concept drifts in data stream mining methods. Int. J. Intell. Syst. Appl. **8**(12), 65 (2016)
31. Mittal, V., Kashyap, I.: An overview of real world applications with concept drifting data streams (2018)

Controlling of Non-minimum Phase System Using Harmony Search Algorithm

Vivek Kumar Jaiswal, Anurag Singh, Shekhar Yadav and Shyam Krishna Nagar

Abstract In this paper, a non-minimum phase system with dead time is controlled by intending the optimized proportional, integral, and derivative controller (PIDC). A system problem is formulated to the non-minimum phase system in which zeroes in the right half-plane (RHP) make system insignificant as delay raises. To enhance the system performance, the factors of the conventional PIDC are optimized by a heuristic algorithm (HA), chosen harmony search (HS). HA, copying the invention of music players, chosen harmony search algorithm (HSA). The HSA looks to acquire a global optimal magnitude of the conventional PIDC and genetic algorithm (GA) in the area portrayed by the regular PID controller and GA. The simulation results demonstrate the transient responses such as settling, rise, peak time, undershoot, and overshoot. Thus to optimized the parameters by minimizing integral square error (ISE) of the given system is improved by the proposed method.

Keywords Non-minimum phase (NMP) system · Conventional PID controller Harmony search algorithm (HSA) · Harmony memory considering rate (HMCR) Pitch adjusting rate (PAR)

V. K. Jaiswal · A. Singh (✉) · S. Yadav
Department of Electrical Engineering, MMMUT, Gorakhpur, India
e-mail: anuragsinghucer123@gmail.com

V. K. Jaiswal
e-mail: vivekjaiswal190@gmail.com

S. Yadav
e-mail: syee@mmmut.ac.in

S. K. Nagar
Department of Electrical Engineering, IIT (BHU), Varanasi, India
e-mail: sknagar.eee@iitbhu.ac.in

L. C. Jain et al. (eds.), *Data and Communication Networks*, Advances in Intelligent Systems and Computing 847, https://doi.org/10.1007/978-981-13-2254-9_16

1 Introduction

In spite of the fact that the advanced control hypothesis has been connected to numerous regions, for example, robots and airplane, and the industry universe are as yet ruled by the crude PIDC. Over 90% of controllers utilized as a part of plant control are yet PIDC. People may ask why the interval is so huge. By looking at the two perspectives, one may discover the appropriate response [1]. For a PIDC outline, no data about the given system is expected, which implies the progression of the given system is thought to be obscure. This really constructs it simple to execute and is more useful. For control configuration classical, i.e., established or modern—present-day control theory, a numerical model of the system is required at the absolute starting point, which implies a large portion of the elements of the system is thought to be known. The achievement of the conventional PID controller acquired utilizing harmony search (HS) algorithm is verified by controlling the NMP system with the delay in time. And basically, the stable system is easy to control than the unstable system. A system having at least one zero in the right half-plane (RHP) of s domain is called NMP stable system, and when the delay in time raises in the system, the scope of controller gain moves toward becoming smaller; in this manner, the response of the system will additionally be decayed. Accordingly, the basic criterion for NMP stable system may not be obtainable.

For controlling the unstable system without or with the delay in time, researchers are working progressively. There are many techniques for controlling unstable system with or without delay, and numerous researchers have watched the traditional PIDC less powerful to enhance the response of the NMP dead-time system [2–4].

So, to enhance the overall response, dead-time or delay-time compensating (DTC) model is used in the system. There are many researchers who have used the Smith predictor (SP) and extensions of SP to control time delay system. In this, SP presence of delay in time is wiped out from the given characteristic equation of the closed-loop (CL) system. From these advantages of SP, a time delay system is converted into without time delay system. Although, disadvantages of SP it cannot be employed to open-loop (OL) unstable system [5–9].

In this paper, the given system is optimized to get a better transient response of the given system using the HSA. Music is a standout among the most fulfilling forms created by human effort. Another HA created by an artificial event establish in music response, in particular, the way toward hunting down better harmony, can be presented.

This paper has been organized as follows. Section 2 exhibits the concise depiction of NMP systems with dead time. In Sect. 3, conventional PIDC is displayed. In Sect. 4, HSA is clarified. Section 5 shows the result and discussion of this paper. Section 6 introduces the conclusion of this paper with related references to this paper.

2 Non-minimum Phase System with Dead Time

In control theory, the nonlinear systems which have at least one system zero in the RHP. A system recognized in this paper is fine-modeled by suitable linear time-invariant (LTI) systems by feasible NMP components and delay in time. For any control system, internal stability is the essential criterion of actual CL system that can be calibrated by testing the observability and controllability gramians of a given system. And G(s) is the transfer function of the system [2]:

$$G(s) = \frac{KN(s)}{D(s)} = \frac{K\mathrm{Nmp}(s)N\mathrm{nmp}(s)}{\mathrm{Ds}(s)\mathrm{Du}(s)}\mathrm{e}^{-sT} \tag{1}$$

where K stands for gain, Nmp(s) is minimum phase, Nnmp(s) is NMP, Du(s) and Ds(s) are unstable and stable polynomials, and T stands for time delay. In NMP systems, beginning response of the system is sluggish because of the presence of RHP zero in the system, and dead time is called the transport lag, and at higher frequency, thundering phase lags along no attenuation. The amplitude is always unity or 0 dB in log amplitude, and phase angle changes linearly along frequency of transport lag [3].

$$|G(j\omega)| = |\cos\omega T - j\sin\omega T| = 1 \tag{2}$$

$$\angle G(j\omega) = -\omega T(\mathrm{rad.}) \tag{3}$$

3 Typical PID Controller

PIDC is widely used in industrial control. The universality of PIDC can be certified most of the way to its strong implementation in a broad variety of working conditions and halfway to its reasonable ease that empowers the designer to work them in coordinate and an essential manner. As the name proposes, PID controller includes three basic terms: proportional, integral, and derivative which are given to get a perfect response [10].

In Fig. 1, $e(t)$ indicates error signal which is the difference between $R(t)$, reference input signal, and $Y(t)$, obtained output signal. The $e(t)$ works as input to PIDC and output of the PIDC control signal, $u(t)$, to be applied to given plant that is equivalent to proportional gain (K_P) times the amplitude of $e(t)$ accumulation to the integral gain (K_I) times the fundamental of $e(t)$ in accumulation to the derivative gain (K_D) times the derivative of $e(t)$.

Thus, time domain demonstrations of $u(t)$:

$$u(t) = K_P e(t) + K_I \int e(t)\mathrm{d}t + K_D \frac{\mathrm{d}e(t)}{\mathrm{d}t} \tag{4}$$

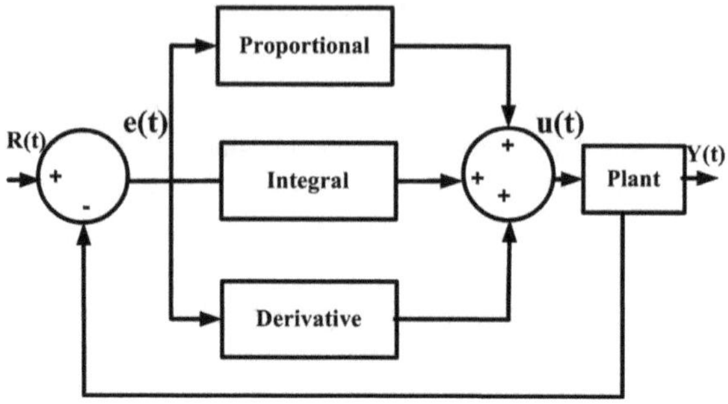

Fig. 1 Typical block diagram of PID controller

A plant on accepting the input $u(t)$ will create a changed output $Y(t)$ which will be again contrasted with the reference signal until the point that the desired level is come to in this way shaping a nearby closed-loop system [11].

4 Description of Harmony Search Algorithm

Presently, the time has come to make an inquiry. Is it conceivable to build up another heuristic algorithm with better response (better results with less iteration) than existing heuristic calculations? This algorithm would fill in as an appealing contrasting option to another effectively settled algorithm [11].

Since the 1970s, numerous heuristic calculations have been created that consolidate standards and haphazardness copying natural event. These procedures or techniques incorporate simulated annealing (SA), physical annealing (PA), evolutionary algorithms (EA), human memory, etc. Consequently, another calculation may likewise be based on a natural event, or in synthetic one. A synthetic event, musical or melodic harmony, can fill in as the mold for concocting another technique. Harmony or music is a standout among the most fulfilling forms created in human being's effort [12]. Another heuristic algorithm is obtained from a synthetic event found in melodic response (for instance, a jazz triad); in particular, the way toward looking for better harmony with less iteration can be presented.

Music concordance is a mix of sounds thought about satisfying from an elegant perspective. There is a peculiar connection between a few sound effects that have distinctive frequencies in nature of harmony. Melodic responses look for an optimum state (awesome harmony) controlled by elegant estimation, and the optimization techniques look for an optimum state (state of global optimum: maximum gain with minimum worth) dictated with objective function assessment. Elegant evaluation is controlled with the arrangement of sound played with connected instruments, and similarly, objective function assessment is dictated by the arrangement of values created by factors variables; for better sound, elegant estimation could be enhanced with a great many practices; similarly, superior objective function assessment could be enhanced by many iterations [13–16]. Short introductions of these perceptions appear inside Table 1.

Harmony search (HS) algorithm procedure can be understood with the help of flowchart in this way:

Step 1: Initializing a harmony memory (HM).
Step 2: Improvise or create a new harmony from step 1.
Step 3: In this step, if the new harmony is superior to anything least harmony in HM, incorporate the new and exclude the least harmony in HM.
Step 4: In this step, if halting norms are not fulfilled, then again follow step 2.

The global optimum solution lies at first in HM. At the point when this is not the situation, keeping in mind the end goal to search global optimum, HS starts a parameter, HMCR, which ranges between 0 and 1. Let a HMCR of 0.90 implies on that subsequent stage, the algorithm picks a variable incentive from HM with 90% probability [17, 18].

Enhancing the results and getting away local optima, the additional choice might be presented. Let aPAR 0.20, it means this algorithm picks an adjacent with 20% probability (lower and upper values by ±10%) [19] (Fig. 2).

The proposed parameters of the conventional PIDC, GA, and HSA are shown in Table 2.

5 Result and Discussion

For controlling the NMP time delay system by optimized PIDC [2], consider a transfer function of actual NMP system with time delay:

Table 1 Comparison of musical and optimization response

Comparing factors	Optimization procedure	Response procedure
Optimum state	Globally optimum	Awesome harmony
Estimated by	Objective function	Elegant norm
Estimated with	Variables values	Pitch of musical instruments
Procedure unit	Every iteration	Every practice

Table 2 Gain of the controllers

Parameters	HAS	GA	PID
KP	4.4295	4.9043	5.1610
KI	0.0073	0.1975	0.2775
KD	20.5130	23.8970	24.0
ISE	9.7605	11.9572	13.2720

Fig. 2 Flowchart of HSA

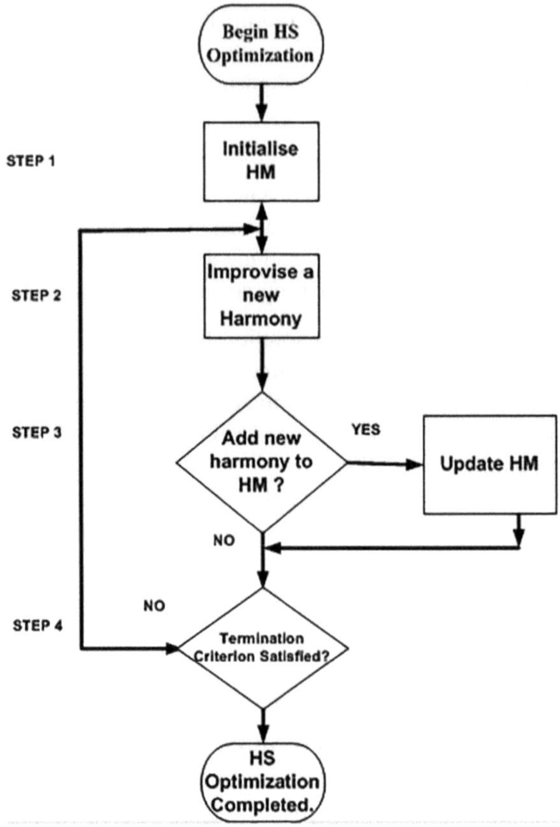

$$G(s) = G(s) = \frac{0.1}{0.05s^4 + 1.25s^3 + 5.2s^2 + 4s} e^{-8s} \tag{5}$$

Step response of actual NMP system is shown in Fig. 3.

Approximation of the system using Pade's:

First of all, the given NMP system transfer function with dead time is approximated with the help of Pade's first-order approximation in MATLAB with command Pade *(G,1)* that produces the approximated transfer function as shown in Eq. 6, and step response of the system is shown in Fig. 4.

$$G(s) = \frac{-0.1s + 0.025}{0.05s^5 + 1.263s^4 + 5.513s^3 + 5.3s^2 + 5} \tag{6}$$

Step response of approximated system using PIDC:

When we used PIDC to approximate a system, we got the result that settling time of system is reduced but the maximum overshoot and undershoot of the system are increased as shown in Fig. 5. These problems are overcome by using evolutionary techniques.

Step response of the system using HSA and GA:

As shown in Fig. 6, comparison of step response of different algorithms can be understood with the help of Table 3.

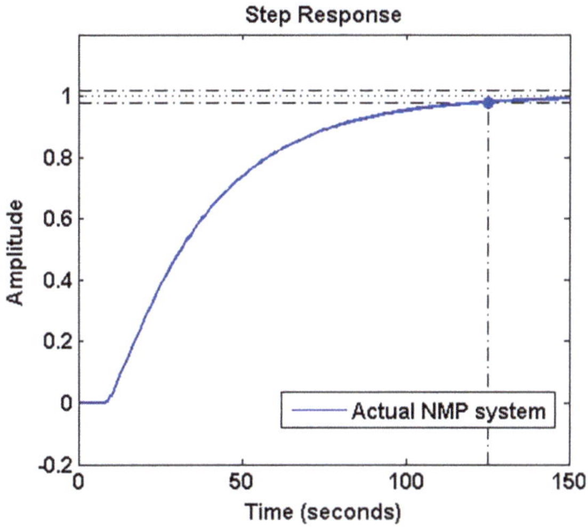

Fig. 3 Actual response of NMP system

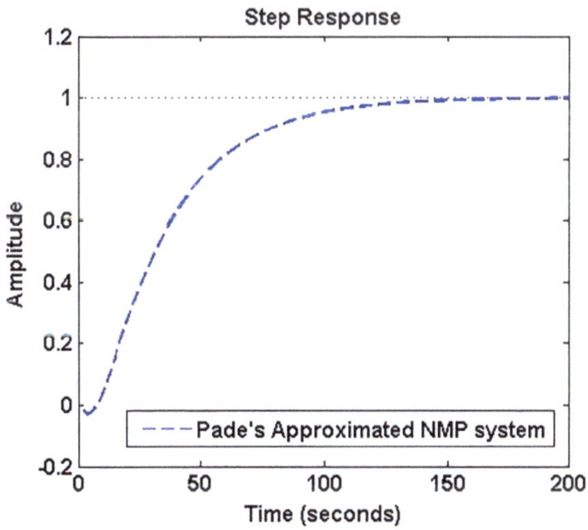

Fig. 4 Pade's approximated response of NMP system

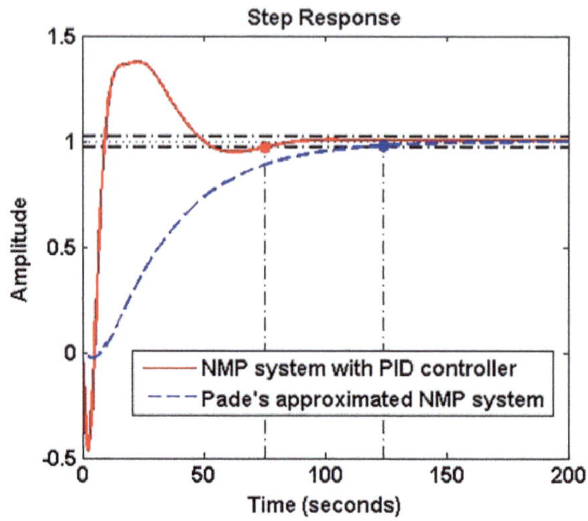

Fig. 5 Step response of NMP system with PIDC

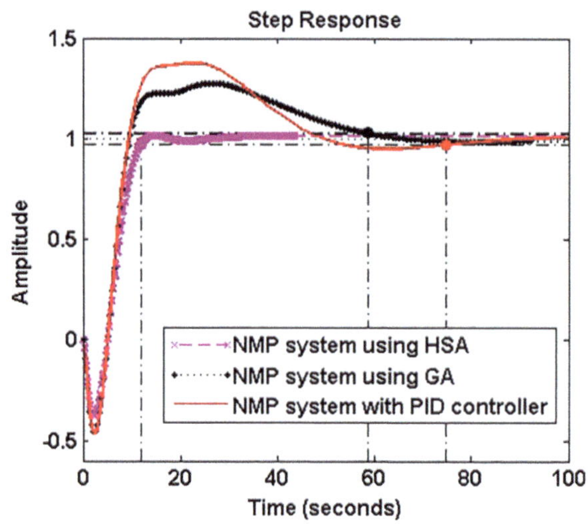

Fig. 6 Comparison of step response of NMP system

6 Conclusion

In this paper, the performance of HSA is considered for optimizing the parameters of PIDC. The proposed PIDC is optimized by HSA to control the NMP system. The given NMP system is approximated with Padé's first-order approximation. The HSA consolidates ISE for optimizing the factors of the PIDC. From the above discussion,

Table 3 Comparison of response specification of different technique

Specification	Approximated system	PIDC	GA	HSA
Peak time	229.8469	22.8672	26.5317	14.5837
Settling time	123.90	74.8433	58.6122	11.7892
Overshoot	0	37.5177	27.7547	1.6629
Undershoot	2.8013	46.6808	45.5332	37.6136
Rise time	65.4610	3.2765	3.4684	5.0763

it has been noticed that HSA gives the global optimum results and enhances the characteristics compared to other evolutionary algorithms like GA and conventional control techniques like PIDC tuned using Ziegler–Nichols method.

References

1. Hagglund, T.: An industrial dead-time compensating PI controller. Control Eng. Pract. **4**(6), 749–756 (1996)
2. Verma, S.K., Yadav, S., Nagar, S.K.: Optimized fractional order PID controller for non-minimum phase system with time delay. In: International Conference on Emerging Trends in Electrical Electronics and Sustainable Energy Systems (ICETEESES), pp. 169–173, IEEE, Sultanpur (2016)
3. Yadav, S., Verma, S.K., Nagar, S.K.: Design of an optimized PID controller for non-minimum phase system with delay. In: 2015 39th National Systems Conference (NSC), IEEE (2015)
4. Palmor, Z.J.: Time-delay compensation Smith predictor and its modifications. In: Levine, W. (ed.) The Control Handbook, pp. 37–48. CRC Press, Boca Raton, FL (1996)
5. Watanabeand, K., Ito, M.: A process-model control for linear systems with delay. IEEE Trans. Autom. Control. **26**(6), 1261–1269 (1981)
6. Ogata, K.: Modern Control Engineering, 4th edn, Prentice Hall, New York (2002)
7. Gu, K., Niculescu, S.I.: Survey on recent results in the stability and control of time-delay systems. J. Dyn. Syst. Meas. Control **125**(2), 158–165 (2003)
8. Richard, J.P.: Time delay systems: an overview of some recent advances and open problems. Automatica **39**(10), 1667–1694 (2003)
9. García, P., Albertos, P.: A new dead-time compensator to control stable and integrating processes with long dead-time. Automatica **44**(4), 1062–1071 (2008)
10. Åströmand, K.J., Hägglund, T.: PID Controllers: Theory, Design, and Tuning, vol. 2, Research Triangle Park, NC: Instrument Society of America (1995)
11. Chidambaram, M., Sree, R.P.: A simple method of tuning PID controllers for integrator/dead-time processes. Comput. Chem. Eng. **27**(2), 211–215(2003)
12. Deb, K.: Multi-objective Optimization Using Evolutionary Algorithms, vol. 16, John Wiley & Sons (2001)
13. Geem, Z.W., Kim, J.H., Loganathan, G.V.: A new heuristic optimization algorithm: harmony search. Simulation **76**(2), 60–68 (2001)
14. Lee, K.S., Geem, Z.W.: A new structural optimization method based on the harmony search algorithm. Comput. Struct. **82**(9–10), 781–798 (2004)
15. Lee, K.S., Geem, Z.W.: A new meta-heuristic algorithm for continuous engineering optimization: harmony search theory and practice. Comput. Methods Appl. Mech. Eng. **194**(36–38), 3902–3933 (2005)

16. Mahdavi, M., Fesanghary, M., Damangir, E.: An improved harmony search algorithm for solving optimization problems. Appl. Math. Comput. **188**(2), 1567–1579 (2007)
17. Geem, Z.W. (ed.): Music-Inspired Harmony Search Algorithm: Theory and Applications, vol. 191, Springer (2009)
18. Yang, X.S.: Harmony search as a metaheuristic algorithm. Music-Inspired Harmony Search Algorithm, pp. 1–14, Springer, Berlin, Heidelberg (2009)
19. Yang, X.S.: A new metaheuristic bat-inspired algorithm. Nature Inspired Cooperative Strategies for Optimization (NICSO 2010), pp. 65–74, Springer, Berlin, Heidelberg (2010)

Framework for Real-World Event Detection Through Online Social Networking Sites

Ritesh Srivastava, M. P. S. Bhatia, Veena Tayal and J. K. Verma

Abstract In recent few years, due to the exponential growth of users on online social networking sites (OSNs), mainly over micro-blogging sites like Twitter, the OSNs now resemble the real world very cohesively. The excess of continuously user-generated online textual data by OSNs that encapsulates almost all verticals of the real world has attracted many researchers who are working in the area of text mining, natural language processing (NLP), machine learning, and data mining. This paper discusses the feasibility of OSNs in detecting real-world events from the horizon of the virtual world formed over OSNs. Moreover, this paper also describes the framework for real-world event detection through online social networking sites.

Keywords Online social network · Event detection · Social network analysis
Data mining · Text mining

1 Introduction

The evolution of Web 2.0 [1] allows users to interact and collaborate with each other on OSNs platform [2]. In recent years, an exponential growth in the users of OSNs has been witnessed. The numbers of users on OSNs are getting double in every five years with about 2.5 billion users this year. A projected result shows that about 39.9%

R. Srivastava (✉) · V. Tayal
CSE Department, FET, MRIIRS, Faridabad, India
e-mail: ritesh21july@gmail.com

V. Tayal
e-mail: veena.mittal06@gmail.com

M. P. S. Bhatia
NSIT, University of Delhi, New Delhi, India
e-mail: mpsbhatia@nsit.ac.in

J. K. Verma
Galgotias University, Greater Noida 201310, India
e-mail: jitendra.verma.in@ieee.org

L. C. Jain et al. (eds.), *Data and Communication Networks*, Advances in Intelligent Systems and Computing 847, https://doi.org/10.1007/978-981-13-2254-9_17

of world population will become OSN users with the end of 2020. The users of micro-blogging sites (like Twitter) actively participate in sharing their opinions on various hot topics (e.g., personalities, products, and events) by posting their comments about the topics. Usually, the comments written by users of micro-blogging sites are short snippets of text that are limited to only a few characters. These short snippets of textual comments over OSN are also termed as online micro-texts [3]. The consistent posting of comments by millions of users of micro-blogging sites and the exponential increase in the number of users on micro-blogging sites have flourished two interesting aspects of social-networking-enabled micro-blogging sites:

(i) The WWW has now become a massive source of online micro-texts. Micro-texts are extremely subjective in nature as they generally contain the positive or negative opinion of users about an entity. Data mining for searching some interesting information from such data has gained the affection of many researchers in the previous years. Sentiment analysis (SA) is one of the most prominent ways for the analysis of this valuable collection of subjective data for making predictions in various events.
(ii) Another interesting facet of OSN is the creation of virtual communities. A virtual community is a set of social entities, especially human beings that are connected to each other on the basis of some common interests on any topics and events over OSN [2].

With the exponential growth of users on OSNs, the OSNs now resemble the corresponding real-world community very cohesively. As a conscience, any event that may be initiated in any of these communities has a significant reaction in both the communities and vice versa. This cohesiveness among the virtual and its corresponding real community has motivated many researchers and data analyst to sense the happenings of the real-world events through the contents of OSN.

Knowing about future has always been fascinating. Predictive analytics is a way by which one can predict the unknown future events. In the predictive analysis, historic and the current data are analyzed to make a prediction, which utilizes many methods such as statistics, data mining, and machine learning. The problem of the predictive analysis can be abstractly classified into two different set of problems.

(i) Making predictions of future by utilizing the current data.
(ii) Making predictions of some attributes of one observation space from another observation space at the same time.

Nowadays, the OSNs offer great opportunities for making predictions about the real-world happenings in both cases. From Fig. 1, it can be easily understood that the degree of cohesion (d) between a real-world community and its corresponding online virtual community formed over OSN is directly proportional to the number of active users on the online virtual community and the number of users commenting about an event occurred in the real world as given in (1) and (2).

$$d \propto \text{Number of active users on Online Virtual Communitiy} \tag{1}$$

$$d \propto \text{Number of users commenting about an event} \tag{2}$$

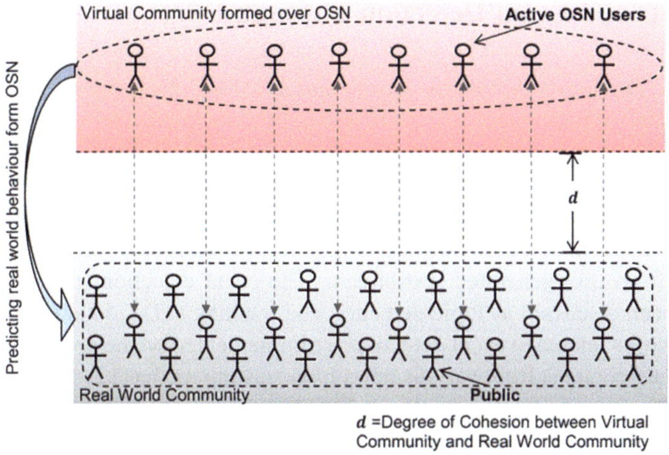

Fig. 1 Degree of cohesion between real world and virtual world of OSN

The consistent posting of comments by millions of users on micro-blogging sites and the exponential increase in the number of users on micro-blogging sites have made the WWW a massive collection of online micro-texts and increased the degree of cohesiveness between the real-world communities and their corresponding online virtual communities, respectively. The availability of massive data on OSN that represents the social and behavioral aspects of a big section of the population provides the immense possibilities to us to observe the public behaviors from the virtual world formed over OSN. Furthermore, it also opens up new opportunities to conduct predictive analysis for the future events. Consequently, an emerging area of research is to make the prediction of real-world happenings and behaviors of people in the real world by analyzing their behaviors in the virtual worlds of OSNs.

The social networking sites have proven their huge power of prediction in predicting the results of the events of real world. Recently, Twitter is utilized in various tasks such as monitoring, predicting, and analyzing the real-world events and activities like breaking news tracking [4], election prediction [5], natural disasters like earthquake [6], and crime, radicalization, and terrorism [7]. Certainly, the text stream mining of Twitter data can substitute the traditional polling [8].

We believe that the real-world activities could be monitored in real time by performing the real-time analysis of contents of micro-blogging site like Twitter. Such type of real-time analysis can enhance the real-time decision-making and an alternative for both types of predictive analysis as mentioned above.

2 Events and Event Detection Through OSN

The Oxford Dictionary defines the word ***event*** as a thing that happens or takes place, especially one of importance. An event is usually associated with time and location. The process of event detection from temporally ordered textual data can be explained as an automatic process for identifying novel events evolved meanwhile in that textual stream [9]. In early years of 2000s, with the evolution of Web 2.0, the enormous use of computer-mediated communications motivated researchers for automatic event detection from user-generated text stream. The event detection from textual data has long been discussed as topic detection and tracking (TDT) [9–11]. Most of the previous works related to event detection are implemented on the conventional news based textual contents from various news broadcasting media [12–14].

The task of event detection in the Twitter data stream can be broadly categorized into two: (i) specified or targeted events and (ii) unspecified (untargeted) events [15]. The targeted event detection is a kind of supervised process in which a sufficient amount of prior information is known in advanced; based on this information, the events are detected. For specified event detection, the prior information may include place, time, domain, description, and features about the event. For example, election in any country is kind of specified event as the date of the election, names of contesting parties as well as the names of the candidates are known in advance for performing the analysis. Conversely, in the case of unspecified event detection process, no clues about the event are known in advance. Moreover, an event can undergo with some sub-events. Sub-events can be defined as those events that occur in between the discussion duration of any event and cause significant impact over the points of discussion. Sub-events generally change the sentiment of the main event significantly. We state such event as sentimental events or sub-events. The task of event detection in the Twitter data stream can be broadly categorized in two: (i) specified or targeted events and (ii) unspecified (untargeted) events (Fig. 2) [15]. The targeted event detection is a kind of supervised process in which a sufficient amount of prior information is known in advance; based on this information, the events are detected. For specified event detection, the prior information may include place, time, domain, description, and features about the event. For example, election in any country is a kind of specified event as the date of the election, names of contesting parties as well as the names of the candidates are known in advance for performing the analysis. Conversely, in the case of unspecified event detection process, no clues about the event are known in advance. Moreover, an event can undergo with some sub-events. Sub-events can be defined as those events that occur in between the discussion duration of any event and cause significant impact over the points of discussion. Sub-events generally change the sentiment of the main event significantly.

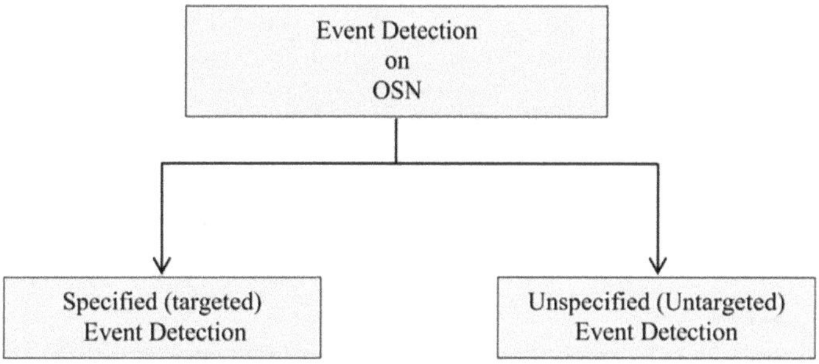

Fig. 2 Types of event detection task on OSN

2.1 Types of Events on OSN

Based on the ways, in which the events evolved in real world and discussed on OSN, we can classify the events on OSN into the following three types:

(i) Periodic events.
(ii) Sudden events.
(iii) Long-duration gradual events.

As depicted in the graphs of Fig. 3, the periodic events are those events that reoccurred periodically on certain pre-decided date and time. Over OSNs, such type of events hugely attracts comments of users on the date or time of event occurrence periodically. The periodic events are generally designated by certain prior information such as a list of frequent keywords and time of occurrence. For example, #FollowFriday is a periodic event that involves discussions on new movie release on every Friday of a week. The periodic events can further be classified on the basis of their recurrence intervals; for example, daily event may be (#GoodMorning), the weekly event may be follow Friday (#FollowFriday), and the yearly event may be any festival (#HappyChrismas). The next type of events is sudden events; such type of events gains the sudden interest of users of OSN in their posts after the occurrence of the events. For example, after the occurrence of an earthquake, the user's posts mentioning the word earthquake increase suddenly. Unlike the periodic events, such events do not offer any prior information concerning the time and the place of events; moreover, in many cases, they do not designate any predefined keywords or hashtags keywords also. Terrorist attacks and riots generally belong to such category of events.

Another essential category of events discussed on OSN is long-duration gradual events. Such kinds of events are specified and designated by the date, place, and other related information such as keywords and terminologies in advance. The discussions about a topic in a long-duration event persist for a long period. The users on OSN have prior knowledge of such kind of events. An example of such kind of events includes an election in any country. Usually, the date of an election is declared in advance. The users of OSN start discussing the election few days prior to the election (e.g., one

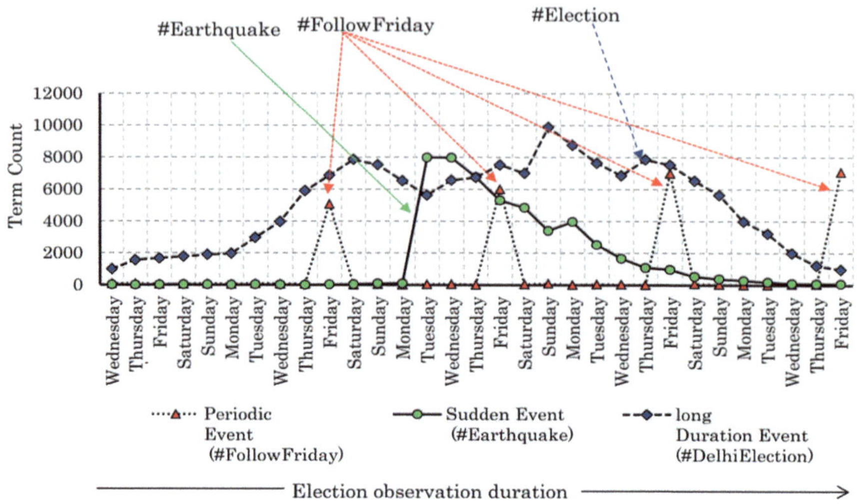

Fig. 3 Types of events on OSN

month) and persist it throughout the election campaign. Some major keywords for an election are comprised of contesting party names, contesting candidates' names, etc.

3 Framework for New Event Detection in OSN

A generic framework for new event detection is text data stream process which is a two-step process as depicted in Fig. 4. The first step is responsible for feature-based signal generation from the input text stream. The second step is responsible for detecting the burst in the signals.

(i) **Signal Generation**: The signal generation depends on features of the incoming text streams. Identification of the best features of the incoming text stream is very crucial for accurate event detection.

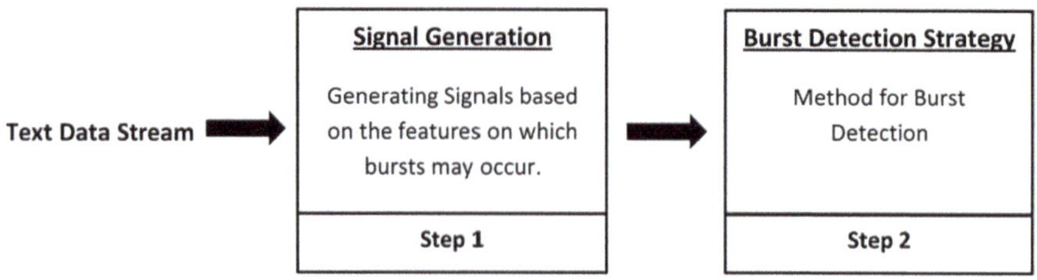

Fig. 4 Model for event detection in text data stream

(ii) **Burst Detection Strategy**: Any unprecedented change in the observed signal generated by signal generator the incoming text streams can be considered as a burst in the feature-based signals. Bursts in the generated signal are the potential indicators of the occurrence of events.

The rest parts of this section describe various strategies widely used for signal generation and burst detection in textual data stream.

3.1 Features for Signal Generation in Text Data Stream

1. **Frequency of words**: The most prominent burst detection methods in text data streams relied on the burst detection based on the examination of the frequencies of words present in the text stream [16–18]. The term frequency and inverse document frequency (TF-IDF) is the most popular method for generating term-frequency-based signal. The TF-IDF can be explained as follows: Let d be a document in a corpus of N documents and let t be a term in documents, then TF-IDF can be calculated by using (3).

$$\text{TF} - IDFt, d = 1 + \log TFt, d \times ID(t) \tag{3}$$

   where $\text{TF}(t, d)$ is the term frequency of t in d, and $\text{IDF}(t)$ is the inverse document frequency, i.e.,$N/n(t)$, where $n(t)$ is the number of documents containing the term t.
   In Twitter data streams, instead of calculating the TF-IDF, the *Term Frequency and Inverse Tweet Frequency* (TF-ITF) has been calculated for a given temporal window. For the real-time event detection, the TF-ITF is monitored periodically and any unprecedented changes in the TF-ITF are recorded as the bursts in the signal.
2. **Platform-Specific Features**: Online social networking sites like Twitter generally offer some specific notations to emphasize the topic of discussion in order to grasp the attention of other users. The notion of hashtags (#) is widely used by almost all OSN these days. Hashtags are generally created by users to indicate event or issue and floated over OSNs for drawing comments of other users over the events. Hashtags are the most prominent features for detecting events via OSNs. Instead of monitoring frequency of all words belonging to the text stream, the monitoring of the frequency of hashtags is more beneficial.
3. **Signal Generation based on online analysis of raw features**: There are certain signal generators which take raw features of the text data streams and process them online for generating the signals, for example, change in the sentiment score and change in domain of discussion.

   a. **Sentiment-Analysis-Based Signal Generation**: The sentiment analysis is defined as the automatic process of determining the sentiment of digitally stored textual documents [19–21]. While performing online sentiment analysis [18, 20, 22, 23] for a specified event, a significant change in the sentiment score with respect to time is a strong indicator of the occurrence of sub-event. Such significant changes can be utilized for tracking the occurrence of sub-events.

b. **Domain-Specific Features**: In text streams, the features represent the domain changes very frequently; for instance, the discussion of users on OSNs may change from the topic politics to sports after any crucial sport result. Observing the domain-specific features and generating the signal accordingly is also very prominent way for accurate event detection. However, such kind of signal generation required online domain classification. During the time of change in the domain of discussion, the underlying data distribution of the text stream also changes significantly.

3.2 Burst Detection in Text Data Stream

Burst detection is a process of detecting high-frequency period of in time series data analysis. The burst detection methods are utilized in variety of ways:

1. Fixed threshold value.
2. Minimum period between two bursts.
3. Duration of bursts.
4. Adaptive threshold parameter.

4 Conclusion

With the exponential growth of users on OSNs, the OSNs now resemble the corresponding real-world community very cohesively. The events occurred in the real world are often discussed on the virtual world of OSNs; hence, the analysis of contents of online OSNs provides immense possibility to track the real-world happening from the horizon of virtual world formed over social network. In keeping the view of the feasibility of OSNs in real-time tracking of the new events through OSNs, this paper discusses the general framework for real-world event detection through online social networking sites.

References

1. Web 2.0: January 2016, cited 2016. Available from: https://en.wikipedia.org/wiki/Web_2.0
2. Kaplan, A.M., Haenlein, M.: Users of the world, unite! The challenges and opportunities of social media. Bus. Horiz. **53**(1), 59–68 (2010)

3. Rosa, K.D., Ellen, J.: Text classification methodologies applied to micro-text in military chat. In: ICMLA'09. International Conference on Machine Learning and Applications. IEEE (2009)
4. Jackoway, A., Samet, H., Sankaranarayanan, J.: Identification of live news events using Twitter. In: Proceedings of the 3rd ACM SIGSPATIAL International Workshop on Location-Based Social Networks. ACM (2011)
5. Srivastava, R., et al.: Analyzing Delhi assembly election 2015 using textual content of social network. In: Proceedings of the Sixth International Conference on Computer and Communication Technology. ACM (2015)
6. Sakaki, T., Okazaki, M., Matsuo, Y.: Earthquake shakes Twitter users: real-time event detection by social sensors. In: Proceedings of the 19th International Conference on World Wide Web. ACM (2010)
7. Weimann, G.: New Terrorism and New Media. Wilson Center Common Labs (2014)
8. O'Connor, B., et al.: From tweets to polls: linking text sentiment to public opinion time series. ICWSM **11**(122–129), 1–2 (2010)
9. Yang, Y., Pierce, T., Carbonell, J.: A study of retrospective and on-line event detection. In: Proceedings of the 21st Annual International ACM SIGIR Conference on Research and Development in Information Retrieval. ACM (1998)
10. Allan, J.: Topic Detection and Tracking: Event-Based Information Organization, vol. 12. Springer Science & Business Media (2012)
11. Allan, J.: Introduction to topic detection and tracking. In: Topic Detection and Tracking, pp. 1–16. Springer, Berlin (2002)
12. AlSumait, L., Barbará, D., Domeniconi, C.: On-line LDA: adaptive topic models for mining text streams with applications to topic detection and tracking. In: Eighth IEEE International Conference on Data Mining. IEEE (2008)
13. Fiscus, J.G., Doddington, G.R.: Topic detection and tracking evaluation overview. In: Topic Detection and Tracking, pp 17–31. Springer, Berlin
14. Brants, T., Chen, F., Farahat, A.: A system for new event detection. In: Proceedings of the 26th Annual International ACM SIGIR Conference on Research and Development in Information Retrieval. ACM (2003)
15. Atefeh, F., Khreich, W.: A survey of techniques for event detection in twitter. Comput. Intell. **31**(1), 132–164 (2015)
16. Cordeiro, M.: Twitter event detection: combining wavelet analysis and topic inference summarization. In: Doctoral Symposium on Informatics Engineering (2012)
17. Bahir, E., Peled, A.: Real-time major events monitoring and alert system through social networks. J. Conting. Crisis Manag. **23**(4), 210–220 (2015)
18. Cui, L., et al.: Topical event detection on Twitter. In: Australasian Database Conference. Springer, Berlin (2016)
19. Srivastava, R., Bhatia, M.: Ensemble methods for sentiment analysis of on-line micro-texts. In: International Conference on Recent Advances and Innovations in Engineering (ICRAIE). IEEE (2016)
20. Srivastava, R., Bhatia, M.: Challenges with sentiment analysis of on-line micro-texts. Int. J. Intell. Syst. Appl. **9**(7), 31 (2017)
21. Srivastava, R., et al.: Exploiting grammatical dependencies for fine-grained opinion mining. In International Conference on Computer and Communication Technology (ICCCT). IEEE (2010)
22. Srivastava, R., Bhatia, M.: Offline versus online sentiment analysis: issues with sentiment analysis of online micro-texts. Int. J. Inf. Retr. Res. (IJIRR) **7**(4), 1–18 (2017)
23. Srivastava, R., Bhatia, M.: Real-time unspecified major sub-events detection in the twitter data stream that cause the change in the sentiment score of the targeted event. Int. J. Inf. Technol. Web Eng. (IJITWE) **12**(4), 1–21 (2017)

Novel Architecture for Internet of Things and Blockchain Technologies

Chetna Dabas and Aniket Dabas

Abstract Diverse companies and platforms are falling madly in love with the blockchain technology in the present scenario. So, it is of due importance that the relations among the different types of blockchains are established. Further, the extraction of capabilities of strengths of diverse blockchains catering to different kind of micro-services is a crucial step toward the realistic blockchain. This issue even became of utmost significance when specifically the researchers are dealing with Internet of things (IoT) purposes. The IoT in general is composed of a wide spectrum of technologies varying from soft real time to hard real time, from stateless through state full, and from unconstrained to constrained. Therefore, for all definite implementations, no definite blueprint of the IoT architecture may be referenced. When it comes to blockchain technology melded with IoT, the existing literature seems to lack a hybrid architecture which contains the best of both these worlds. This research paper proposes a hybrid architecture for IoT and blockchain along with the evaluation of this architecture by implementing an application for the same. The application is for a common user using these technologies which enable the user to select among diverse cryptocurrencies efficiently. The evaluation of the proposed architecture is carried out by implementing blockchain on IoT devices. The results retrieved as a part of this work are encouraging in terms of execution times.

Keywords Internet of things · Blockchain · Radio frequency identification IoT architecture

C. Dabas (✉)
Department of CSE, Jaypee Institute of Information Technology, Noida, India
e-mail: chetna.dabas@jiit.ac.in

A. Dabas
Erricson India Global Services Pvt. Ltd., Noida, India
e-mail: aniket.dabas@erricson.com

L. C. Jain et al. (eds.), *Data and Communication Networks*, Advances in Intelligent Systems and Computing 847, https://doi.org/10.1007/978-981-13-2254-9_18

1 Introduction

In almost all the information industries across the globe, the concept of Internet of things is prevalent in its wider most form. It has deeply impacted the life of a common man and has caught the attention of academia, industries, and the governments worldwide. An Internet of things (IoT) [1] is an advanced and extending technology having its basis grounded on the well-known Internet technology. There lie immense possibilities of expansion and scalability with IoT. The IoT includes any kind of objects or things connected through wired or wireless technologies involving orientation tracing, intelligent recognition, equipments like Radio Frequency Identification (RFID) [2, 3] tags and readers, numerous kinds of sensors, GPS, laser scanners to name a few. Further, the IoT can be well connected to cloud computing [4–9] technology to make the matter more complicated to address [10].

There are various initial developments in the direction of the Internet of things which have paid special attention to the product life cycle applications and RFID infrastructures in Business-to-Business (B2B) logistics but still the IoT is farther to go and make even a common man capable to effectively access as well as contribute rich, secure, and cost-effective information about locations and things.

Due to the heterogeneous environment of the IoT scenarios, there is a lack of precise and generic architecture which can govern the majority of applications in diverse scenarios. Apart from this, there exist lots of constraints in such an environment which hampers the performance of applications and increases the cost involved in developing methods to assist fast and reliable transactions across the network. Further, there come security issues embedded with such heterogeneous and scalable environments where IoT is connected to the cloud as well.

Blockchain [11–13] can perform a crucial role in such a scenario to suggest reliable and secure solutions with the Internet of things [1].

Specifically, the blockchain is considered as a distributed and public ledger. The transactions in blockchain are kept in a linked list consisting of blocks. This chain expands upon the addition of new blocks which are added continuously on the Internet. Various cryptography and distributed algorithms have been utilized for maintaining the security and consistency of the ledger. Blockchain comes with advantages of persistence, decentralization, auditability, and anonymity which enable the improvement of the efficiency and saves time.

Specifically, mentioned in [13] paper, the blockchain holds great potential in constructing the future Internet systems to come in due years.

In this work, a hybrid IoT architecture is proposed, the possible applications of the proposed architecture are suggested, and the architectural elements of the proposed IoT architecture are discussed. Further, the integrated view of the proposed hybrid IoT architecture composed of technologies like RFID, cloud computing, and blockchain is presented. Apart from this, IoT data analytics is performed along with result assessment.

The rest of the research paper is ordered in the way mentioned ahead. The related work is presented in Sect. 2. Section 3 presents the proposed Internet of things hybrid architecture. This section also contains subsections which contain the possible applications for the proposed architecture, the architectural elements of the proposed IoT architecture, and the integrated view of the proposed architecture. In Sect. 4, authors give a detailed explanation of our proposed work. Section 4 presents the IoT data analytics along with result assessment. Section 5 presents open challenges and future vision. The research paper concludes with Sect. 5 which presents the conclusions of the work and summary is discussed as well.

2 Related Work

This related work section describes the existing literature on various existing architectures of IoT-based systems along with the possibilities of the existing technologies like cloud computing, Radio Frequency Identification, and blockchain along with IoT.

In [10], authors have discussed architectural aspects of integrating cloud and RFID for the scalability. The authors have also talked about the possibilities of kinds of applications that can be running on this integrated architecture. The authors of this paper have revealed the blending possibilities of the two important technologies, namely cloud and RFID.

The authors of paper [14] have given a cloud-centric vision for IoT scenarios. This paper also makes a point that with the present existence of a variety of numerous RFID tags and other wireless sensor technologies and sensor equipments, IoT is believed to be the forthcoming revolutionary technology which is supposed to transform the present Internet to future Internet which is fully integrated.

In the paper [15], the authors present a basic architecture of IoT with three layers, namely perception layer, network layer, and application layer. This paper lacks the workflow of these layers in detail.

The authors of research work [16] have tabulated the security issues in IoT. Further, they have presented the role of blockchain in solving a huge number of security issues associated with IoT and hence revealing the significance of blockchain in IoT scenarios.

In [17], the research work by the authors discussed the future architecture for the IoT. This paper discusses the definitions and existing developments in the area, along with key requirements. This paper also discusses a proposal for the future IoT. Apart from this, in both user-centric and business-centric environments, this paper examines the usability by stakeholders.

The authors of the paper [18] discuss the significance of IoT in day-to-day life, present the architecture of IoT, the protocols for IoT, and further discuss possible IoT applications.

The research work in [19] presents a framework for IoT architectures. The framework proposed by the authors is designed for the applications which are service-based and in smart cities. The data input for this framework is retrieved from data sources which form data cloud.

In [20], the authors have proposed the classification along with a comparison of the blockchains and the blockchain-based systems. This paper claims that the presented work helps in assisting the assessment and design procedure of blockchains on the existing architectures.

The authors of [21] highlight the existing trends and generic architecture of IoT composed up of application layer, middleware layer, perception layer, business layer along with the future application. The authors of this paper specifically stress on the fact that IoT is gaining popularity across diverse fields like industry, academia to name a few. Further to this, the authors convey that IoT bears the capability to bring benefits to professionals and even to individuals.

In [22], the authors of this paper have proposed specific media-aware traffic security architecture. This architecture depends on the given traffic classification for catering various multimedia services.

In the existing literature, so far no work has been noted on the integration of all these technologies like blockchain, RFID, cloud computing in a single architecture melded with IoT.

The authors of this paper took this step further in the proposed research work as follows.

The research aim of this proposed work is to offer fusion architectures with Internet of things and blockchain technologies, while utilizing Radio Frequency Identification and cloud computing perception altogether. The vision of this proposed research work is to provide data security, management, analytics, and implications to a common man while keeping low cost and optimum performance.

3 The Proposed Internet of Things Hybrid Architecture

System blockchains are the inventions of the present era that have revolutionized the way one distributes personal information over the cloud and Internet in secured manner. Blockchain has been of great importance to the daily personal (both textual and image) data playing its crucial role on social Web sites, etc., and hence in the commercial aspect. RFID has on the other hand revolutionized the association of unique identities and tracking to things, places, animals, and even human beings. The hybrid use of RFID and blockchain leverages a benefit of distributed ledger which fortifies both security and transparency.

Aryanvika is a self-improved RFID and blockchain system and method which has the ability to boost the performance and decline the cost of the data analyzed in optimum ways in which data can be handled.

The idea was first conceived by the authors when they were busy observing their surroundings to seek an innovative design architecture on how to use their creativity for the enhancement of the society. As a consequence, the attention of the authors fell on the day-to-day problem of storing the magnanimous amount of data generated by photographs taken by people every day all over the world and not being able to store them properly due to issues like lack of security. Then the authors thought why not to think of some innovative way of solving this problem.

This idea is not sold; neither has it gone through any official process. No prototypes have been made. Authors do have a hard copy of our calculations and workouts and researches that made us finalize the features for the system and method.

3.1 Example Scenarios of Applications of the Proposed IOT Architecture

In this section, the authors explain with the help of a few examples the possible applications that may be used on the proposed architecture.

- For instance, suppose a person in India was working on a technology 'X' and had his personal photograph album for it stored on his personal cloud at his home. He also wants to associate a unique identification to his photographs for security reasons. Suppose he now moves to a different country, let's say Australia, for a job and wants to show his secure photographs to his new colleagues. This can be easily accomplished with the proposed hybrid IoT architecture which he might have constructed in his home back in India.
- Another interesting example of this scenario is related to mining of user choice-based cryptocurrencies pertaining to scalable, efficient, and cost-effective way in a heterogeneous IoT environment. The proposed architecture delivers a significant performance in a cost-effective manner in such a scenario.

3.2 Elements of the Proposed IoT Architecture

The proposed architecture in the present research work is described in Fig. 1 in abstract as a system model. It makes use of blockchain, Radio Frequency Identification, and cloud technologies. This system model is primarily composed of four layers as is self-explanatory in Fig. 1. The authors have named it as Aryanvika.

The detailed workflows of the proposed hybrid IoT architecture are presented in the subsections ahead.

3.2.1 Workflow of Layer 1

Layer 1 corresponds to the virtual private network on the cloud. It is composed of three heterogeneous devices, namely EndUser Device 1 (laptop), EndUser Device 2 (smartphone), and EndUser Device 3 (raspberry pi 3). These heterogeneous devices may be scaled up or down according to the requirements. Further, these devices have their own respective user interfaces (may be same or different). With the help of these

Fig. 1 System model

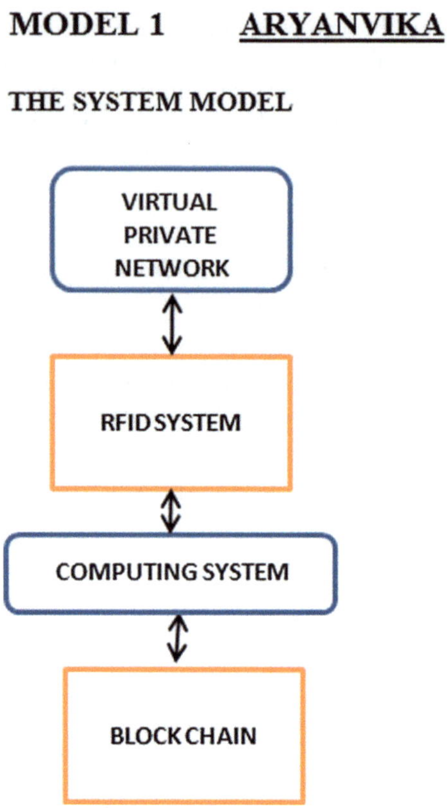

user interfaces, the user can see or update the data on these devices. These end user devices form a virtual private network. This layer 1 is further connected to layer 2 of the proposed hybrid IoT architecture which is explained in the next subsection. The system model of layer 1 is presented in Fig. 2.

3.2.2 Workflow of Layer 2

Layer 2 as depicted in Fig. 3 takes its input from layer 1. The interface of this layer interacts with the layer 1 as well as the RFID reader. There exists a sound possibility of using more than one RFID reader in this layer of the architecture. The RFID reader offers a unique identification of the asset linked to another interface which is linked to layer 3 of this architecture. The output of layer 2 is fed as an input to layer 3.

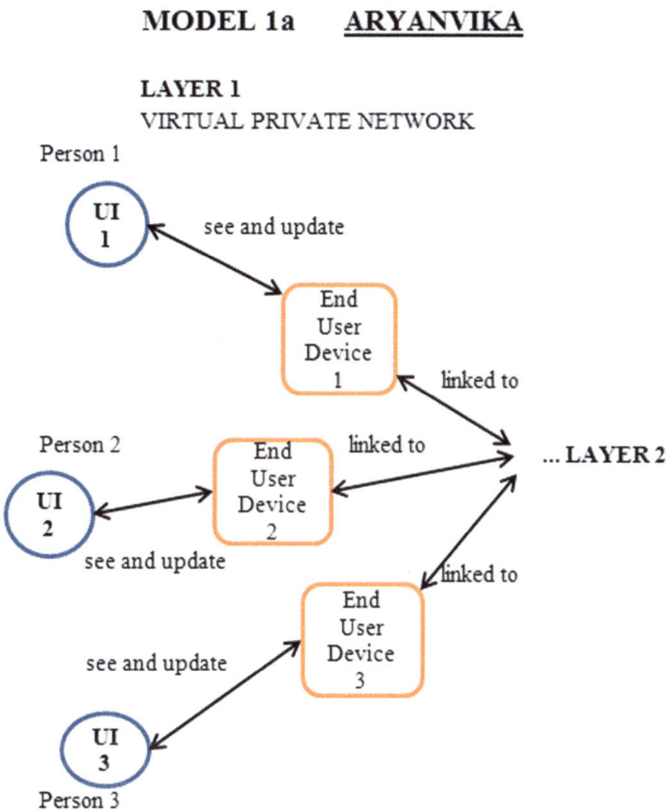

Fig. 2 System model: layer 1

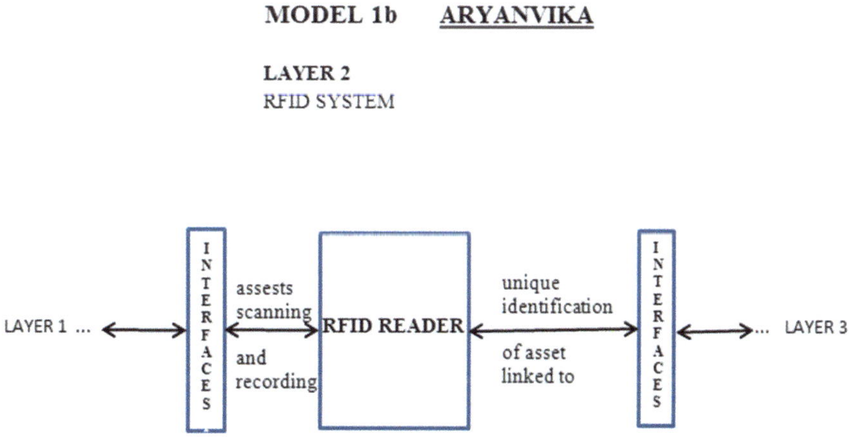

Fig. 3 System model: layer 2

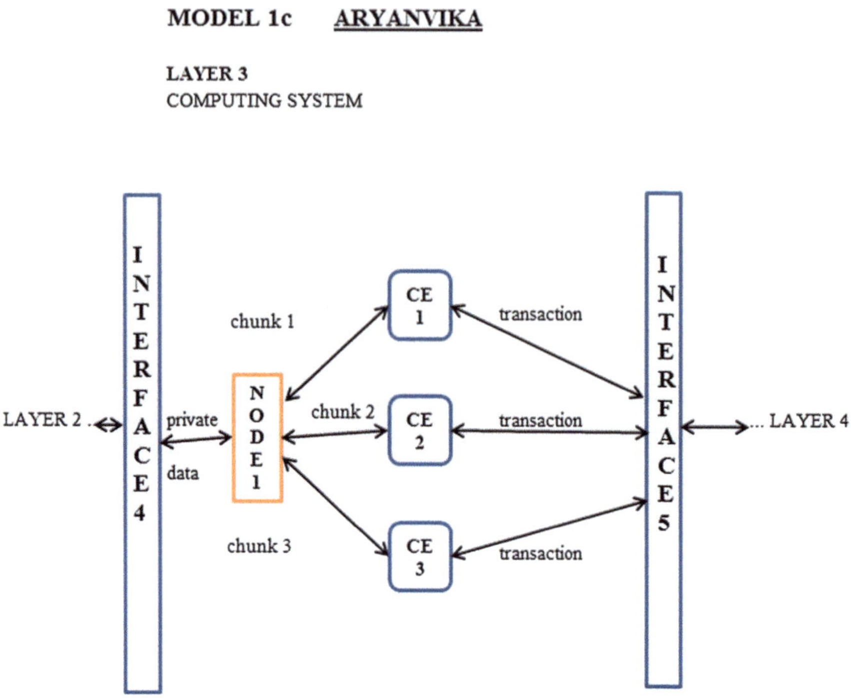

Fig. 4 System model: layer 3

3.2.3 Workflow of Layer 3

Layer 3 of the proposed novel architecture is shown in Fig. 4. This layer takes its input from layer 2 and passes on its output to layer 4. In layer 3, there exists a single node (to save cost) although there exists a strong possibility to scale it to multiple nodes depending on the requirements. There exist three computational elements in the present architecture (which may be extended to n in numbers). By this layer of the architecture, a transaction is initiated and layer 4 is invoked.

3.2.4 Workflow of Layer 4

Layer 4 corresponding to the proposed hybrid IoT architecture is presented in Fig. 5. This layer is composed of the blockchain. The data in the block of the chain may be like the time stamp, date hash, and link as is clearly indicated in the figure. The data block is added securely and uniquely with tracking details.

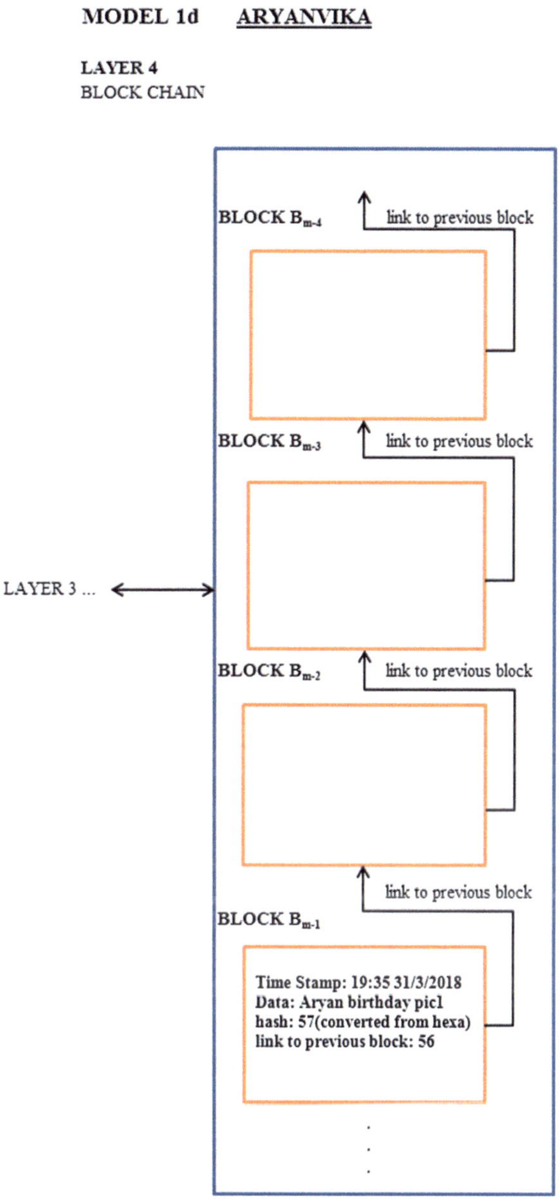

Fig. 5 System model: layer 4

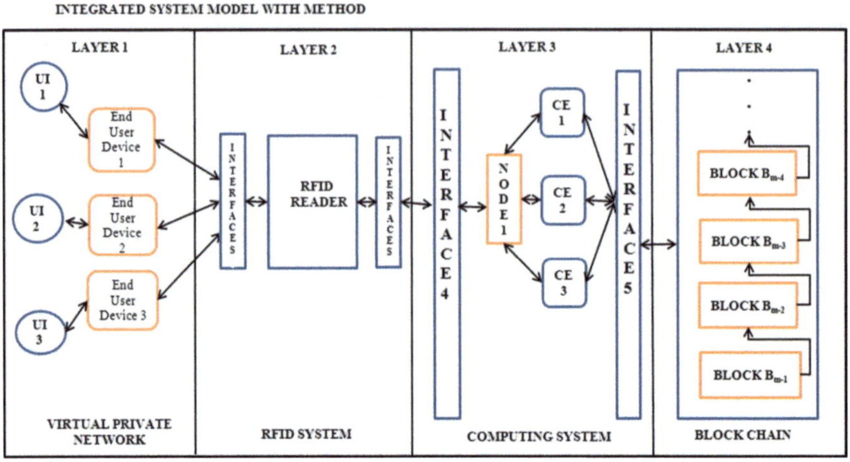

Fig. 6 Proposed hybrid Internet of things architecture (with blockchain and RFID technologies)

3.3 *Integrated View of Hybrid Internet of Things Architecture*

Here, the authors describe the integrated functioning of the proposed hybrid architecture of the Internet of things. The pictorial view of the proposed novel IoT architecture is presented in Fig. 6.

The functioning of the proposed IoT architecture is described by the following four major components or layers:

(1) *Virtual Private Network (VPN): Accessing Data*
Whenever a request from the user comes, a data file is generated locally on the virtual private network (local cloud) from one of the heterogeneous devices, for example, from EndUser Device 1 (laptop) or EndUser Device 2 (smartphone) or EndUser Device 3 (raspberry pi 3). These devices may be further generalized to n numbers. The data file may be in a different format depending on the source. There exists a Python script to convert the requested file in the desired format.
(2) *RFID System: Reading Data*
The RFID reader is connected to the VPN and reads data from the same. The Python script is used to connect the RFID reader to the VPN and reading the data and associating a unique identification to the data. This ensures the uniqueness of the desired data. There can be a single RFID system or a collection of RFID devices. In case of more than one RFID reader, there will be a necessity of incorporation of the RFID anti-collision protocols [23].
(3) *Computing system: Processing Data*
This data from the RFID reader is further fed to a cluster of small fast processing computers at a single node (there is a possibility of multiple nodes connected at

this point consisting of n number of computational elements depending on the horizontal scalability parameters) which contains execution of Python scripts.

(4) *Blockchain: Managing Data*
The concepts of proofs are written according to the requirements, the specific or generic algorithms are up and running on the computational nodes, and the data is appended as a block in the chain in the most reliable and cost-effective manner.

4 Proposed Architecture Assessment and Results

This section discusses the experimental setup and assumptions made while executing the designed architecture, and then, results are displayed corresponding to the successful execution of the proposed hybrid IoT architecture. The application under consideration is written in Python and aims to offer a user the best-rated cryptocurrency among the four considered in the present scenario. The four cryptocurrencies considered in this work are BTC, Etherum, Ripple, and Zcash.

4.1 Environment of Execution

The environment under consideration is as discussed in the proposed hybrid architecture for IoT under Sect. 3 of this research paper.

4.2 The Scenario and Application Under Consideration

A user of the proposed architecture has migrated to some other location, and his IoT architecture setup exists at his home in India. The user who is temporarily migrated to some other location now wishes to offer a new business proposal to his new organization where he wishes to seek the old mined data kept on his cloud at his home. He wishes to offer an application which makes use of all the technologies like RFID, cloud, blockchain, and IoT to make secure transaction and offer his new organization a secure, unique, and reliable solution.

4.3 Execution of the Application

The data in layer of the architecture was successfully mined with the help of designed Python scripts. Each of the data center storage had one of the four different types of files, namely low priority type-1, high priority type-1, low priority type-2, and

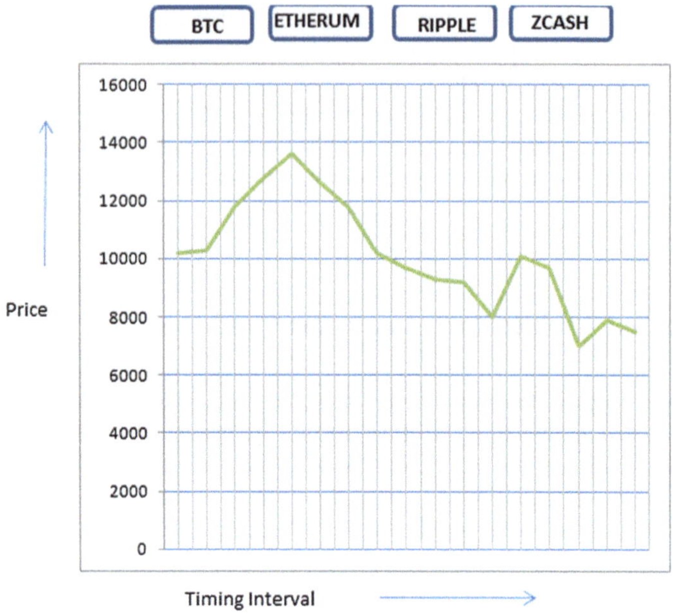

Fig. 7 Timing interval versus price of various cryptocurrencies

high priority type-2. For simulating the configuration, the authors used three end user devices, a cluster of three graphical processing units, one RC522 RFID reader, specially designed user interfaces, one raspberry pi 3 module with Wi-Fi module, one Samsung smartphone, and one HP laptop (core i3). The results were successfully obtained at the user interface of a different node as depicted in Fig. 7.

5 Open Challenges and Future Visions

The challenges associated with the proposed work are to bring the instances of the proposed architectures to life for amplification of the performance of the hybrid IoT system and also to reduce the cost of data handled to a prime level while keeping into consideration the heterogeneous devices utilized in Internet of things. More challenges associated with the proposed research work are multi-technologies awareness and integration capabilities along with the blockchain vulnerabilities. The future vision for the proposed work imbibes the possibilities of implementation of the proposed architectures with polyglot programming and polyglot approaches.

6 Summary and Conclusions

The prime idea of this research work is that the proposed IoT architecture would require no external characteristics to govern the cost or the performance of the hybrid system. Further, the instances resulting from implications of the proposed architecture would possess the self-capability for analyzing the performance of the private data which is in the system and hence set the cost, timing to list a few among crucial parameters. Further, there lie intensive market opportunities for the proposed research work. This is due to the fact that one can make discrete Aryanvika kind of system and methods for diverse stretches across the globe, catering to their own needs of preserving data from heterogeneous sources in a secure and cost-effective manner. Aryanvika is a huge research, business, and expansion opportunity.

References

1. Huckle, S., Bhattacharya, R., White, M., Beloff, N.: Internet of things, blockchain and shared economy applications. Procedia Comput. Sci. **98**, 461–466 (2016)
2. Molnar, D., Wagner, D.: Privacy and security in library RFID: issues, practices, and architectures. In: Proceedings of the 11th ACM Conference on Computer and Communications Security, pp. 210–219. ACM, Oct 2004
3. Bolic, M., Simplot-Ryl, D., Stojmenovic, I. (eds.): RFID systems: research trends and challenges. Wiley, New York (2010)
4. Buyya, R., Yeo, C.S., Venugopal, S., Broberg, J., Brandic, I.: Cloud computing and emerging IT platforms: vision, hype, and reality for delivering computing as the 5th utility. Future Gener. Comput. Syst. **25**(6), 599–616 (2009)
5. Moreno-Vozmediano, R., Montero, R.S., Llorente, I.M.: IaaS cloud architecture: from virtualized datacenters to federated cloud infrastructures. Computer **45**(12), 65–72 (2012)
6. Zhu, J.: Cloud computing technologies and applications. In: Handbook of Cloud Computing, pp. 21–45. Springer, Boston (2010)
7. Al-Dhuraibi, Y., Paraiso, F., Djarallah, N., Merle, P.: Elasticity in cloud computing: state of the art and research challenges. IEEE Trans. Serv. Comput. **11**(2), 430–447 (2018)
8. Malik, S.U.R., Khan, S.U., Ewen, S.J., Tziritas, N., Kolodziej, J., Zomaya, A.Y., Malluhi, Q.M.: Performance analysis of data intensive cloud systems based on data management and replication: a survey. Distrib. Parallel Databases **34**(2), 179–215 (2016)
9. Kapoor, S., Dabas, C.: Cluster based load balancing in cloud computing. In: Eighth International Conference on Contemporary Computing (IC3), pp. 76–81. IEEE, Aug 2015
10. Dabas, C., Gupta, J.P.: A cloud computing architecture framework for scalable RFID. In: Proceedings of the International MultiConference of Engineers and Computer Scientists, vol. 1, Mar 2010
11. Zheng, Z., Xie, S., Dai, H.N., Wang, H.: Blockchain challenges and opportunities: a survey. Work Paper (2016)
12. Cachin, C.: Architecture of the hyperledger blockchain fabric. In: Workshop on Distributed Cryptocurrencies and Consensus Ledgers, vol. 310, July 2016
13. Zheng, Z., Xie, S., Dai, H., Chen, X., Wang, H.: An overview of blockchain technology: architecture, consensus, and future trends. In: IEEE International Congress on Big Data (BigData Congress), pp. 557–564. IEEE, June 2017
14. Gubbi, J., Buyya, R., Marusic, S., Palaniswami, M.: Internet of things (IoT): a vision, architectural elements, and future directions. Future Gener. Comput. Syst. **29**(7), 1645–1660 (2013)

15. Yang, Z., Yue, Y., Yang, Y., Peng, Y., Wang, X., Liu, W.: Study and application on the architecture and key technologies for IOT. In: International Conference on Multimedia Technology (ICMT), pp. 747–751. IEEE, July 2011
16. Khan, M.A., Salah, K.: IoT security: review, blockchain solutions, and open challenges. Future Gener. Comput. Syst. **82**, 395–411 (2018)
17. Uckelmann, D., Harrison, M., Michahelles, F.: An architectural approach towards the future internet of things. In: Architecting the Internet of Things, pp. 1–24. Springer, Berlin (2011)
18. Kraijak, S., Tuwanut, P.: A survey on IoT architectures, protocols, applications, security, privacy, real-world implementation and future trends (2015)
19. Al-Fagih, A.E., Al-Turjman, F.M., Alsalih, W.M., Hassanein, H.S.: A priced public sensing framework for heterogeneous IoT architectures. IEEE Trans. Emerg. Top. Comput. **1**(1), 133–147 (2013)
20. Xu, X., Weber, I., Staples, M., Zhu, L., Bosch, J., Bass, L., Pautasso, C., Rimba, P.: A taxonomy of blockchain-based systems for architecture design. In: IEEE International Conference on Software Architecture (ICSA), pp. 243–252. IEEE, Apr 2017
21. Khan, R., Khan, S.U., Zaheer, R., & Khan, S.: Future internet: the internet of things architecture, possible applications and key challenges. In: 10th International Conference on Frontiers of Information Technology (FIT), pp. 257–260. IEEE, Dec, 2012
22. Zhou, L., Chao, H.C. Multimedia traffic security architecture for the internet of things. IEEE Netw. **25**(3) (2011)
23. Dabas, A.C., Balhara, B.M., Gupta, C.J.P.: CDMA based anti-collision deterministic algorithm for RFID tags. Int. J. Recent Trends Eng. **1**(1), 603 (2009)

Homomorphism Between Fuzzy Set-Valued Information Systems

Waseem Ahmed, M. M. Sufyan Beg and Tanvir Ahmad

Abstract Communication between various information systems (IS) has been an urgent issue that needs to be discussed in the granular world. Compatible and consistent homomorphism is an important analytical tool to study communication among various IS. Fuzzy set-valued information systems (FSIS) are those IS that contain fuzzy set-values for some attribute. This paper aims to discuss important properties related to communication between FSIS. Fuzzy relation mapping from the perspective of FSIS is discussed. The proposed approach proved that feature selection and other properties of the original FSIS and the corresponding image FSIS are assured under consistent and compatible homomorphism. Finally, a real-life example demonstrated the utilization of the proposed work.

Keywords Fuzzy set-valued information systems · Homomorphism
Feature selection · Fuzzy relation mapping

1 Introduction

Rough set theory (RST) is an influential soft computing tool in the area of information technology and has attracted much attention and interest [1–5]. Continuous or real-valued IS cannot be handled directly by the traditional rough set because it requires discretization of real attributes before feature selection, so to handle this, fuzzy rough set (FRS) is introduced [2]. In FRS, fuzzy similarity relation is used to handle real-valued attributes. RST begins with a single-valued information system

W. Ahmed (✉) · T. Ahmad
Department of Computer Engineering, Jamia Millia Islamia, New Delhi, India
e-mail: waseem.ahmed86@gmail.com

T. Ahmad
e-mail: tahmad2@jmi.ac.in

M. M. S. Beg
Department of Computer Engineering, Aligarh Muslim University, Aligarh, India
e-mail: mmsbeg@eecs.berkeley.edu

L. C. Jain et al. (eds.), *Data and Communication Networks*, Advances in Intelligent Systems and Computing 847, https://doi.org/10.1007/978-981-13-2254-9_19

[2–5]; however, in real-life situations, some attribute values may contain set-values, resulting in the formation of SIS [6, 7]. Dai and Tian [6] investigated FRS model for SIS. SIS studied so far contains crisp set-values [6, 7] and does not consider the case of fuzzy set-valued attributes. To handle these types of attributes, a FSIS was introduced [1].

In today's Information era, information system (IS) is the realm of information technology, and the major concern here is communication between IS. Due to different nature of IS, it becomes necessary to transfer information within these IS [3, 5]. This motivates us to examine the communication among various IS's. In this work, important properties related to communication between FSIS using FRS model are discussed. Communication means transformation of information between IS, alongside keeping unblemished its prime properties and functions.

From mathematical viewpoint, mapping is an efficient way to communicate between IS by analyzing their properties [5]. Homomorphism mapping plays a vital role to accomplish this task [3, 5]. In this paper, important properties related to communication between FSIS under FRS model are discussed.

The proposed approach proved that feature selection and other properties of the original FSIS and the corresponding image FSIS are assured under consistent and compatible homomorphism.

The structure of the remaining paper is defined below:

A brief description about FSIS is given in Sect. 2. In Sect. 3, fuzzy relation mapping for FSIS is formalized. Section 4 discusses the notion of homomorphism between FSIS. In Sect. 5, the conclusion is outlined.

2 Preliminaries

2.1 Introduction to FSIS

Definition 1 A FSIS = (S, C, V, f) is an IS that contains real or fuzzy set-values for some of the attribute [1].

Where S and C are set of sample and set of fuzzy multivalued attributes, respectively. f: $S \times C \to V$, where for $m \in S$, $c \in C, f(\text{m, c}) \in V_c$ assigns fuzzy set-values to samples. Also, V_c is the set of pairs $(x, \mu_{Vc}(\text{x}))$ such that $\mu_{Vc}(x)$ is the membership value of x in V_c between [0, 1].

Example 2 FSIS is given in Table 1. Here, $S = \{m_1, m_2, m_3, m_4\}$ is a set of objects; $C = \{R, W, S\}$ is a set of attributes; V = {French, German, English} is domain values. For ease, French, German, and English are denoted by f, g, and e, respectively.

If suppose, 'c' is an attribute 'Reading'= $\{f, g, e\}$. Then $c(m) = \{(f, 0.8), (g, 0.6), (e, 0.9)\}$ illustrates that reading abilities of 'm' in French, German, and English are 0.8, 0.6, and 0.9, respectively.

Table 1 An example of FSIS

FSIS	Reading (R)	Writing (W)	Speaking (S)
m_1	{(*f*, 0.8), (*g*, 0.6), (*e*, 0.9)}	{(*f*, 0.3), (*g*, 0.7), (*e*, 0.5)}	{(*f*, 0.6), (*g*, 0.9), (*e*, 0.7)}
m_2	{(*f*, 0.8), (*g*, 0.6), (*e*, 0.9)}	{(*f*, 0.3), (*g*, 0.7), (*e*, 0.5)}	{(*f*, 0.6), (*g*, 0.9), (*e*, 0.7)}
m_3	{(*f*, 1), (*g*, 0.3), (*e*, 0)}	{(*f*, 0.8), (*g*, 0.3), (*e*, 0)}	{(*f*, 0.9), (*g*, 0.6), (*e*, 0.3)}
m_4	{(*f*, 0.5), (*g*, 0.9), (*e*, 0.4)}	{(*f*, 0.7), (*g*, 0.7), (*e*, 0)}	{(*f*, 0.5), (*g*, 0.8), (*e*, 0.7)}

2.2 *Fuzzy Similarity Relation for FSIS*

Definition 3 For FSIS = (S, C) and $a \in C$ and $m, n \in S$, the fuzzy similarity relation R_a is defined as:

$$\mu_{R_a}(m,n) = \frac{\sum \inf(a(m), a(n))}{\sum \sup(a(m), a(n))}$$

3 Fuzzy Relation Mapping for FSIS

A definition of fuzzy relation mapping to communicate between FSIS using Zadeh's extension principle is given below:

For two universal sets S_1 and S_2, let $R(S_1 \times S_1)$ and $R(S_2 \times S_2)$ represent classes of all fuzzy binary relation on S_1 and S_2, respectively.

Definition 4 Let f: $S_1 \rightarrow S_2$ be a mapping. 'f' generate a mapping from $R(S_1 \times S_1)$ to $R(S_2 \times S_2)$ as:

$$f(R)(z_1, z_2) = \begin{cases} \sup_{m_1 \in f^{-1}(z_1)} \sup_{m_2 \in f^{-1}(z_2)} R(m_1, m_2), & (z_1, z_2) \in f(S_1) \times f(S_1) \\ 0, & (z_1, z_2) \notin f(S_1) \times f(S_1) \end{cases}$$

3.1 *Consistent Functions*

Definition 5 Let S_1 and S_2 are universe of discourse, f: $S_1 \rightarrow S_2$ is their mapping from S_1 to S_2, and B, B_1, $B_2 \in R(S_1 \times S_1)$. Let $[a]_f = \{b \in U\text{:} f(a) = f(b)\}$;

For any $m_1, m_2 \in [a]_f$, and $z_1, z_2 \in [b]_f$, we say 'f' is compatible with relation B, if $B(m_1, z_1) = B(m_2, z_2)$.

For any a, $b \in S_1$, if any of the condition holds:

(1) $B_1(m_1, z_1) \le B_2(m_1, z_1)$ for any $(m_1, z_1) \in [a]_f \times [b]_f$
(2) $B_1(m_1, z_1) \ge B_2(m_1, z_1)$ for any $(m_1, z_1) \in [a]_f \times [b]_f$, then the mapping '$f$' is consistent to relation B_1 and B_2.

Theorem 6 *Let f: $S_1 \to S_2$, B, B_1, $B_2 \in R(S_1 \times S_1)$; then, we have the following:*

(1) $f(B_1 \cup B_2) = f(B_1) \cup f(B_2)$
(2) $f(B_1 \cap B_2) \subseteq f(B_1) \cap f(B_2)$; if '$f$' is a consistent mapping, then they are equal.

Proof (1) For any $z_1, z_2 \in S_2$

$$\begin{aligned} f(B_1 \cup B_2)(z_1, z_2) &= \sup_{m_1 \in f^{-1}(z_1)} \sup_{m_2 \in f^{-1}(z_2)} (B_1 \cup B_2)(m_1, m_2) \\ &= \sup_{m_1 \in f^{-1}(z_1)} \sup_{m_2 \in f^{-1}(z_2)} (B_1(m_1, m_2) \cup B_2(m_1, m_2)) \\ &= (f(B_1) \cup f(B_2))(z_1, z_2). \end{aligned}$$

(2)

$$\begin{aligned} f(B_1 \cap B_2)(z_1, z_2) &= \sup_{m_1 \in f^{-1}(z_1)} \sup_{m_2 \in f^{-1}(z_2)} (B_1 \cap B_2)(m_1, m_2) \\ &= \sup_{m_1 \in f^{-1}(z_1)} \sup_{m_2 \in f^{-1}(z_2)} (B_1(m_1, m_2) \cap B_2(m_1, m_2)) \\ &\le \sup_{m_1 \in f^{-1}(z_1)} \sup_{m_2 \in f^{-1}(z_2)} B_1(m_1, m_2) \cap \sup_{m_1 \in f^{-1}(z_1)} \sup_{m_2 \in f^{-1}(z_2)} B_2(m_1, m_2) \\ &= (f(B_1) \cap f(B_2))(z_1, z_2). \end{aligned}$$

Now, it will be proved that the equality holds, if the mapping 'f' is consistent mapping:

Now, since 'f' is consistent mapping to B_1 and B_2, it follows from Definition 5 that it satisfies one of the following conditions:

1. $B_1(m_1, m_2) \le B_2(m_1, m_2)$
2. $B_1(m_1, m_2) \ge B_2(m_1, m_2)$.

For any $(m_1, m_2) \in f^{-1}(z_1) \times f^{-1}(z_2)$
For case 1,

$$\begin{aligned} f(B_1 \cap B_2)(z_1, z_2) &= \sup_{m_1 \in f^{-1}(z_1)} \sup_{m_2 \in f^{-1}(z_2)} (B_1 \cap B_2)(m_1, m_2) \\ &= \sup_{m_1 \in f^{-1}(z_1)} \sup_{m_2 \in f^{-1}(z_2)} (B_1(m_1, m_2) \cap B_2(m_1, m_2)) \\ &= \sup_{m_1 \in f^{-1}(z_1)} \sup_{m_2 \in f^{-1}(z_2)} B_1(m_1, m_2) \\ &= f(B_1)(z_1, z_2) \end{aligned}$$

Now taking RHS,

$$\begin{aligned} (f(B_1) \cap f(B_2)(z_1, z_2) &= f(B_1)(z_1, z_2) \cap f(B_2)(z_1, z_2) \\ &= (\sup_{m_1 \in f^{-1}(z_1)} \sup_{m_2 \in f^{-1}(z_2)} (B_1(m_1, m_2)) \end{aligned}$$

$$\cap\left(\sup_{m_1\in f^{-1}(z_1)}\sup_{m_2\in f^{-1}(z_2)} B_2(m_1, m_2)\right)$$
$$= f(B_1)(z_1,\ z_2)$$

Hence $f(B_1 \cap B_2) = f(B_1) \cap f(B_2)$.

Example 7 (continued to example 2) Let $S_1 = \{m_1, m_2, m_3, m_4\}$ and $S_2 = \{z_1, z_2, z_3\}$ f_2: $S_1 \rightarrow S_2$ is a mapping from S_1 to S_2 and defined as: $f_2(m_1) = f_2(m_2) = z_1$; $f_2\ (m_3) = z_2; f_2(m_4) = z_3$;

Fuzzy similarity relations for attributes reading (R), writing (W), and speaking (S) given in Table 1 are calculated using Definition 3 and shown in Table 2. ($R_R \cap R_W \cap R_S$) is given in Table 3.

Images of R, W, and S using 'f_2' are computed and presented in Table 4. $f_2(R_R) \cap f_2(R_W) \cap f_2(R_S)$ and $f_2(R_R \cap R_W \cap R_S)$ are given in Table 5.

We find that $f_2(R_R \cap R_W \cap R_S) = f_2(R_R) \cap f(R_W) \cap f(R_S)$, since mapping f_2 is a compatible mapping.

4 Homomorphism Between FSIS

This section focuses on the concept of homomorphism and discusses important properties of FSIS using homomorphism.

Definition 9 Let S_1 and S_2 are two universes, f: $S_1 \rightarrow S_2$ be its mapping from S_1 to S_2. Assume $\boldsymbol{R} = \{B_1, B_2, \ldots, B_n\}$ is collection of fuzzy relations on S_1; then $f(\boldsymbol{R}) = \{f(B_1), f(B_2), \ldots, f(B_n)\}$. Then $(S_1, \boldsymbol{R})$ is termed as FSRIS, and the corresponding $(V_1, f(\boldsymbol{R}))$ is its induced FSRIS [5].

Definition 9 Let $(S_1, \boldsymbol{R})$ be a FSRISs and 'f' is a function mapping. We define homomorphism using 'f' satisfying certain conditions as follows [5]:

(1) $\forall\ B_i, B_j \in \boldsymbol{R}$, if '$f$' is consistent with every B_i and B_j, then we say that 'f' is consistent homomorphism.
(2) $\forall\ B_i \in \boldsymbol{R}$, *if is* compatible with each B_i, then we say that 'f' is compatible homomorphism.

Definition 10 Let $(S_1, \boldsymbol{R})$ be FSRISs and $P \subseteq \boldsymbol{R}$ satisfies the following:

(1) $\cap P = \cap R$
(2) $\forall B_i \in P, \cap P \subset \cap(P - \{B_i\})$.

Then we say P is a reduct of $\boldsymbol{R}$

Theorem 11 *Let $(S_1, \boldsymbol{R})$ be FSRISs and 'f' be a consistent homomorphism mapping from S_1 to S_2. $P \subseteq \boldsymbol{R}$ is a reduct of $\boldsymbol{R}$ only when $f(P)$ is reduct of $f(\boldsymbol{R})$ and vice versa.*

Table 2 Similarity relations for attribute R, W, and S

R_R	m_1	m_2	m_3	m_4	R_W	m_1	m_2	m_3	m_4	R_S	m_1	m_2	m_3	m_4
m_1	1	1	0.4	0.6	m_1	1	1	0.3	0.5	m_1	1	1	0.6	0.9
m_2	1	1	0.4	0.6	m_2	1	1	0.3	0.5	m_2	1	1	0.6	0.9
m_3	0.4	0.4	1	0.3	m_3	0.3	0.3	1	0.7	m_3	0.6	0.6	1	0.6
m_4	0.6	0.6	0.3	1	m_4	0.5	0.5	0.7	1	m_4	0.9	0.9	0.6	1

Table 3 Relation for $(R_R \cap R_W \cap R_S)$

$(R_R \cap R_W \cap R_S)$	m_1	m_2	m_3	m_4
m_1	1	1	0.3	0.5
m_2	1	1	0.3	0.5
m_3	0.3	0.3	1	0.3
m_4	0.5	0.5	0.3	1

Table 4 Images of *R*, *W*, and *S* using f_2

$f_2\,(R_R)$	z_1	z_2	z_3	$f_2\,(R_W)$	z_1	z_2	z_3	$f_2\,(R_S)$	z_1	z_2	z_3
z_1	1	0.4	0.6	z_1	1	0.3	0.5	z_1	1	0.6	0.9
z_2	0.4	1	0.3	z_2	0.3	1	0.7	z_2	0.6	1	0.6
z_3	0.6	0.3	1	z_3	0.5	0.7	1	z_3	0.9	0.6	1

Table 5 Relation for $f_2(R_R) \cap f(R_W) \cap f(R_S)$ and $f_2(R_R \cap R_W \cap R_S)$

$f_2(R_R \cap R_W \cap R_S)$	z_1	z_2	z_3	$f_2(R_R) \cap f_2(R_W) \cap f_2(R_S)$	z_1	z_2	z_3
z_1	1	0.3	0.5	z_1	1	0.3	0.5
z_2	0.3	1	0.3	z_2	0.3	1	0.3
z_3	0.5	0.3	1	z_3	0.5	0.3	1

Proof $\Rightarrow$ Let us suppose, *P* be reduct of ***R***. Therefore, $\cap P \neq \cap(P - \{B_i\})$.

Then, there must be $m_1, m_2 \in S_1$, such that $\cap P(m_1, m_2) < \cap\,(P - \{B_i\})(m_1, m_2)$, which implies,

$$
\begin{aligned}
& f(\cap(P - \{B_i\}))(f(m_1), f(m_2)) \\
&= \sup_{z_1 \in f^{-1} f(m_1))} \sup_{z_2 \in f^{-1} f(m_2))} \cap(P - \{B_i\})\,(z_1, z_2) \\
&> \sup_{z_1 \in f^{-1} f(m_1))} \sup_{z_2 \in f^{-1} f(m_2))} \cap P\,(z_1, z_2) \\
&= f(\cap P)(f(m_1), f(m_2)) = f(\cap R)(f(m_1), f(m_2))
\end{aligned}
$$

Now, $\cap P = \cap \boldsymbol{R}$. Hence, $f(\cap P) = f(\cap \boldsymbol{R})$
Using Theorem 6, $\cap f(P) = \cap f(\boldsymbol{R})$.
Assume, $f(P)$ is not reduct of $f(\boldsymbol{R})$, $\exists\, B_i \in P$ such that $\cap(f(P) - f(B_i)) = \cap f(P)$.
Since, $f(P) - f(R_i) = f(P - \{B_i\})$; therefore, $\cap f(P - \{B_i\}) = \cap f(P) = \cap f(\boldsymbol{R})$
Again, by Theorem 6, $f(\cap(P - B_i)) = f(\cap \boldsymbol{R})$
which is a conflict to the assumption that $f(P)$ is not reduct of $f(\boldsymbol{R})$.
$\therefore f(P)$ is not reduct of $f(\boldsymbol{R})$.

Example 12 (continued to example 7) Let $S_1 = \{m_1, m_2, m_3, m_4\}$ and $S_2 = \{z_1, z_2, z_3\}$

It is evident from Tables 2 and 4 that $\{R_R, R_W\}$ is reduct of '**R**', if and only if $\{f_2(R_R), f_2(R_W)\}$ is a reduct of $f_2(\boldsymbol{R})$.

5 Conclusion

This paper aims to discuss important properties related to communication between FSIS using FRS model. The definition of fuzzy relation mapping from the perspective of FSIS is discussed. The proposed approach proved that feature selection and other properties of the original FSIS and the corresponding image FSIS are assured in the case of consistent and compatible homomorphism.

References

1. Ahmed, W., Beg, M.S., & Ahmad, T.: Entropy based feature selection for fuzzy set-valued information systems. 3D Res. **9**(2), 19 (2018)
2. Dubois, D., Prade, H.: Rough fuzzy sets and fuzzy rough sets. Int. J. Gen. Syst. **17**, 191–208 (1990)
3. Grzymala-Busse, J.W., Sedelow, W.A.: On rough sets, and information system homomorphism. Bull. Pol. Acad. Sci. Tech. Sci. **36**(3–4), 233–239 (1988)
4. Pawlak, Z.: Rough Sets. Int. J. Comput. Inf. Sci. **11**, 341–356 (1982)
5. Tsang, E.C.C., Wang, C., Chen, D., Wu, C., Hu, Q.: Communication between information systems using fuzzy rough sets. IEEE Trans Fuzzy Syst. **21**(3), 527–540 (2013)
6. Dai, J., Tian, H.: Fuzzy rough set model for set-valued data. Fuzzy Sets Syst. **229**, 54–68 (2013)
7. Guan, Y.Y., Wang, H.K.: Set-valued information systems. Inf. Sci. **176**(17), 2507–2525 (2006)

A Novel Approach for Predicting the Outcome of Request in RAOP Dataset

Amreen Ahmad, Tanvir Ahmad and Abhishek Bhatt

Abstract In today's era, online social communities such as Q&A sites are widely used for asking favors, so it would be beneficial to formulate a technique that would help in predicting the success of the response. The objective of the paper is to enhance the accuracy of prediction of the success of altruistic request that follows the same approach as used by ADJ (Proceedings of AAAI International Conference on Web and Social Media, ICWSM, 2014 [1]. Three more features are proposed, i.e., topic, role, and centrality in addition to the features proposed by ADJ [1]) to capture user's interaction in the past and topic effect on the prediction of response. We also propose a graph-based success prediction (GSP) model that uses feature weights and uses the underlying graph structure for the propagation to predict the outcome of a request. Experiments were conducted on the RAOP dataset which belongs to sub-community of Reddit.com using GSP, and it outperformed ADJ and other baseline methods using limited training data.

Keywords Success prediction · Altruistic request · Social interactions

1 Introduction

In today's era of the World Wide Web, online social communities are being widely used for the noble cause. These social communities are often used by needy people for donations or help. Recently, researchers have focussed on studying the underlying hidden factors that increase the probability of a request being responded [2] and help in the promotion of such request. Some key factors were identified by existing

A. Ahmad (✉) · T. Ahmad · A. Bhatt
Department of Computer Engineering, Jamia Millia Islamia, New Delhi, India
e-mail: amreen.ahmad10@gmail.com

T. Ahmad
e-mail: tahmad2@jmi.ac.in

A. Bhatt
e-mail: abhishek1995bhatt@gmail.com

L. C. Jain et al. (eds.), *Data and Communication Networks*, Advances in Intelligent Systems and Computing 847, https://doi.org/10.1007/978-981-13-2254-9_20

researchers that included social interaction between the giver and user, the scale of the request, and whether the giver reciprocates. ADJ [1] work focussed on uncovering the effect of the linguistic factor on the success prediction of request given a situation in which the giver is not receiving any reward. It was also proved that the linguistic factor that included politeness, evidentiality, and narrative structure had a great effect on decision making and is strongly correlated with the success of requests.

The paper aims to improve the prediction accuracy of the altruistic request, and it follows the same approach as proposed by ADJ [1]. ADJ work suffered from the following limitations: (i) It required a large dataset for prediction; (ii) it did not take into consideration user interaction in the history that could be one of the important reason for the success of the request. The central assumption of the paper is that inclusion of three extra features, i.e., topic, centrality, and role, along with temporal, social, and textual features would prove to be effective in boosting the accuracy of the success prediction of the request. A graph-based success prediction (GSP) model is proposed to predict the outcome of the request. Finally, for evaluation of GSP model, ADJ, LR, and SVM are used.

The remaining part is structured as follows: The dataset and the proposed features are explained in detail in Sect. 2. Next, we proceed with the explanation of the proposed graph-based success prediction model in Sect. 3. We proceed with an analysis of results obtained from evaluation described in Sect. 4 followed by conclusion discussed in Sect. 5.

2 Dataset and Features

A brief description of the dataset and features selection is given below:

2.1 Request Dataset

The pizza request dataset consists of 5728 pizza requests and 1.87 million relevant posts that are used for computing feature values.

2.2 Features

A detailed description of the three proposed features, i.e., topic, role, and centrality, is discussed.

1. Centrality features: We capture user interaction that takes place in Reddit.com through comments so that their effect can be analyzed which would affect the success of a request.

2. Role features: The role of a user in communities can be measured in the following ways:
   (a) Core node: Individuals who are highly active in a network community imply that they have high TPR, i.e., triangle participation ratio.
   (b) Bridging effect: Some individuals involved in the network community play the role of transmitting information. RawComm is used to measure the bridging effect [4] of a node.
3. Topic features: It is assumed that the success of altruistic request is highly dependent on the topic to which it belongs. For modeling latent topics, the following ways are used:
   (a) Bag-of-Word: The words with AD (adverb), VA (adjective), VV (verb), NR (proper noun), and NN (noun) are considered as BOW features.
   (b) N-gram: Recursive feature elimination technique [5] is used for the identification of discriminative and important features of bi-gram and tri-gram.

3 Graph-Based Success Prediction (GSP) Model

Graph-based success prediction (GSP) model contains two phases:

3.1 Graph Construction

In the underlying graph, request text is represented as nodes and feature correlation between nodes is computed using probabilistic approach that is used to predict the outcome of unseen request. The graph is used to map the interaction between request nodes denoted by $G=(V;E)$. Let $V = U_{train} \cup U_{test}$, where U_{train} is a set of training nodes (t_r) and U_{test} is a set of testing nodes (t_{st}). Let node u represents testing request node, i.e., $u \in U_{test}$, and v represent training request node, i.e., $v \in U_{train}$. Now there are two steps: (a) Connect every node $u \in U_{test}$ to the top-$\lambda 1$ similar node $v \in U_{train}$ and (b) connect every $u \in U_{test}$ to the top-$\lambda 2$ similar nodes u_w such that $u \neq u_w$ and $u_w \in U_{test}$, where $\lambda 1 = k \times |U_{train}|$ and $\lambda 2 = k \times |U_{test}|$, k is a parameter $\in [0,1]$. In the proposed approach, the value of k is varied to determine its efficiency and effectiveness. The probability of each request node for each label sc(x) (for node x) can either be 1 (successful) or 0 (unsuccessful) denoted by $P_{sc}(x)$. Firstly, the probabilistic scores for trained request nodes are set. For $v \in U_{train}$, let $P_1(v) = 1$ and $P_0(v) = 0$ if sc(v) = 1 else $P_1(v) = 0$ and $P_0(v) = 1$. Also for $u \in U_{test}$, let $P_1(u) = 0$ and $P_0(u) = 0$.

The edge of graph G denotes the correlation between request texts based on their features. The process of edge creation is explained further. Let us assume that there is a certain feature f_t where f_t represents one of the features of nodes, then feature-based request correlation factor $F_c(n_1; n_2)$ is computed for nodes $(n1; n2) \in E$ that can be derived from difference in features $F_c(n_1; n_2) = \|\text{ft}(n_1) - \text{ft}(n_2)\|$. For a given set of features denoted by $F = f_t$ (where $t = 1; \ldots; z$) and z is the count of features, feature-aware correlation factor can be computed using feature-based request correlation factor and is given as follows:

$$F_arc\left(n_{1,}n_2\right) = \exp\left(-\sum_{t=1}^{z} w_t^2 \times F_c\left(n_1, n_2\right)\right) \tag{1}$$

where w_t denotes the weight of respective feature f_t. The $F_arc(n_1; n_2)$ is edge weight which can be denoted as $w_c(n_1; n_2)$ for edge $(n_1; n_2) \leftarrow \in E$ in the graph G.

3.2 *Learning-Based Propagational Optimization*

It works in two steps:

1. The probabilistic values of $P_1(u)$ and $P_0(u)$ are inferred from $P_1(v)$ and $P_0(v)$.
2. The probabilistic values of $P_1(u)$ and $P_0(u)$ can also be derived from u's neighboring nodes. Let the degree of node u be denoted as du, then $P_1(u)$ and $P_0(u)$ can be derived using the given approach that enables us to get optimal set of edge weights in the graph:

$$P_{sc}(u) = \frac{1}{d_u} \sum_{(u,y)\in E} w_c(u, y) \times P_{sc}(y) \tag{2}$$

where y denotes other nodes(both training and testing dataset) present in the graph G.

Furthermore, learning and optimizing w_t can enable us to compute similar probabilities for nodes in the close vicinity of the testing node. In this model, a heuristic approach is proposed to learn the feature weights w_t based on entropy of the probabilities and the probabilities are fine-tuned toward more accurate results. E_p denotes the probability of success that is given as:

$$E_p(u) = \frac{1}{U_{test}} \times - \sum_{sc\in(0,1)} P_{sc}(u)(1 - P_{sc}(u)) \log(1 - P_{sc}(u)) \tag{3}$$

where U_{test} denotes the count of requests in testing dataset of graph G and sc denotes success label.

Learning Optimization It is assumed that success probabilities computed would be of higher confidence if they are achieving lower entropy values. Gradient descent algorithm is used to minimize the entropy w.r.t w_t given by:

$$\delta E_{\mathrm{p}}(u) = \frac{1}{U_{\mathrm{test}}} \times \sum_{y \in (U_{\mathrm{test}})} \frac{(\log(1 - P(y))}{P(y)} \times \delta P(y) \tag{4}$$

It is further differentiated using chain rule, and the output is:

$$\delta w_{\mathrm{t}} = 2 \times w_{\mathrm{c}}(n_1 \times n_2) \times F_{\mathrm{c}}(n_1 \times n_2) \times w_{\mathrm{t}} \tag{5}$$

Prediction Likewise, the feature weights w_{t} are updated with each iteration where $w_{\mathrm{t}} = w_{\mathrm{t}} - \delta w_{\mathrm{t}}$. These updated feature weights w_{t} are used to derive new edge weights w_{c} $(n_1; n_2)$ using Eq. (1) and can be further used for the derivation of success probabilities of testing request $P_1(y)$ and $P_0(y)$ from the underlying graph as follows:

$$\max_{P_{\mathrm{sc}}} u = 2 \times w_{\mathrm{c}}(n_1 \times n_2) \times F_{\mathrm{c}}(n_1 \times n_2) \times w_{\mathrm{t}}. \tag{6}$$

The next step proceeds with the updation of E_{p} using Eq. (3), and the procedure is repeated iteratively till E_{p} converges. The success status of y nodes where $y \in U_{\mathrm{test}}$ can be derived using success probability is obtained from Eq. (5).

4 Evaluation

The t_{r} data is varied in the ratio of 30%:70% and 70%:30% for analysis. Initially, k is set to 0.1 and is varied to assess the efficiency of GSP model using different percentages of t_{r} data. The performance of GSP is evaluated using AUC score and accuracy, where accuracy $= \frac{CPR}{t_{\mathrm{s}}}$ where CPR is the count of requests that are correctly predicted and t_{s} is the sum total of testing data. The percentage of t_{s} and t_{r} data is selected, and the experiment is repeated up to 200 times. Finally, average of AUC score and average of accuracy score is computed and used for analysis purpose.

4.1 Competitive Methods

The proposed GSP model is compared with ADJ work [1] and other competitive classifiers such as SVM [6] and LR [3]. It can be seen from the obtained results shown in Figs. 1 and 2 that GSP model outperforms ADJ and other competitive rivals for every case. As can be seen from Fig. 2a that the ADJ approach achieved 0.67 AUC score with 70% t_{r} data while 0.81 AUC score and accuracy score was achieved by the proposed GSP, with same amount of training data. This is mainly because of Fig. 3a, b reports the AUC score and execution time for different values of k. As can be seen from Fig. 3a, when the value of k increases, AUC score also increases since it increases the size of the graph G. In Fig. 3b, although good results are achieved by increasing k value, alongside execution time also increases that adds

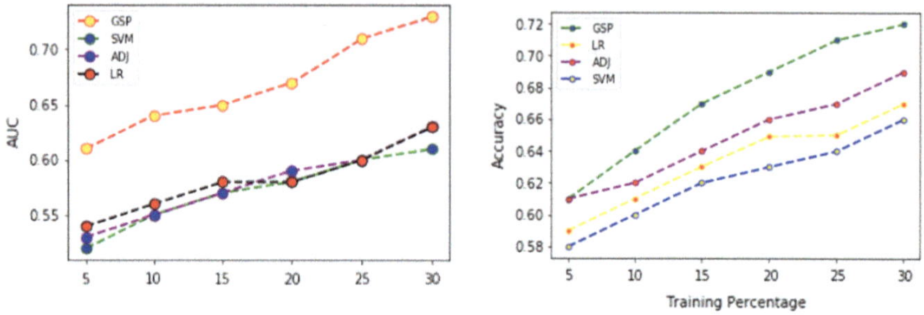

(a) AUC score (b) Accuracy score

Fig. 1 AUC and accuracy score when training data percentage is varied from 5 to 30%

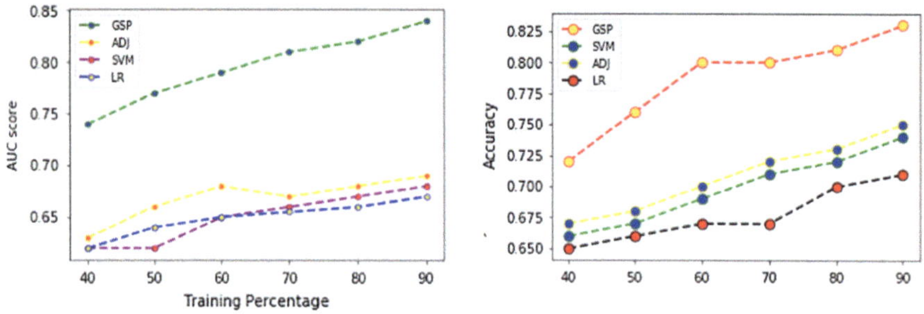

(a) AUC score (b) Accuracy score

Fig. 2 AUC and accuracy score when training data percentage is varied from 40 to 90%

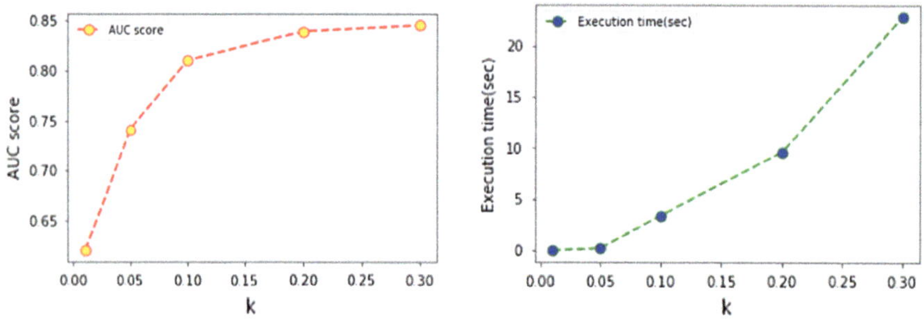

(a) AUC score for different values of k (b) Execution time for different values of k

Fig. 3 AUC score and execution time values obtained under the GSP approach the incorporation of three extra features, i.e., centrality, role, and topic, that played a major role in the success prediction

to its disadvantage. But for $k = 0.10$, GSP takes 4 s to execute and it is feasible for practical use, so we set $k = 0.10$.

5 Conclusion

The proposed work focusses on collaborative knowledge obtained from sites Reddit.com and evaluates the efficiency of proposed GSP model for the prediction of real-world events. Considering the work done by ADJ as the base, a new methodology is proposed to predict the outcome of the request. By considering three more features, i.e., topic, role, and centrality apart from those proposed by ADJ and using graph-based success prediction (GSP) model, 0.81 AUC score was obtained. It was observed that the proposed GSP model has outperformed other classifiers and has shown an improvement over ADJ in terms of AUC score with limited training data.

References

Althoff, T., Danescu-Niculescu-Mizil, C., Jurafsky, D. (ADJ): How to ask for a favor: a case study on the success of altruistic request. In: Proceedings of AAAI International Conference on Web and Social Media, ICWSM (2014)

Backstrom, L., Huttenlocher, D., Kleinberg, J., Lan, X.: Group formation in large social networks: membership, growth, and evolution. In: Proceedings of the ACM SIGKDD International Conference on Knowledge Discovery and Data Mining, KDD, pp. 44–54 (2006)

Freedman, D.A.: Statistical Models: Theory and Practice, pp. 128 (Cambridge University Press, 2009)

Scripps, J., Tan, P.N., Esfahanian, A.H.: Exploration of link structure and community-based node roles in network analysis. In: Proceedings of the IEEE International Conference on Data Mining, ICDM, pp. 649–654 (2007)

Ugander, J., Backstrom, L., Marlow, C., Kleinberg, J.: Structural diversity in social contagion. Proc. Nat. Acad. Sci. U.S.A. (PNAS) **109**(16), 5962–5966 (2012)

Vapnik, V.: Support-vector networks. Mach. Learn. **20**(3), 273297 (1995)

A Soft Computing Methodology for Estimation and Forecasting of Daily Global Solar Radiation (DGSR)

J. Christy Martina and T. Amudha

Abstract Energy is one of the most crucial building blocks of economic development. The energy sector in India has a rapid growth in recent years. The factors that took the nation to a very acute energy crisis are increase in population, transportation, urbanization, industrialization, high standard of living and fast depleting fossil fuels. Energy demand problems have increased all over the world, and renewable energy sources are more crucial to solve these problems. Solar energy is a stepping stone to satisfy the growing energy demands in India and across the globe. In this research work, Coimbatore location was considered for daily global solar radiation (DGSR) analysis and the meteorological parameters used for this research are minimum air temperature, global solar radiation, maximum air temperature, sunshine hours, wind speed, mean air temperature, extraterrestrial radiation, average atmospheric pressure, average precipitation and relative humidity. Statistical models and artificial intelligence computational technique with ANFIS, i.e. adaptive neuro-fuzzy inference system, were used for forecasting and estimation of DGSR. The results of ANFIS model were found to be the best fit for forecasting and estimation of DGSR in any region.

Keywords Solar energy · Estimation and forecasting of solar radiation
Statistical models · ANFIS · ANFIS chaotic time series prediction

1 Introduction

Renewable energy is abundant in India and used as an alternative source for electricity generation. Solar energy, biomass, hydro and wind are the renewable energy sources that offer sufficient supply of clean energy, economic prosperity, social and envi-

J. Christy Martina (✉) · T. Amudha
Bharathiar University, Coimbatore, India
e-mail: martinachristy16@gmail.com

T. Amudha
e-mail: amudhaswamynathan@buc.edu.in

L. C. Jain et al. (eds.), *Data and Communication Networks*, Advances in Intelligent Systems and Computing 847, https://doi.org/10.1007/978-981-13-2254-9_21

ronmental benefits [1]. Indian energy sector approach towards meeting the energy demands has focussed on resource exploration and exploitation, capacity additions, creating awareness on the potential and promoting use of renewable energy. Solar energy is adequate and free resource of energy that plays a vital role in meeting the energy demands in India and across the globe. India has rich solar resources. There are 250–325 bright days in a year in India. Most of the regions in India receive 4.7 kWh/m^2/day of solar radiation [2]. The main objective of this research work is to forecast and estimate DGSR for solving the energy demand in India and across the globe.

In this research work, Coimbatore region was considered for daily global solar radiation (DGSR) analysis. Climatological data including daily solar radiation, maximum air temperature, relative humidity, minimum air temperature, extraterrestrial radiation, mean air temperature, wind speed, average atmospheric pressure, average precipitation, day length and sunshine hours were used for estimation and forecasting of DGSR. The data sets for 11 years (2005–2015) were collected from Agro Climatic Research Centre (TNAU), Coimbatore, and NASA near real-time data. Statistical methods were used to develop models using the temperature, sunshine and meteorological-based parameters. A soft computing technique, ANFIS model, was developed for the estimation of DGSR based on available climatological parameters. The developed ANFIS model results were compared with the best statistical models. The historical data sets of daily observed solar radiation from January 2005 to December 2015 were used to forecast solar radiation using ANFIS chaotic time series prediction method. Using the daily observed solar radiation in Coimbatore region, the corresponding daily solar radiation from January 2016 to December 2026 was calculated. The newly developed ANFIS models were found to be the best models and are suitable for forecasting and estimation of DGSR in any region of the world.

2 Estimation and Forecasting of Solar Radiation

Solar radiation estimation analysis helped to understand the need for solar energy and identify the best models to estimate DGSR. Coimbatore region in Tamil Nadu was considered for this research work. Statistical models and a soft computing technique, ANFIS method, were used for estimation of DGSR using the available climatological parameters. ANFIS chaotic time series prediction method was used for forecasting DGSR. In this research work, existent statistical models were categorized into three types: temperature, sunshine and climatological parameter-based models. Nine statistical models were used in which they include one temperature model, four sunshine models and four different climatological parameter-based models to evaluate the performance of DGSR estimation. The statistical indices such as mean absolute error (MAE), RMSE root-mean-square error, determination coefficient (R^2) and mean absolute relative error (MARE) were used to find the best-fit model for implementing solar radiation estimation techniques. The statistical models mentioned in Table 1 were found as the best models on the basis of investigations on DGSR esti-

Table 1 Statistical models

S. No	Type	Model	Equation
1	Temperature based	Hunt et al. [13]	$Z = j(T_{\max} - T_{\min})^{0.5} Z_0 + k$
2	Sunshine based	Ogelman et al. [14]	$Z/Z_0 = j + k(\mathrm{SH/SH_0}) + l(\mathrm{SH/SH_0})^2$
3		Bahel et al. [15]	$Z/Z_0 = j + k(\mathrm{SH/SH_0}) + l(\mathrm{SH/SH_0})^2 + m(\mathrm{SH/SH_0})^3$
4		Newland [16]	$Z/Z_0 = j + k(\mathrm{SH/SH_0}) + l\log(\mathrm{SH/SH_0})$
5		Bakirci [17]	$Z/Z_0 = j + k(\mathrm{SH/SH_0}) + l\exp(\mathrm{SH/SH_0})$
6	Meteorological based	Swartman et al. [18]	$Z/Z_0 = j + k(\mathrm{SH/SH_0}) + l\mathrm{RH}$
7		Abdalla [19]	$Z/Z_0 = j + k(\mathrm{SH/SH_0}) + lT + m\mathrm{RH}$
8		Ododo et al. model 1 [20]	$Z/Z_0 = j(\mathrm{SH/SH_0})^k T_{\max}^l \mathrm{RH}^m$
9		Ododo et al. model 2 [20]	$Z/Z_0 = j + k(\mathrm{SH/SH_0}) + lT_{\max} + m\mathrm{RH} + nT_{\max}(\mathrm{SH/SH_0})$

mation models where 'Z' is the global solar radiation, 'Z_0' is the extraterrestrial solar radiation, 'SII' is thc actual sunshinc hour, '$\mathrm{SH_0}$' is thc maximum possible sunshine duration/day length, 'RII' is the relativc humidity, 'T' is thc daily mcan air tcmperature, '$T_{\max}$' is the daily maximum air temperature, '$T_{\min}$' is the daily minimum air temperature and 'j, k, l, m, n' are empirical coefficients.

3 Adaptive Neuro-Fuzzy Inference System (ANFIS)

ANFIS technique is an integration of ANN knowledge representation, learning abilities, implication capabilities of fuzzy logic into a single system that drives to achieve desired performance by self-modifying their membership function. Fuzzy logic is used in ANFIS data learning technique. It transforms the given input into a target output. ANN learning ability is capable of creating interrelationship with the input and output variables. The input and output parameters can be related using If-Then sentences which are called as rules of fuzzy system. The salient features of ANFIS are used to manage complex decision-making systems successfully, easy to implement, and former human knowledge is not required. Figure 1 shows the ANFIS architecture of Sugeno fuzzy model which consists of two If-Then fuzzy rules [3].

RULE 1: *If r is C1 and s is D1 Then k1 = t1r + u1s + v1*
RULE 2: *If r is C2 and s is D2 Then k2 = t2r + u2s + v2*

Here, '*r*' and '*s*' are considered as inputs; 'C_i and D_i' are the fuzzy sets; 'k_i' are the outputs in the fuzzy region specified by the rules of fuzzy; and the design parameters are 't_i, u_i, v_i'. ANFIS architecture has five layers, namely input fuzzification, second layer is the application of fuzzy operators, third layer is the application method, fourth layer is the aggregation of output, fifth is defuzzification layer [4]. Each layer is given as follows:

Layer one is called as the input fuzzification layer in which all the inputs pass through membership functions.

$$\mathrm{OP}_{1,i} = \mu C_i(r); \quad i = 1, 2 \tag{1}$$

$$\mathrm{OP}_{1,i} = \mu D_i(s); \quad i = 3, 4 \tag{2}$$

where '*r*' and '*s*' are the inputs to node '*i*', and 'C_i' and 'D_i' are the linguistic labels such as high, low associated with this node function. $\mu C_i(r)$ and $\mu D_i(s)$ can adopt any fuzzy membership function. Triangular, Pi-shaped, Gaussian are some of the input membership functions, and there are two output membership functions such as constant and linear. The nodes are fixed in layer two and represented as a circle labelled Π which multiplies the incoming signals for fuzzification of inputs using the AND operator [5]. Thus, rules weight is achieved in this layer. Following is the output of this layer,

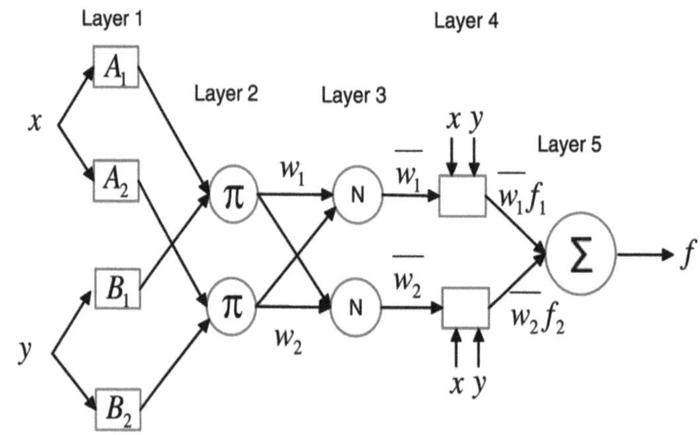

Fig. 1 ANFIS architecture with two rules, two inputs and one output [3]

$$\mathrm{OP}_{2,i} = w_i = \mu C_i(r).\mu D_i(s); \quad i = 1, 2 \tag{3}$$

Every node in layer three consists of a circle node labelled as N specifies that they play a normalization role to the firing strengths from the preceding layer. Following is the output of this layer,

$$\mathrm{OP}_{3,i} = \bar{w}_i = \frac{w_i}{w_1 + w_2}; \quad i = 1, 2 \tag{4}$$

Defuzzification is the fourth layer. In this every node, it is an adaptive node, with consequent parameters that include t_i, u_i and v_i. Given below is the output of this layer

$$\mathrm{OP}_{4,i} = \bar{w}_i k_i = \bar{w}_i(t_i r + u_i s + v_i); \quad i = 1, 2 \tag{5}$$

In layer five, single node is labelled as Σ. Summation of all incoming signals is performed in this layer in which it represents the overall output of the ANFIS system which is given below

$$\mathrm{OP}_{5,i} = \sum \bar{w}_i k_i = \frac{\sum w_i k_i}{\sum w_i k_i}; \quad i = 1, 2 \tag{6}$$

3.1 Solar Radiation Estimation Using ANFIS

Besides the statistical models, artificial intelligence ANFIS technique is one of the powerful tools to estimate DGSR using the climatological parameters. The learning procedure of ANFIS is similar to neural networks [6]. In the present work, ANFIS model was built using grid partition method with back-propagation learning algorithm for estimation of GSR. The ANFIS GUI editor was used to load both the training and checking data sets. The climatological parameters including observed solar radiation, maximum air temperature, extraterrestrial radiation, wind speed, minimum air temperature, precipitation, relative humidity, sunshine hours, atmospheric pressure and day length were used as input parameters for training and checking the data. Gaussian curve membership function with three numbers of membership function and linear type of output was used to generate FIS structure. Then the back-propagation optimization method was used to train the data set. The number of training epochs (1000) and the training error tolerance (0) were set to stop the training process. Checking process was done after training the FIS and plotting the checking data against the FIS output. Finally, the average testing error for training and checking data was obtained by applying the checking data to the model. The best model was obtained based on the RMSE error rate.

ANFIS Chaotic Time Series Prediction for Forecasting Solar Radiation: Time series prediction is one of the most important prediction methods that refer to the past recorded observations of variables. The values of the variables are measured at specified intervals of time and analysed to obtain the underlying relationships between observations and finding out a descriptive example [7]. This model is applied to extrapolate future time series. The present work was performed with ANFIS chaotic time series prediction modelling on 11 years of meteorological data during the period 2005–2015 for Coimbatore, Tamil Nadu. In time series prediction, Ti represents the known values of the time series up to the point in time and to predict the future value at some point, $Ti+P$ are used. The prediction can be done by creating a mapping from DP sample data points, sampled every Δ units in time, x (Ti − (DP − 1) Δ), ..., x ($Ti - \Delta$), x (Ti), to a predicted future value x=($Ti+P$) [8]. To predict the solar radiation, DP is set to *4* and $\Delta = P = 3$. For each Ti, the input training data for ANFIS is a four-column vector of the form, $\omega(Ti)=[x(Ti-9), x(Ti-6), x(Ti-3), x(Ti)]$. The output training data that corresponds to the trajectory prediction is represented as $\delta\ (Ti)=x\ (Ti+3)$. For each Ti, ranging in values from 12 to 4013, the training input–output data is a structure whose first component is the four-dimensional input ω, and whose second component is the output δ. There are 4017 input–output data values. The first 2812 data values were used for training the ANFIS model, while the others were used as checking data for validating the identified fuzzy model.

4 Implementation Results and Discussion

In this research work, estimation and forecasting of solar radiation for Coimbatore region were implemented successfully with real-time data collected for the period 2005–2015. MATLAB and Microsoft Office Excel 2007 were used for implementation. Microsoft Office Excel 2007 was used to compute statistical equation for the estimation of DGSR, and statistical indices including RMSE, MARE, MAE, correlation coefficient using the formulae. Soft computing technique, ANFIS, was implemented using MATLAB R 2010 a. ANFIS GUI editor was used to train, test and plot the data. RMSE error rate was obtained in order to find the best model for the estimation of DGSR. Best empirical equations based on sunshine, temperature and climatological parameters were identified and implemented for estimation of DGSR. ANFIS models were developed using different combinations of available meteorological parameters for estimation of daily GSR. Statistical evaluation was performed using performance measures RMSE, MAE, MARE, R^2, and the obtained results showed that there was a good agreement between the observed and estimated value by Bahel et al. model and ANFIS_M11 model. ANFIS chaotic time series prediction method was implemented for forecasting of solar radiation for the period 2015–2026. The developed model is suitable for the regions with similar climatic condition and unavailability of recorded DGSR data.

4.1 Performance Measures for Evaluation of Statistical Models

In this research work, statistical indices such as mean absolute error (MAE), root-mean-square error (RMSE), determination coefficient (R^2) and mean absolute relative error (MARE) were used to find the best-fit model for implementing solar radiation estimation techniques. The following are the formulas used to find the statistical indices [9].

$$\mathrm{RMSE} = \sqrt{\frac{\sum_{i=1}^{N} (H_{i,c} - H_{i,m})^2}{N}} \tag{7}$$

$$\mathrm{MAE} = \frac{1}{N}\sum_{i=1}^{N} \left| H_{i,c} - H_{i,m} \right| \tag{8}$$

$$\mathrm{MARE} = \frac{1}{N}\sum_{i=1}^{N} \left| \frac{H_{i,c-H_{i,m}}}{H_{i,m}} \right| * 100\% \tag{9}$$

$$R^2 = \left(\frac{\sum_{i=1}^{N}\left(H_{i,c} - H_{c,\mathrm{avg}}\right)\left(H_{i,m} - H_{m,\mathrm{avg}}\right)}{\sqrt{\left[\sum_{i=1}^{N}\left(H_{i,c} - H_{c,\mathrm{avg}}\right)^2\right]\left[\left[\sum_{i=1}^{N}\left(H_{i,m} - H_{m,\mathrm{avg}}\right)^2\right]\right]}} \right)^2 \tag{10}$$

where '$H_{i,c}$' represents the *i*th estimated daily solar radiation, '$H_{i,m}$' is the *i*th observed daily solar radiation, '$H_{c,\mathrm{avg}}$' is the mean of estimated values, and 'N' is the number of daily solar radiation values. The lower values of *MAE*, *MARE and RMSE* indicate good precision of the model, and higher value of R^2 indicates the best model. R^2 takes the value between 0 and 1. The performance measures of empirical equation were obtained by Eqs. (7–10). Performance analysis of the statistical models implemented in this research work is shown in Table 2.

The results showed that there was a good agreement between the estimated and observed value by Bahel et al. model with lower value of *RMSE 1.096*, *MAE 0.821*, *MARE 20.56%* and higher value of R^2 *0.969*. Thus, Bahel et al. sunshine-based model proved to be best fit among other statistical models for estimating DGSR.

Estimation and Forecasting of GSR Using ANFIS: Meteorological parameters data sets with 4017 samples were used to build ANFIS model in which 2812 were used for training and 1205 for testing the model. Grid partition with back-propagation learning algorithm and Sugeno fuzzy inference system was used. Three membership functions for each input were assigned, such that rules were framed based on the number of input parameters. ANFIS model was trained effectively, and testing was performed using the data samples that were not involved in the training process. Statistical performance measures such as *RMSE*, *MAE*, *MARE* and R^2 were calculated to compare the estimated and predicted values in which the best fit of the model was obtained. Fuzzy inference system (FIS) was generated using grid partition with

Table 2 Performance evaluation of statistical models

Model	RMSE	MAE	MARE (%)	R^2
Hunt et al. model	1.626	1.334	34.08	0.955
Ogelman et al. model	1.101	0.821	20.55	0.969
Bahel et al. model	1.096	0.821	20.56	0.969
Newland model	1.123	0.821	21.62	0.968
Bakirci model	2.088	0.859	21.28	0.882
Swartman et al. Model	1.276	0.971	25.28	0.969
Abdalla model	1.264	0.959	24.95	0.969
Ododo et al. model 1	2.957	2.753	60.77	0.962
Ododo et al. model 2	1.318	1.017	26.54	0.969

back-propagation optimization method. Each input parameters were assigned with three membership functions and a linear type of output. The proposed ANFIS model structure is shown in Fig. 2.

Gaussian curve membership function was identified as the best membership function based on evaluation of temperature-based model during the year 2009. Average testing error of training and checking was obtained to find the best fit of the membership function. The meteorological parameters for Coimbatore region were used as inputs to train the ANFIS model. Various combinations of meteorological parameters were used to train and test the ANFIS models. Statistical indices of ANFIS model

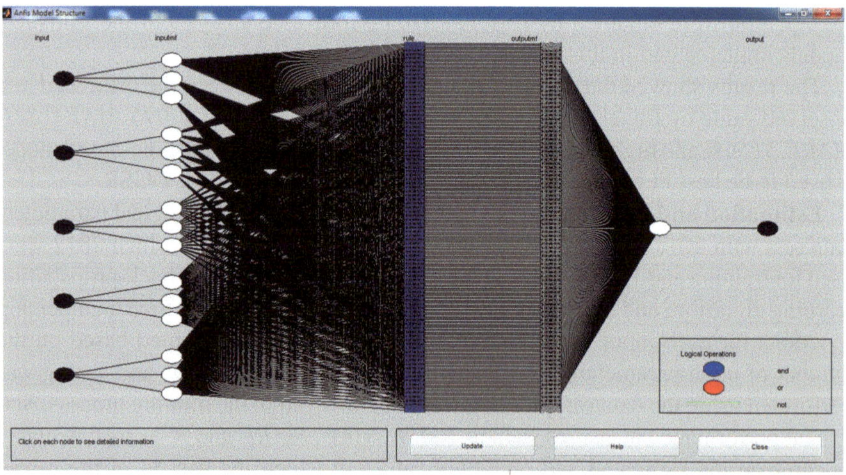

Fig. 2 Structure of ANFIS model

were used to find the best model for estimation of DGSR. The results of statistical analysis of ANFIS models proved that *ANFIS_M11* with available climatological parameters such as relative humidity RH, bright sunshine hours S, day length S_0, mean air temperature T, precipitation P and extraterrestrial radiation H_0 gave better result than other models. ANFIS_M11 model obtained lower value of *RMSE 0.722*, *MAE 0.545*, *MARE 13.62%* and higher value of R^2 *0.981*. Therefore, *ANFIS_M11* model showed good performance and can be used for estimation of DGSR across the globe with similar climatic condition with available meteorological parameters.

The performance comparison of statistical and ANFIS model using statistical performance measures given in Table 3 showed that *ANFIS_M11* model outperformed the *Bahel* et al. statistical model. Thus, *ANFIS_M11* model proved to be the best model for estimating daily GSR among other models. Figure 3 shows comparison graph of daily observed GSR, statistical and ANFIS_M11 model.

ANFIS chaotic time series prediction method was used for forecasting daily global solar radiation. Eleven years of observed solar radiation during the period 2005–2015 for Coimbatore region were given as input parameter in ANFIS chaotic time series modelling. Data sets of 2812 × 1 for training and 1205 × 1 for checking FIS were used. This model was able to predict DGSR for the period 2016–2026. The developed model was validated against actual DGSR for the period January 2016 to October 2017. The results of this model showed better performance for forecasting DGSR in Coimbatore region, and thus, the proposed model can be implemented easily for

Table 3 Performance comparison of statistical and ANFIS model using statistical performance measures

Model	RMSE	MAE	MARE (%)	R^2
Bahel et al. model	1.096	0.821	20.56	0.969
ANFIS_M11 model	0.722	0.545	13.62	0.981

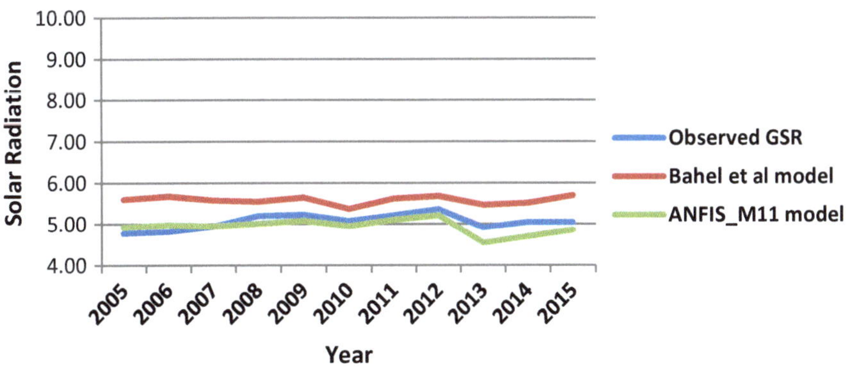

Fig. 3 Comparison graph of daily observed GSR, statistical and ANFIS_M11 model

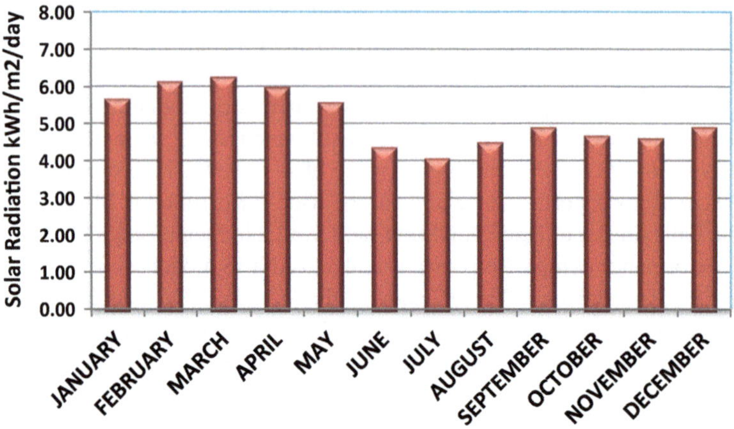

Fig. 4 Monthly average solar radiation from 2005 to 2026

any other region. Tamil Nadu has rich source of solar energy. There are around 300 clear bright days that receive high solar radiation. This shows that the state has higher potential for solar power generation. Coimbatore region in Tamil Nadu was considered for this research work has annual average of 5.06 kWh/m^2/day solar radiation [10]. The days are bright and sunny in the month of February to April and receive rainfall in the months June to August. The climate is warm and humid during September to November and winter from December to January. Figure 4 shows the predicted and observed monthly average solar radiation for the period 2005–2026.

5 Conclusion and Future Enhancements

India is in the world's third place to produce largest electricity and fourth in largest consumer of electricity. The conventional electricity generation using coal, oil and gas has a high impact on environment. The pollutants added to the environment by these methods affect the ecological balance of plants, animals and humans [11]. Thus, there is an earnest requirement for other ecological friendly resources. These resources are green source of energy including solar, wind, hydro, biomass and geothermal, and among these, solar is readily available during the day, abundant, clean and free source of energy. Solar radiation forecasting acts as a primary key to ensure the efficient use of energy that can help in better planning and distribution of energy in photovoltaic system. Solar radiation value is not available for all locations, and accurate forecasting is very important in solving energy problems. The major goal is to motivate the local governments to implement the green source of energy with efficiency measures. Coimbatore City is also one among 48 cities identified for developing as solar cities, up to 50 lakhs per city is provided for the development of the project [12]. Every year the generation of electricity by solar increases with higher demand. This research work was done to depict the importance of solar energy and utilize them in a better way to meet energy demands with pollution free environment in future. The developed ANFIS model is well suitable and can be adopted to estimate DGSR with available meteorological parameter for any location. This research work can be extended in the field of analysing the performance of power production in photovoltaic systems based on DGSR. The present work results are best fit for estimating DGSR with available meteorological parameters over Coimbatore region. In future, satellite images can be used as an alternative to ground-based measuring networks to minimize the cost for installing expensive measuring equipment's and for environments with inadequate meteorological data. Further, investigation of the relationship between solar radiation and climatological parameters is recommended to minimize the input parameters with more accurate results.

References

1. https://en.reset.org/knowledge/renewable-energy-environmentally-friendly-and-low-cost-energy-inexhaustible-sources
2. Dwivedi, A., Bari, A., Dwivedi, G.: Scope and application of solar thermal energy in India—a review. Int. J. Eng. Res. Technol. **6**(3), 315–322 (2013)
3. Al-Hmouz, A., Shen, J., Al-Hmouz, R., Yan, J.: Modeling and simulation of an adaptive neuro-fuzzy inference system (ANFIS) for mobile learning. IEEE Trans. Learn. Technol. **5**(3) (2012)
4. Mehdizadeh, S., Behmanesh, J., Khalili, K.: Comparison of artificial intelligence methods and empirical equations to estimate daily solar radiation. J. Atmos. Solar Terr. Phys. **146**, 215–227 (2016)
5. Sumithira, T.R., Nirmal Kumar, A.: Prediction of monthly global solar radiation using adaptive neuro fuzzy inference system (ANFIS) technique over the state of Tamilnadu (India): a comparative study. Appl. Solar Energy **48**(2), 140–145 (2012)
6. http://radio.feld.cvut.cz/matlab/toolbox/fuzzy/fuzzyt32.html
7. https://arxiv.org/ftp/arxiv/papers/1302/1302.6613.pdf
8. https://in.mathworks.com/help/fuzzy/examples/chaotic-time-series-prediction.html
9. Krishnaiah, T., Srinivasa Rao, S., Madhumurthy, K., Reddy, K.S.: Neural network approach for modelling global solar radiation. J. Appl. Sci. Res. **3**(10), 1105–1111 (2007)
10. http://shodhganga.inflibnet.ac.in/bitstream/10603/143959/11/11_chapter%203.pdf
11. https://beeindia.gov.in/sites/default/files/1Ch1.pdf
12. https://www.slideshare.net/vigneshsekaran520/systematic-roadmap-approach-on-solar-city
13. Hunt, L.A., Kuchar, L., Swanton, C.J.: Estimation of solar radiation for use in crop modelling. Agric. For. Meteorol. **91**, 293–300 (1998)
14. Ogelman, H., Ecevit, A., Tasdemiroglu, : A new method for estimating solar radiation from bright sunshine data. Sol. Energy **33**(6), 619–625 (1984)
15. Bahel, V., Bakhsh, H., Srinivasan, R.: A correlation for estimation of global solar radiation. Energy **12**(2), 131–135 (1987)
16. Newland, F.J.: A study of solar radiation models for the coastal region of South China. Sol. Energy **43**(4), 227–235 (1989)
17. Bakirci, K.: Correlations for estimation of daily global solar radiation with hours of bright sunshine in Turkey. Energy **34**, 485–501 (2009)

18. Swartman, R.K., Ogunlade, O.: Solar radiation estimates from common parameters. In: Solar Energy Conference, Tempe, Arizona (1967)
19. Abdalla, Y.A.G.: New correlations of global solar radiation with meteorological parameters for Bahrain. Int. J. Solar Energy **16**, 111–120 (1994)
20. Ododo, J.C., Sulaiman, A.T., Aidan, J., Yuguda, M.M., Ogbu, F.A.: The importance of maximum air temperature in the parameterisation of solar radiation in Nigeria. Renew. Energy **6**(7), 751–763 (1995)

A Shapley Value Approach for Community Detection in Social Network

Amreen Ahmad, Tanvir Ahmad, Abhishek Bhatt and Sadaf Siddiqui

Abstract The increasing popularity of online social networks (OSNs) has attracted significant attention from the research community. One of the significant tasks of analyzing structure is to detect communities within these OSN. Nodes within a community tend to be densely connected in comparison with nodes outside the community. Community detection is a challenging task in the field of social networks (SNs). Graph model is used to represent OSN where the nodes are representative of actors and the edges represent the interactions between these actors. A novel community detection algorithm named SCD is proposed in this paper. In the underlying graph, nodes are modeled as rational players where each player wants to maximize its Shapley value in terms of interactions. Extensive experiments are conducted on real-world datasets to establish the efficiency of the proposed approach.

Keywords Online social network · Community detection · Shapley value

1 Introduction

Community detection in graphs is gaining serious attention from researchers since it has many applications in different fields such as recognizing functional groups in metabolic networks [1], discovering friendship network in social network [2], and discovering the relationship between topics in World Wide Web [3]. In addition, the

A. Ahmad (✉) · T. Ahmad · A. Bhatt · S. Siddiqui
Department of Computer Engineering, Jamia Millia Islamia, New Delhi, India
e-mail: amreen.ahmad10@gmail.com

T. Ahmad
e-mail: tahmad2@jmi.ac.in

A. Bhatt
e-mail: abhishek1995bhatt@gmail.com

S. Siddiqui
e-mail: sadafsiddiqui1992@gmail.com

L. C. Jain et al. (eds.), *Data and Communication Networks*, Advances in Intelligent Systems and Computing 847, https://doi.org/10.1007/978-981-13-2254-9_22

performance of recommender system can be enhanced with the detection of related customers in purchase network.

Graph model is adopted to model social networks where actors are represented by nodes and edges represent the interactions such as financial exchange, common interest, and friendship between these nodes. A lot of interesting properties is exhibited by social networks, to be mentioned as community structure, power-law degree distributions, and network transitivity. A community can be defined as a cluster in which members within a group are likely to be closely knitted as compared to members outside the group.

Since last decade researchers have proposed a considerable amount of algorithms to detect communities in a social network. The limitation of classical community detection algorithms is time complexity that often fails in case of large-scale social networks. Clauset [2] presented a greedy approach to detect considered every node as a community, and then, two nodes were merged only when they had maximum modularity value.

Researchers [4] have analyzed that user's interactions play a significant role in detecting the communities within the SN since users interact with a limited number of their social group. Hence, analysis of community structure within these SNs should be based on activity instead of the relationship. Weighted links can be used to capture activity in these SNs. The proposed research focuses on static community detection approach.

1.1 Contributions of the Paper

In the proposed work, a novel clustering algorithm is proposed to detect communities in real-world social networks. The work can be summarized as:

1. A Shapley value-based clustering algorithm is proposed for real-world networks. Graph model is adopted to model nodes as rational players that want to maximize its Shapley value in terms of profit (where profit represents interactions).
2. The efficiency of the proposed work is established by comparing it with other competitive methods.

The remaining part is structured in the following way: Sect. 1 proceeds with the discussion of related work followed by Sect. 2 that gives a detail explanation of the proposed methodology. We proceed with an analysis of results obtained from the evaluation discussed in Sect. 3 followed by conclusion discussed in Sect. 4.

2 Proposed Methodology

This section gives a detailed explanation about the procedure for clustering algorithm SCD and is divided into the following modules:

1. In the underlying graph, salient nodes are determined by using SPIN algorithm [5] for every node.
2. The salient nodes identified are used to determine the clusters in the underlying graph discussed in Algorithm 1.

2.1 Clustering Algorithm (SCD)

It is a four-step algorithm that discusses the creation of clusters in the underlying graph. A clustering algorithm is given in a pseudocode form in algorithm 1, and the explanation is given as: (i) Initially, the graph model is adopted to represent actors as nodes and edges representing the interaction between these actors. (ii) Next, we proceed with the identification of salient nodes that is determined by

Algorithm 1 Clustering Algorithm

Also let Φ[] contain the Shapley values of all nodes in decreasing order
Let k represent the number of salient nodes that is given as input
procedure GETCLUSTERS(G, Φ)
Let n represent the nodes in graph G
S = GETSALIENTNODES(G, Φ, k)
Let *Sa* represent the set of SalientNodes that are smallest subset of S such that it gives a coverage of 100% in the underlying graph G
for every node in G different from S_a **do**

  Compute shortest path to all S_a
  closestSalientNode = Identify the closest SalientNode

end for
for every node in G **do**

  newGraph = The node belongs to the community of its S_a

end for
return *newGraph*
end procedure

SPIN algorithm. (iii) Finally, the task remains to identify clusters of the graph. In algorithm 1, first, we have to identify those set of *SalientNodes* that are the smallest subset of S such that it gives a coverage of 100% in the underlying graph G and name it as *Sa*. Then for every non-salient node, its closest salient node is identified. After the closest salient node of a node is identified, the non-salient node belongs to the community of its closest salient node. In the last, n clusters (clusters representing community) are given as output (where n is the number of *SalientNodes*).

3 Evaluation

3.1 Dataset

1. Zachary karate club [6]—This dataset represents a friendship network of karate club consisting of 34 nodes and 74 edges where some nodes are influential than others present in the network. This is used as a benchmark dataset for detecting communities within a network.
2. Dolphin [7]—This dataset is an unweighted and undirected network that represents frequent interaction among 62 dolphins and is divided into two communities due to some reasons.
3. Football network [2]—This dataset is a network of football teams split into independent teams and 11 conferences.
4. US political books(UPB)[1] —This dataset represents a network for US political books constituting 105 books, an interaction network of primary school students, and a collaborative network of 241 physicians.

3.2 Evaluation Metrics

1. Modularity: It is one of the well-known metrics for measuring the quality of clusters in social networks (SNs). Modularity can be defined in the following way:

$$Q_m = \sum_{c=1}^{q} \left[\frac{e_C}{e} - \left(\frac{d_C}{2e} \right)^2 \right] \tag{1}$$

where q denotes the number of detected communities, e_C represents the count of edges in community C, e denotes the aggregate count of edges in the graph G, and d_C represents the aggregate sum of the degrees of nodes in community C.

2. Clustering coefficient: It is a metric that measures the extent to which nodes tend to form clusters. It is a local measure, and it can be defined in the following way for a node:

$$C_{\text{coeff(i)}} = \frac{E_i}{dg_i(dg_i - 1)} \tag{2}$$

where $C_{\text{coeff}(i)}$ denotes the clustering coefficient for node i, dg_i computes degree of node i, and E_i is count of edges between neighbors of node i.

[1]www.orgnet.com.

Table 1 Performance comparison based on Qm

Algorithms	Dataset			
	Zachary	Dolphin	Football	UPB
CC-GA	0.420	0.529	0.594	0.527
Newman	0.381	0.465	0.524	–
GN	0.401	0.519	0.599	–
CNM	0.0.381	0.495	0.577	–
SCD	**0.51**	**0.73**	**0.90**	**0.70**

Table 2 Performance comparison based on $C_{coeff\ (i)}$

Algorithms	Dataset			
	Zachary	Dolphin	Football	UPB
Bi-section K means	0.56	0.21	0.40	–
SCD	**0.56**	**0.26**	**0.60**	–

3.3 Comparative Study and Analysis

Experiments are performed on four real-world datasets to analyze the efficiency of the SCD. The results are further analyzed by comparing SCD with well-known methods: (i) CC-GA [8], (ii) Newman [2], (iii) GN algorithm [2], (iv) CNM algorithm [9], and (v) bi-section k means [10]. The modularity and clustering coefficient values for different algorithms are presented in Tables 1 and 2. As can be seen from Tables 1 and 2, the modularity and clustering coefficient values of the proposed SCD community detection algorithm are better than the other competitive methods. The reason can be inferred as the incorporation of information diffusion concept along with Shapely value has lead to the creation of good quality clusters. Hence, we arrive at a conclusion that the proposed SCD is better than other algorithms.

4 Conclusion

A Shapley value-based approach for detecting communities in the underlying network is proposed. For every node in the graph, SPIN algorithm is used to identify salient nodes. Then, the clustering algorithm is applied to determine communities.

The results obtained prove the efficiency of the proposed approach over other algorithms. The future work remains to extend it to dynamic networks on large real-world networks.

References

1. Chen, J., Yuan, B.: Detecting functional modules in the yeast protein–protein interaction network. Bioinformatics **22**(18), 2283 (2006)
2. Girvan, M., Newman, M.E.J.: Community structure in social and biological networks. Proc. Natl. Acad. Sci. **99**(12), 7821–7826 (2002)
3. Flake, G.W., Lawrence, S.R., Giles, C.L., Coetzee, F.M.: Self-organization and identification of web communities. IEEE Comput. **35**, 66–71 (2002)
4. Viswanath, B., Mislove, A., Cha, M., Gummadi, K.P.: On the evolution of user interaction in facebook. In: Proceedings of the 2nd ACM workshop on Online Social Networks, pp. 37–42. ACM, New York (2009)
5. Narayanam, R., Narahari, Y.: A shapley value-based approach to discover influential nodes in social networks. IEEE Trans. Autom. Sci. Eng., pp. 1–18 (2011)
6. Zachary, W.: An information flow model for conflict and fission in small groups. J. Anthropol. Res. **33**, 452–473 (1977)
7. Lusseau, D.: The emergent properties of a dolphin social network. Proc. Natl. Acad. Sci. **270**(Suppl. no. 0962-8452), S186–8 (2003)
8. Saida, A., Abbasi, R.A., Maqbool, O., Daudb, A., Aljohani, N.R.: CC-GA: a clustering coefficient based genetic algorithm for detecting communities in social networks. Appl. Soft Comput., pp. 59–70 (2018)
9. Moore, C., Newman, M.E.J.: Phys. Rev. E Stat. Phys. Plasmas Fluids Relat. Interdiscip. Top. **61**, 5678–5682 (2000)
10. Raju, E., Hameed, M.A., Sravanthi, K.: Detecting communities in social networks using unnormalized spectral clustering incorporated with bisecting K-means. In: Proceedings of 2015 IEEE International Conference on Electrical, Computer and Communication Technologies (ICECCT) (2015)

Multiple Face Detection Using Hybrid Features with SVM Classifier

Sandeep Kumar, Sukhwinder Singh and Jagdish Kumar

Abstract Nowadays, multiple face detection (MFD) and extraction play an important role in face identification for various applications. In the proposed algorithm, Support Vector Machine (SVM) has been used for multiple face detection, and Discrete Wavelet Transform (DWT), Edge Histogram (EH), and Auto-correlogram (AC) are used for feature extraction. The proposed methodology worked on two different database i.e. Carnegie Mellon University (CMU) and BAO database for MFD. In this research paper, the proposed methodology gives a better result than the existing technique. Finally, our accuracy raised up to 90% approximately.

Keywords Face detection · SVM · DWT · Edge Histogram · Correlogram
Median filter

1 Introduction

In face detection, the different type of facial expression recognition (FER) will play a major role to identify the human face, for example, ID, security framework, human asset information, and participation administration. The initial step to be taken for recognizing the face is face detection to find a face in a picture. Face detection is an innovation to decide whether a static picture incorporates a face and recognizes it. The face detection accuracy depends on the various kinds of the parameters such as the size of the pictures, horizontal or vertical turning, profiles and front faces, facial appearances, and light conditions, which cause reduced recognition rate. In

S. Kumar (✉) · S. Singh
Department of Electronics & Communication, PEC University of Technology, Chandigarh, India
e-mail: er.sandeepsahratia@gmail.com

S. Singh
e-mail: sukhwindersingh@pec.ac.in

J. Kumar
Department of Electrical Engineering, PEC University of Technology, Chandigarh, India
e-mail: jagdishkumar@pec.ac.in

L. C. Jain et al. (eds.), *Data and Communication Networks*, Advances in Intelligent Systems and Computing 847, https://doi.org/10.1007/978-981-13-2254-9_23

like manner, as a result of the trouble in and the significance of FD, this territory of research is viewed as a free segment instead of a preparatory advance of FR. As of late, the regions of FD application have kept expanding [1–3]. Picture resizing will be performed in this exploration. The viability of the FD calculation is influenced by the measure of the information picture [1, 4]. A huge picture size will devour all the more handling time yet will keep up the greater part of the imperative data and give a better outcome, while a little picture will devour quick time yet may cause low execution as the face data. Consequently, this calculation begins with a pre-preparing stage where pictures will be resized to a settled size. This progression can upgrade the execution of a standard handling time. The way toward expanding the pixels' size will enhance the location exactness. In any case, decreasing the pixels' size will accelerate the recognition time; however, the precision will be disintegrated [5, 6].

2 Background

Sr. No.	Authors' name	Year	Remarks
1	Zhou et al. [7]	2018	• Identity-specific localized metric learning technique to identify the real-world multiple faces • Three datasets used: class 4, class 8, and class 16 • Comparing the result with Euclidean metric, LMNN, ITML, etc.
2	Shah et al. [6]	2017	• Threefold classification used • Six natural expressions used for multiclass data • Time-consuming process during training
3	Mahajan et al. [8]	2017	• Quahog feature used for robust face detection • The proposed methodology gives huge changes on contortive images
4	Nasr et al. [9]	2017	• Viola–Jones methodology used for face detection • SURF feature is extracted with Bag-of-Words technique • More accurate and fast recognition
5	Tao et al. [10]	2017	• Proposed a locality-sensitive SVM with kernel combination for feature extraction • Multiple-cascaded neural network to joint for local features

(continued)

(continued)

Sr. No.	Authors' name	Year	Remarks
6	Julina et al. [11]	2017	• Histogram of gradient for feature extraction and SVM for classification • Database used AT&T due to its various challenges and poses • Accuracy improved and got the minimal false-positive rate
7	Naik et al. [12]	2015	• Used a fusion RGB and depth image technique to identify the real-time face • Time process of face recognition reduces • Accuracy also improved with speed
8	Yadav et al. [13]	2015	• Skin color segmentation and facial feature for face detection • Increase the accuracy • Two datasets used: BAO and CMU
9	Raghavendra et al. [14]	2014	• Light-field camera used for face recognition in the real world • Data collected by the conventional camera also • In future, the author wants to work on parallax effect and 3D reconstruction

3 Problem Statement

The fundamental issue to be explained is to actualize a calculation for MFD in a picture. At a first look, the undertaking of MFD may not appear to be so overpowering particularly considering how simple it is tackled by a human. However, there is a conspicuous difference to how troublesome it really is to make a PC effectively unravel this errand.

4 Proposed Methodology

The proposed methodology contains various steps, i.e., face detector, noise removal, and feature extraction and classification which are explained in Fig. 1.

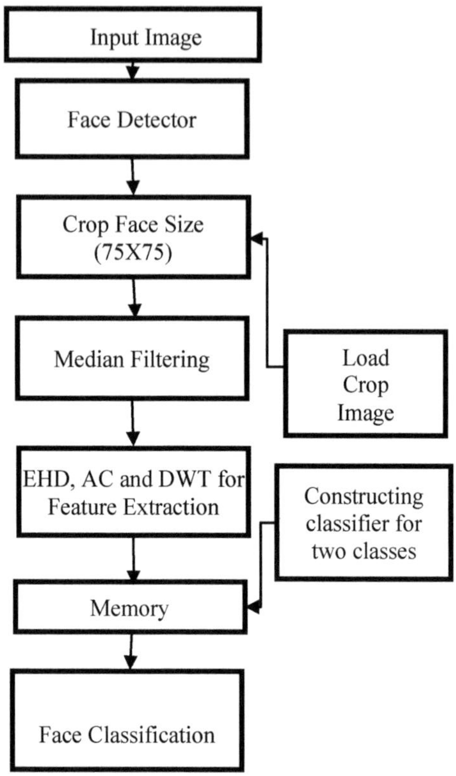

Fig. 1 Block diagram of proposed methodology

4.1 *Face Detection (FD)*

The essential rule of the Viola–Jones FD calculation is to check the locator commonly through a similar picture—each time with another size. A picture contains at least one faces clearly an inordinate substantial measure of the assessed sub-windows would even now be negatives (non-faces). Each phase of the course was prepared to utilize a positive set, a negative set, and for execution estimation an assessment set. For each stage, the positive set and the assessment set were the same (informational collection A) while the negative set was particularly intended for precisely that stage. As depicted before false positives are favored over false negatives, and since the AdaBoost calculation goes for limiting false negatives it needs a touch of tweaking. In their paper, Viola–Jones depicts the methodology that step by step brings down the classifier limit of a phase to the point that a given execution necessity is met [15]. Figure 2 shows (a) original image, (b) face-detected image using Cascade classifier, and (c) face-registered image in the folder.

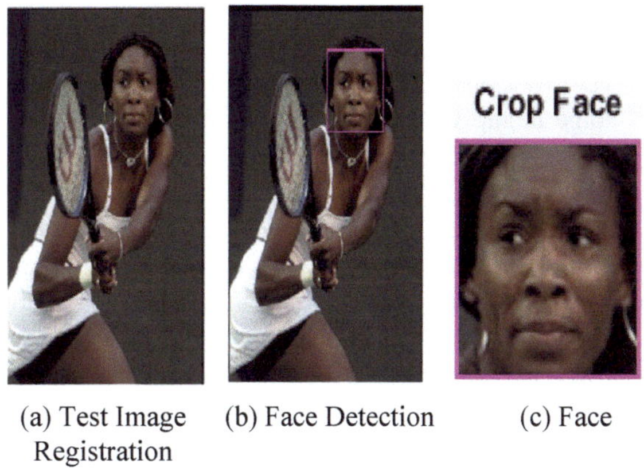

Fig. 2 **a** Original image, **b** detected face, and **c** crop detected image

4.2 Feature Extraction

- **Edge Histogram Descriptor**: Different types of edges will provide the complete information regarding the human face. To investigate this component, one path is to utilize histogram. Edge Histogram Descriptor (EHD) gives information about neighborhood pixels of edges over the whole picture. EHD essentially speaks to five sorts of edges in a neighborhood which is known as sub-picture. Figure 3 shows the five sorts of edges and all the edges are used to process the EHD. To register 80-receptacle neighborhood histogram, the primary picture is partitioned into 4 × 4 equivalent estimated non-covering pieces. The nearby histogram is registered for each of these sub-pictures for each edge sort [16].

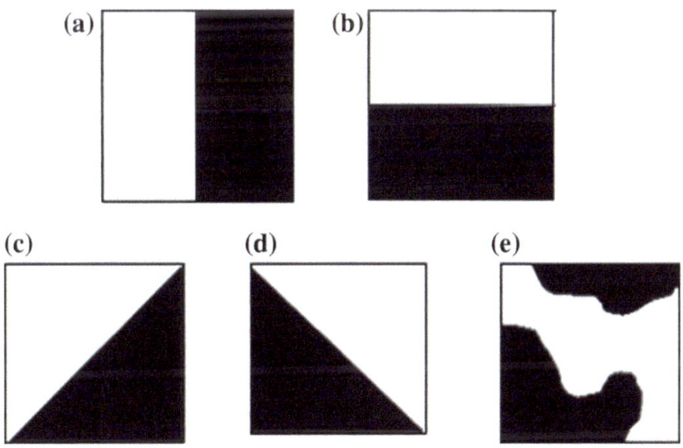

Fig. 3 Various types of edges: **a** vertical, **b** horizontal, **c** 45°, **d** 135°, and **e** non-directional

- **Discrete Wavelet Features**: The 2D-DWT can be characterized effectively as far as the 1D-Haar. 2D-DWT comprises two stages. The primary stage is applying the 1D-DWT on the 2D input signal in row-/column-based order; while, the second stage is applying the 1D-DWT on the yield of the main stage in column-/row-based order. The 2D-DWT can be characterized in view of a falling–sifting plan. In such a case, it can be characterized as a two-phase separating process i.e. low pass as well as high pass, and each channel is trailed by another two channels. Figure 3 demonstrates a case of an info picture and its relating DWT up to level 2. It additionally demonstrates the relating yields from the channels [17] (Figs. 4, 5, 6, and 7).
- **Correlogram Color Descriptor**: Smith and Chang [18] gave a proposed methodology to remove the regions of similar color (or dark level) in the pictures by requantizing the color space of the pictures. In our approach, we utilize this standard to diminish the quantity of the levels G in the pictures. This is finished by requantizing the pictures previously processing of the Correlograms. Utilizing this technique, we accomplish two critical advantages: (1) The decline of G helps the computational cost, and (2) quantization of the picture into sets of comparative levels sums up the picture content, which enhances the recovery, comes about.

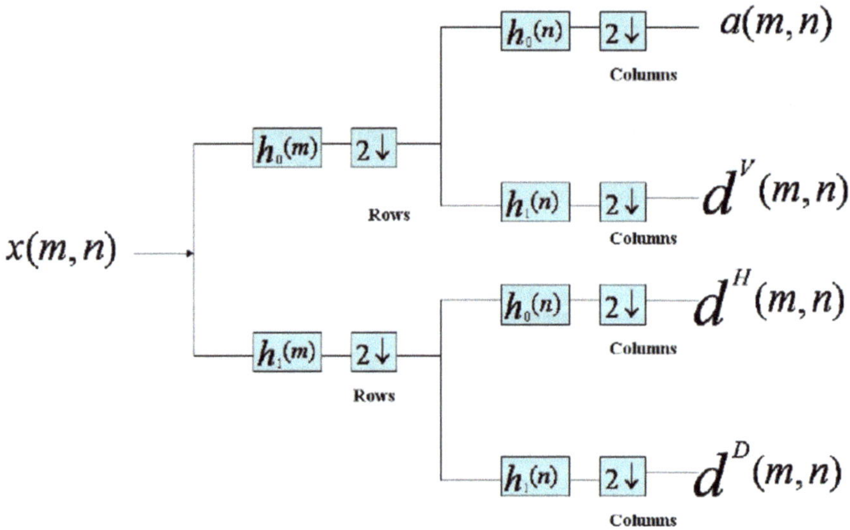

Fig. 4 2D-DWT scheme

Algorithm I

Step 1: Bit by bit brings down the threshold in proper strides until the point when a given genuine positive rate is accomplished.
Step 2: Run the negative cases through the course utilizing the threshold decided in 1.If false positive rate is above 50 %:
Step 3 (a): Increment the measure of highlights by a proper number and go back to 1.Else:
Step 3 (b): Utilize the present threshold and measure of highlights for this stage. Create another negative preparing set and another stage containing an indistinguishable measure of highlights from the present stage. Presently concentrating on the new stage go back to 1.

Fig. 5 Algorithm for preprocessing steps

Fig. 6 Algorithm for median filter steps

Algorithm II

Input: Face Crop ImageFI(Width [w] and Height [h]), window W (size (Sm*Sn)))
Output: Filtered Image Out

```
SetW[Sm * Sn]
x := (Sm / 2)
y := (Sn / 2)
for x from x to image w - x
for y from y to image h - y
i = 0
forfx from 0 to Sm
forfv from 0 to Sn
W[i] := FI[x + fx - x][y + fy - y]
i := i + 1
sort entries in W[]
Out[x][y] := W[Sm * Sn / 2]
```

4.3 Classification Using SVM

SVM is one of the popular and more accurate classifiers for image classification and recognition of the face. The SVM classifier is more accurate for two classes, but error will increase for multiclass classification; then multiclass SVM classifier will reduce the problems which were occurring in binary [19]. SVM is based on hyper-plane, and this hyper-plane gives the information between two classes; numerous hyper-planes

Algorithm III for Feature Extraction

Step 1: After removing the noise image is converted into 4 sub images DWT and each sub images are high as well as low frequency coefficients.

Step 2: Images are further split with DWT to get the more accurate data in limited no of features. In the proposed methodology, decomposition is done up to 4 levels.

Step 3: All the 4 level pixels of sub images gives additional as well as more accurate information.

Step 4: The features are extracted from images by mathematical formula of mean(μ), and standard deviation(σ).

$$\mu = \sum_{i=1}^{N} y_i \, , \ \sigma = \sqrt{\sum_{i=1}^{N} (y_i - \mu)}$$

Y: pixels value of image, N: size of image

Step 5: For every filtered image, we have to calculate the histogram using EHD with the help of different type of edges i.e. vertical edges, horizontal edges, 45° diagonal, 135° diagonal and non directional. Sobel operator used to determine the max value of gradient.

$$[\mathrm{ma, o}] = \max\,(I_{out}, [],3)$$

I_{out} : Filtered image.

e_{dim} : Edges of images determined by canny method. Then multiply both:

$$\mathrm{out} = (\mathrm{i} \times e_{dim})$$

out : Filtered image, e_{dim} : canny illustration

$$e_{hist} = \mathrm{hist}(I_{out}(:,5)$$

e_{hist} : Histogram of image

Step 6: Assume [D] represents a group of static distances image, I is represented the AC for close couple at a distance d.

$$\delta_{g_i, g_j}^{(d)}(I) = Pr_{p_1 \in I_{g_i}, p_2 \in I} \left| p_1 \in I_{g_j} | p_1 - p_2 = d \right|$$

$$\beta_g^{(d)}(I) = \delta_{g,g}^{(d)}(I)$$

p1 and p2 are the probability of the pixels, d: distance between each other, are of a comparative level gi. The partition measure between the histograms, AC, and it is the L1-standard that is computationally light system and used as a part of [13] and [14].

Fig. 7 Algorithm for feature extraction steps

can be attracted to isolate the two classes of training information, and SVM chooses ideal hyper-plane. The hyper-plane linear equation can be represented as:

$$hy + w = 0$$

y	denotes hyper-plane point,
h	represents the hyper-plane,
w	represents the bias,
$(\lvert w\rvert/\lVert h\rVert)$	hyper-plane is perpendicular to origin,
$\lVert h\rVert$	Euclidean mean.

5 Result Analyses

Two databases, specifically BAO [20] database and CMU [21] database, are utilized to assess the proposed framework. The BAO database consists of 274 pictures with single countenance and various faces in the diverse foundation. Another face database from the CMU contains 60 pictures in the database with various sorts of pictures: Four side-view pictures and three frontal pictures are utilized. Every one of the reenactments is executed on this PC by designing programming, MATLAB 14b keep running on Windows 8 Home Edition OS. The proposed approach was assessed by utilizing two comprehensively utilized face databases. The proposed approach was assessed by utilizing two comprehensively utilized face databases. In the analysis, proposed methodology gives better results on BAO database [22] as well as CMU-1 database. SVM classifier is used to isolate multiple or single face in a given image. A total of 110 features are used for verification and threefold cross-validation technique used to find the correct detection. Finally, this methodology gives better result as compared to existing algorithm as shown in Tables 1 and 2.

- Calculate detection rate (CDR) and false-positive rate (FPR) of the system using this formula:

$$\mathrm{CDR}(\mathrm{Re\,call}) = \mathrm{TP}/\mathrm{TP} + \mathrm{FN}$$

Table 1 Accuracy comparison between the proposed methodology and existing approach on BAO database

Method	CDR (%)	FPR (%)
Reference	84.2	4.2
Proposed	89.09	10.90

Table 2 Result obtained using the proposed system for CMU-1 database

Method	CDR (%)	FPR (%)	Time (s)
Proposed	88.67	11.32	3.56

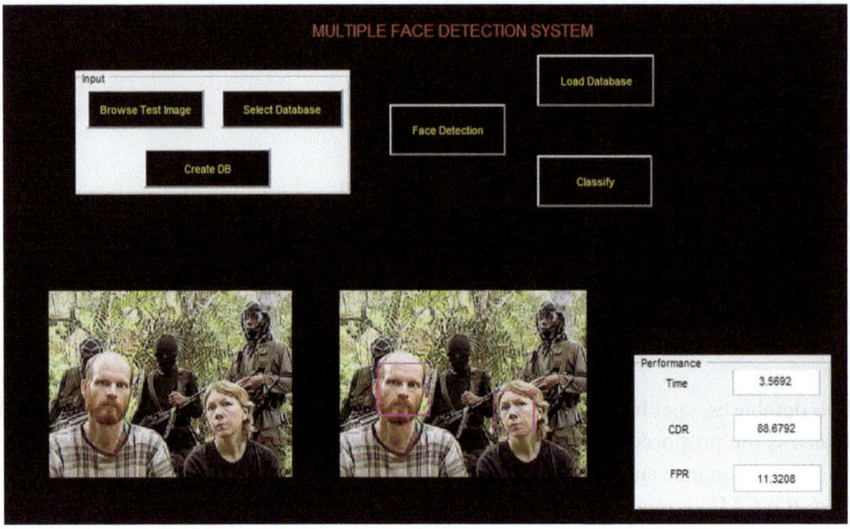

Fig. 8 GUI of proposed system on CMU dataset, **a** test image, **b** face detection

$$\text{FPR(MissRate)} = 1 - \text{CDR}$$

FN represents the false negative, and FP is false positive;

TP represents true positive, and TN is true negative.

According to Table 1, the accuracy or Correct Detection Rate of the proposed method is much higher than that of reference system on BAO dataset. Experimental results show that detection rate can reach approximate 90%, whereas Ref gives 84.2%. This methodology applies on CMU database also, and accuracy achieved 88.67% with fast computational time as shown in Table 2 (Figs. 8, 9, 10, and 11).

6 Conclusion

In this research, we focused on a multiple face detection technique which uses the Correlogram, DWT, and EH for face recognition. The median filter is used for removing the noise in given image. For MFD, the combination of DWT, EH and Correlogram of a face is used as the feature vector and an SVM classifier is trained to classify the input face as multiple or single. The proposed algorithm has been tested on the standard BAO and CMU datasets, and the accuracy reached up to 90%. From the experimentation, we noted that the SVM classifier recognizes the facial images more accurately from the previous classifier. We got an average recognition accuracy of 89% for SVM.

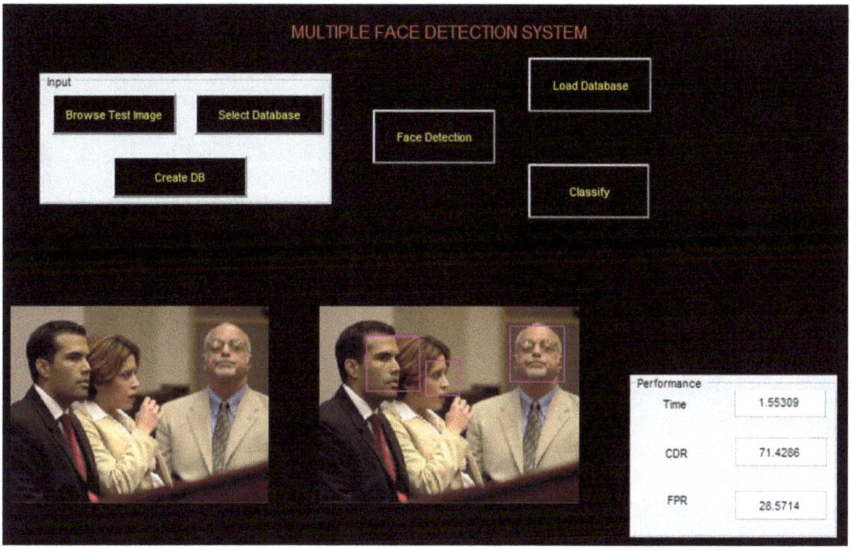

Fig. 9 GUI of proposed system on CMU dataset, **a** test image, **b** face detection

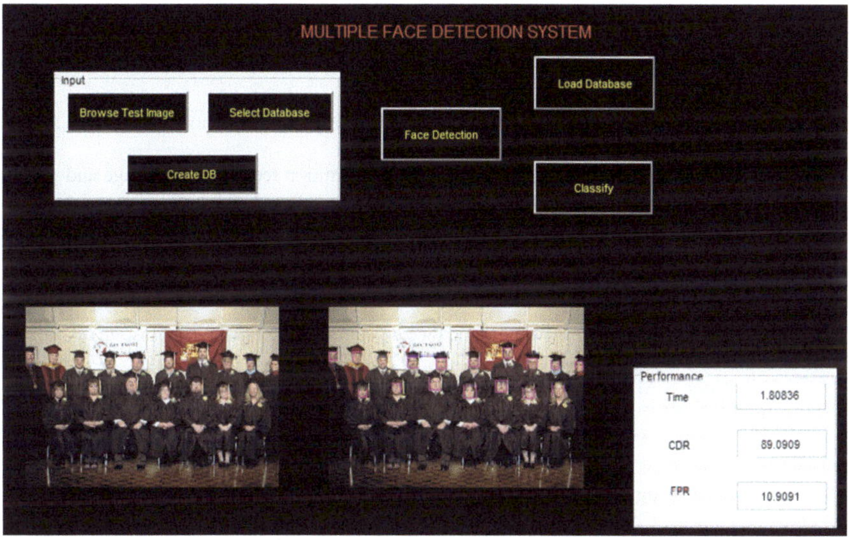

Fig. 10 GUI of proposed system on BAO dataset, **a** test image, **b** face detection

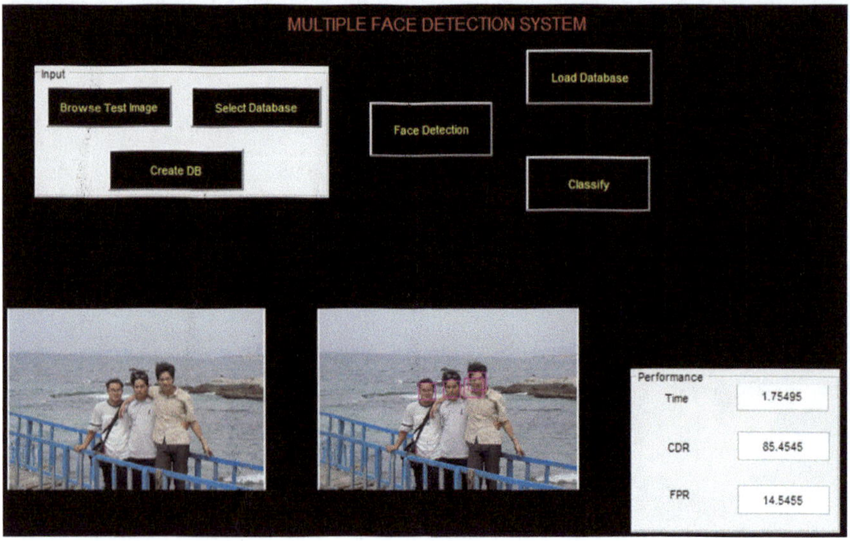

Fig. 11 GUI of proposed system on BAO dataset, **a** test image, **b** face detection

References

1. Kang, S., Choi, B., Jo, D.: Faces detection method based on skin color modeling. J. Syst. Archit. **64**, 100–109 (2016)
2. Kumar, S., Singh, S., Kumar, J.: A study on face recognition techniques with age and gender classification. In: IEEE International Conference on Computing, Communication and Automation (ICCCA), 5, 6 May 2017
3. Kumar, S., Singh, S., Kumar, J.: A comparative study on face spoofing attacks. In: IEEE International Conference on Computing, Communication and Automation (ICCCA), 5, 6 May 2017
4. Kumar, S., Singh, S., Kumar, J.: Automatic face detection using a genetic algorithm for various challenges. Int. J. Sci. Res. Mod. Educ. **2**(1), 197–203 (2017)
5. See, Y.C., Noor, N.M., Lai, A.C.: Hybrid face detection with skin segmentation and edge detection. In: 2013 IEEE International Conference on Signal and Image Processing Applications (ICSIPA), pp. 406–411 (2013)
6. Shah, J.H., Sharif, M., Yasmin, M., Fernandes, S.L.: Facial expressions classification and false label reduction using LDA and threefold SVM. Pattern Recognit. Lett. 1–17 (2017)
7. Zhou, X., Jin, K., Chen, Q., Xu, M., Shang, Y.: Multiple face tracking and recognition with identity-specific localized metric learning. Pattern Recognit. **75**, 41–50 (2018)
8. Mahajan, J.A., Paithane, A.N.: Face detection on distorted images by using quality HOG features. In: 2017 International Conference on Inventive Communication and Computational Technologies (ICICCT), pp. 439–444 (2017)
9. Nasr, S., Bouallegue, K., Shoaib, M., Mekki, H.: Face recognition system using a bag of features and multi-class SVM for robot applications. In: 2017 International Conference on Control, Automation and Diagnosis (ICCAD), pp. 263–268 (2017)
10. Tao, Q.Q., Zhan, S., Li, X.H., Kurihara, T.: Robust face detection using local CNN and SVM based on kernel combination. Neurocomputing **211**, 98–105 (2016)

11. Julina, J., Kulandai J., Sree Sharmila, T.: Facial recognition using a histogram of gradients and support vector machines. In: IEEE International Conference on Computer, Communication and Signal Processing (ICCCSP), pp. 1–5 (2017)
12. Naik, N., Rathna, G.N.: Robust Real-Time Face Recognition and Tracking on GPU Using a Fusion of RGB and Depth Image. arXiv preprint. arXiv:1504.01883 (2015)
13. Yadav, S., Nain, N.: Fast face detection based on skin segmentation and facial features. In: 2015 11th International Conference on Signal-Image Technology & Internet-Based Systems (SITIS), pp. 663–668. IEEE (2015)
14. Ramachandra, R., Yang, B., Raja, K.B., Busch, C.: A new perspective—face recognition with the light-field camera. In: IEEE International Conference on Biometrics (ICB), pp. 1–8. IEEE (2013)
15. Jensen, O.H.: Implementing the Viola-Jones face detection algorithm. Master's thesis, Technical University of Denmark, DTU, DK-2800 Kgs. Lyngby, Denmark (2008)
16. Chaudhary, M.D., Upadhyay, A.B.: Integrating shape and edge histogram descriptor with stationary wavelet transform for effective content-based image retrieval. In: 2014 International Conference on Circuit, Power and Computing Technologies (ICCPCT), pp. 1522–1527. IEEE (2014)
17. Masselos, K., Andreopoulos, Y., Stouraitis, T.: Performance comparison of two-dimensional discrete wavelet transforms computation schedules on a VLIW digital signal processor. IEE Proc. Vis. Image Signal Process. **153**(2), 173–180 (2006)
18. Smith, J.R., Chang, S.F.: Tools and techniques for color image retrieval. In: Storage and Retrieval for Still Image and Video Databases IV, vol. 2670, pp. 426–438. International Society for Optics and Photonics, Washington (1996)
19. Yang, S., Bebis, G., Chu, Y., Zhao, L.: Effective face recognition using a bag of features with additive kernels. J. Electron. Imag. **25**(1), 013025 (2016)
20. https://facedetection.com/datasets/
21. http://db.cs.edu/
22. Huang, Z.C., Chan, P.P., Ng, W.W., Yeung, D.S.: Content-based image retrieval using color moment and Gabor texture feature. In: 2010 International Conference on Machine Learning and Cybernetics (ICMLC), vol. 2, pp. 719–724. IEEE (2010)

A Brief Overview on Generative Adversarial Networks

Raj Patel

Abstract Generative models dwell into a research domain, which incorporates the use of mathematical models by which machines could mimic and generate new data. Generative models are used in tasks such as image or music generation as well as in capturing and mimicking certain features of a dataset. One such generative model is generative adversarial network, which makes use of a generator and discriminator in order to perform such generative tasks. It does this by maintaining a min–max relationship between the two models. This study gives a brief overview of generative adversarial networks its optimization function and metrics as well as its implementation and working. The study also incorporates a case study, which provides a clear illustration of the actual training, which is to be performed. The performance and the results are sampled and very well compared. The study also gives insights regarding the various applications and the variations of generative adversarial network.

Keywords Deep learning · Generative adversarial networks · GAN
Neural networks

1 Introduction

Advancement in deep learning and knowledge-based AI has facilitated computers in performing complex tasks that were perceived only to be possible by humans. Whether it would be image recognition, understanding natural language, speech recognition or to learn and play games, computers are capable of performing such tasks with the highest level of accuracy. This has been possible by incorporating several algorithms and mathematical models. One such set of tasks is generative tasks, in which new data is generated, similar to existing data. This generated data can be used as generative output for generative tasks or in places having missing data,

R. Patel (✉)
Department of Information Technology, Dwarkadas J. Sanghvi College of Engineering, Vile Parle, Mumbai, India
e-mail: raj1471997@gmail.com

L. C. Jain et al. (eds.), *Data and Communication Networks*, Advances in Intelligent Systems and Computing 847, https://doi.org/10.1007/978-981-13-2254-9_24

which may be useful for other machine learning tasks. The models that facilitate such generative tasks are known as generative models. One such generative model is generative adversarial network.

Ian Goodfellow introduced generative adversarial network or GAN in 2014 [1]. GANs work on a zero-sum game framework [2]. It consists of a generator and a discriminator, and the generator performs the task of generating new items, whereas the discriminator performs the task of distinguishing between real and fake data samples. The data samples that generated by the generator would be fake since it is trying to mimic the real data. The main goal of the generator is to make these data items as real or as true to the original data items as possible in order to fool the discriminator. On the other hand, the goal of the discriminator is to flag these fake data items apart from the original ones, as best as possible. So here, the discriminator works as an adversary or judge, judging the real and the fake data items.

1.1 Background

GANs were not the first to tackle the challenge in performing generative tasks, and there have been several studies made over the years. Variational autoencoders (VAE) and restricted Boltzmann machine (RBM) both provide models and optimization techniques, which incorporated can perform generative tasks. VAEs are the probabilistic graphical model whose explicit goal is latent modeling and marginalizing out certain variables [3]. In the study (Pu [4]), the authors develop an autoencoder to model images, as well as to associate labels or captions in a semi-supervised setting. The study employs a convolutional neural network (CNN) inside the encoder for the posterior distribution of the parameters of the image generative model [4]. RBM works on the principle of unsupervised feature extractor, and in the process, it learns the joint probability, which facilitates generation of new data. Ruslan and Geoffrey (Salakhutdinov [5]) formulate a new algorithm with multiple layers of hidden random variables known as deep Boltzmann machines to learn good generative models. Both VAE and RBM use Markov chains for learning. GANs avoid the use of Markov chains due to its high computational cost [6].

Section 2 of this study will discuss the optimization function and its metrics. Sections 3 and 4 will dwell on the implementation and working of GAN with the help of a case study. Section 5 will discuss its various applications and different variations of the GAN, and at last, Sect. 6 will summarize and conclude the study.

2 Objective Function and Metrics

The optimization function for training GAN is represented using a set of two functions G and D in a min–max relationship where G is the generative function and D is the discriminative function. The relationship established is illustrated in Eq. (1).

$$\min_G \max_D V(D, G) = E_{x \sim p_{\text{data}}(x)}\left[\log(D(x))\right] + E_{z \sim p_{\text{data}}(z)}\left[\log(1 - D(G(z)))\right] \quad (1)$$

- z represents the initial random noise or input given to the generator.
- x represents the actual data from the dataset to be fed to the discriminator.
- $G(z)$ represents the output of the generator. G is a differentiable function, which tries to map the input z based on its distribution p_g.
- $D(x)$ is the discriminating function which gives out a probabilistic value on how likely the data item x represents or belongs from the actual dataset.
- $D(G(z))$ suggests the probability on how likely the output of the generator is similar to the actual data. This value can be interpreted as the similarity index for the generator.

The discriminator is trained in order to maximize the probability of assigning the correct label to both training examples and the fake samples generated by the generator [2]. Simultaneously, generator is trained in order to minimize $\log(1 - D(G(z)))$ [1]. Error estimation of both the models is a form of log loss function. Initially, when G is poor, D can reject samples with high confidence because they are clearly different from the training data. In this case, $\log(1 - D(G(z)))$ saturates [1]. Rather than training G to minimize $\log(1 - D(G(z)))$, we can train G to maximize $\log(D(G(z)))$. In other words, D and G play the following two-player minimax game with value function $V(G, D)$ [1]. Ideally, both error estimation of both the models should converge to zero but in a minimax relationship, such convergence will result in minimum possible loss value of either of the model.

3 Case Study: Introduction and Implementation

Working of GAN can be inferred with consideration of a case study. This case study implements a deep convolutional GAN, which is a variation of GAN. Deep convolutional GAN (DCGAN) consists of two primary networks: a deconvolutional neural network as a generator for generating images and a convolutional neural network as a discriminator for recognizing images. The output of the generator will be an image, and the output of the discriminator will be single value suggesting the probability of high likely the generated image from the generator is to the real dataset.

The convolutional neural network or the discriminator will start with an input layer to take in the input as an image. The deconvolutional neural network or the generator takes in input as random noise. During the training, the weights of both the networks are updated so that the generator gets better at deluding the discriminator and the discriminator gets better at identifying fake data.

3.1 Dataset

The dataset used in the case study is the CelebFaces Attributes Dataset (CelebA) [7]. It is a large-scale face attributes dataset with more than 200 K celebrity images, each with 40 attribute annotations. The images in this dataset cover large pose variations and background clutter. From the exploratory visualization, different visual variations are observed in the dataset. These variations range from:

1. Varying gender and age
2. Varying color complexity and ethnicity
3. Varying hairs and face positions.

These above-mentioned variations are clear in Fig. 1.

3.2 Architecture and Implementation

The architecture for the case study is based upon the one proposed in the DCGAN paper (Radford [8]). The final implemented architecture is illustrated in Fig. 2. This architecture will generate RGB images of size 64 * 64. The generator consists of an input layer to take in the noise as an input the next follows a set of reshape layers and convolutional layers with rectified linear nit (ReLu) activation and upsampling layers in between. The final layer is a tanh activation layer. The upsampling layers increase the height of the image and hence increasing the resolution after each convolutional.

Fig. 1 Image samples from the dataset

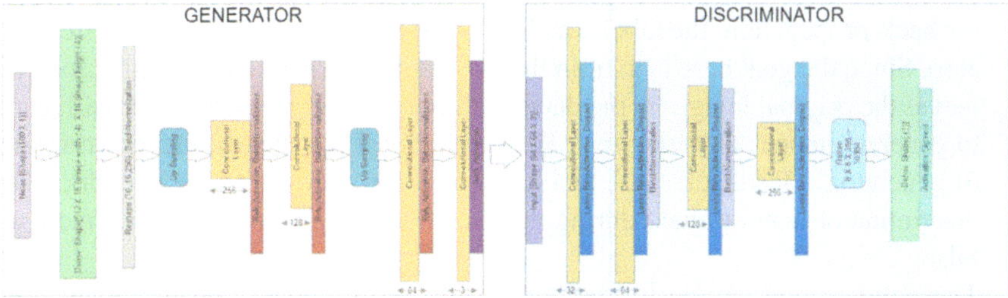

Fig. 2 Architectural implementation of DCGAN

The discriminator is quite reverse of that of the generator. It starts with an input layer, which takes in an image of size 64 * 64 * 3; it then follows with a set of convolutional layers with increasing number of filters. Between the convolutional layers are leaky rectified linear unit (leaky ReLu) activation layers along with batch normalization and dropout. The last layers are the flatten layer followed by the dense layer. Batch normalization helps to prevent arbitrary large weights in the intermediate layers as batch normalization normalizes the intermediate layers, thus helping to converge well. Dropout layers are quite effective in preventing the discriminator from overfitting during training. This forces the network from learning redundant information.

3.3 Training and Refinement

The training process of DCGAN is as follows:

1. A random noise generator is used to generate random noise. This random noise acts as input for the generator. On given input, the generator outputs an image and this generated image is used to train the discriminator in distinguishing the fake images from the real ones.
2. The generated image is given as input to the discriminator along with a real image from the dataset. The label attached to the generated image is zero since it was generated by the generator; hence, it does not follow any similarity with the dataset, whereas the label attached to the image from the dataset is one since it is the real image. The goal here is to train the discriminator to distinguish between real and the fake image as best as possible. Training is performed by back-propagation, by updating the weights of each layer based on the error function. At this stage, weights of discriminator are only updated and this can be done by setting generator's trainable argument as false.
3. The stage that is left is to train the generator and for training the generator. Similar to step two generators generates a new image by feeding it with random noise. This generated image is again fed into the discriminator, which outputs the probabilistic value of how likely the generated image is to the dataset. However,

for back-propagation, the label attached to the generated image is one instead of zero. Since the goal here is to train the generator to generate images that look as real as the original images in the dataset; thus, eventually the generator will learn to generate images that look real. Hence, the label attached is one. For training of generator, a feedback cycle facilitates updating weights of the network. The discriminator is not trained during this step by setting its trainable argument as false.
4. These three stages are performed indefinitely until the loss of each model converges to the minimum. After each step, the generator becomes stronger at generating images identical to the dataset and the discriminator becomes stronger at distinguishing these images. Moreover, in the end, two very strong models are obtained which train itself by trying to outperform each other.

Several hyperparameters were tuned in order to obtain adequate results. The hyperparameters set for each model were as follows:

Hyperparameters set for the generator:

1. Kernel size: 4. (To set the filter size of convolutional layers to {4, 4} and to perform fast convolutions on the image.)
2. Momentum: 0.8. (To avoid the problem of falling in local minima whenever one of the models falls to its minimum value.)

Hyperparameters set for the discriminator:

1. Kernel size: 4. (To set the filter size of convolutional layers to {4, 4} and to perform fast convolutions on the image.)
2. Strides: 2. (In order to jump two pixels at a time.)
3. Dropout: 0.25 between first two layers and 0.4 between the rests and thus prevention from overfitting.

4 Case Study: Validation and Results

The training was conducted with a batch size of 32 samples. Initially, the learning rate was kept at 0.02 and the model was tested for 10,000 steps/epochs. The mean loss of generator and discriminator was found to be 2.4 and 3.43 after the training, which was quite higher than expected. This was probably due to high learning rate. Next, the learning rate was reduced for one-tenth of its value, which resulted in a reduction of the losses.

In the final implementation, the model was trained for 100,000 steps/epochs with a learning rate of 0.002 (Note: The training was done on a GPU-based workstation resulting in total training time of 20 h). The losses of each model over the period are illustrated in Fig. 3; the losses of generator and discriminator are plotted on the Y-axis and the number of steps on X-axis.

Initially, the loss from the generator is quite high averaging around 1.13 until 20,000 steps but it lowers down as the model trains for more time. On the other hand,

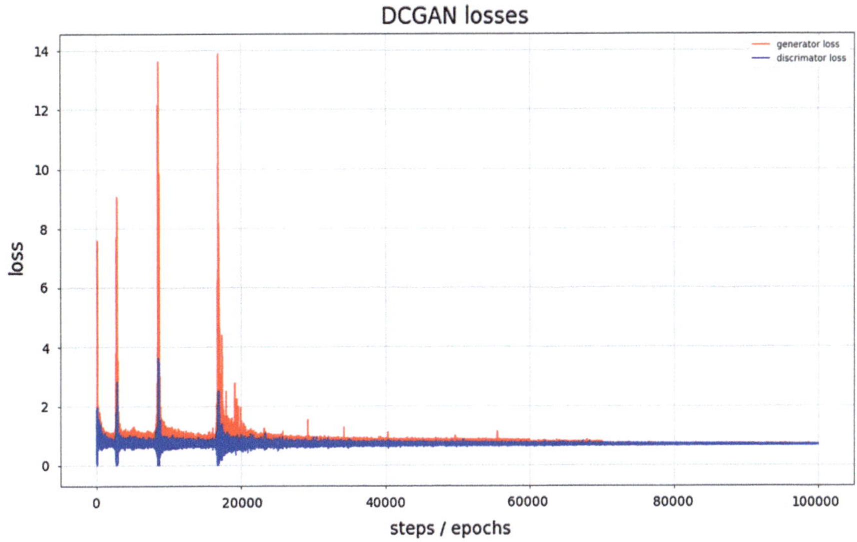

Fig. 3 Loss of generator and discriminator over each step

the loss from the discriminator is quite low as compared to the generator. The above figure illustrates the interesting relationship between generator and discriminator in GAN. In which, whenever the loss of one model goes down the loss of other model goes up. This is the min–max relationship, which is to be seen in GAN where one network tries to outperform the other this is visible near 20,000 epochs where discriminator loss is close to zero and generator loss is at its highest at around 14. However, as the training continuous the two models begin to maintain an equilibrium, which is clear after 40,000–60,000 epoch and after 70,000 epoch. The losses of the generator and discriminator seem to be the same value, but having a closer inspection still, the two models are still trying to balance each other where a decrease in one model's loss leads to increase in the other. This can be observed in Fig. 4.

The average loss of the generator during 100,000 epochs was 0.844, and the average loss of discriminator was 0.707. The average accuracy of the discriminator was 48%, which seems to be an appropriate value for a well-trained GAN.

4.1 Results from the Generator Overtime

The outputs obtained from the GAN can be summarized from Fig. 5. Over time, an improvement in the quality of images generated by the generator can be concluded. Visually comparing the generator is able to capture the variations that were observed in the original dataset and reflect these variations in the produced images; a variety of different images having various hair color, face complexity, background, and

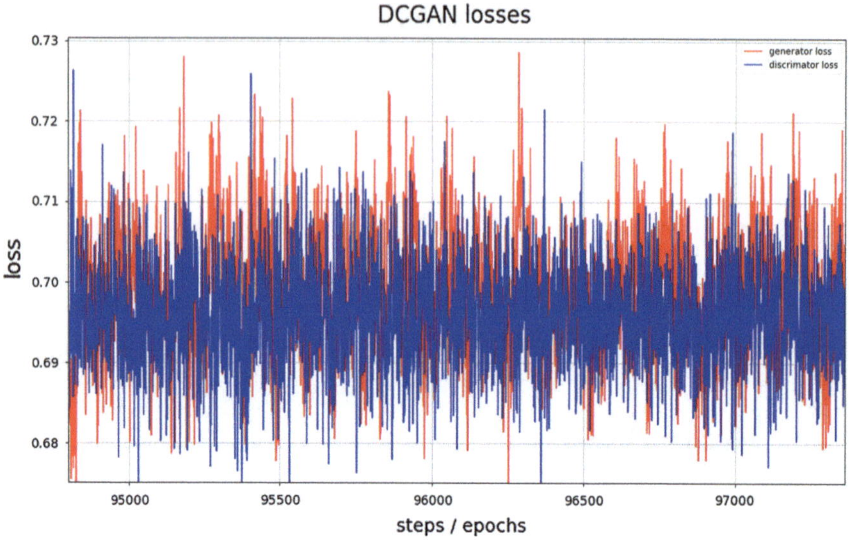

Fig. 4 Zoomed in image of the graph between 95,000 and 100,000 steps

accessories. Figure 5 shows the results obtained over time starting from the first epoch to each horizontal slice displaying the output after every 20,000 steps. Some of the problems that can be seen in the images are:

(1) Blurriness caused in the images due to low resolution.
(2) A minor case of morphing seen in the images likely caused due to less training.

5 Types and Applications of GAN

The case study dwelled into the implementation of the DCGAN architecture which is an essential architecture used for image generation. It yields good quality images with proper training. Some of the other application of GANs are in music and video generation, in image-to-image translation as shown by pix2pix and in text to image translation. Apart from performing generative tasks, GANs can be used in tasks such as anomaly or outlier detection. Quite often, this type of tasks requires huge number of data to make proper distinctions, which in most cases is not available. GANs solve this problem by making use of the generator in generating new data by which the discriminator could take advantage in training. GAN can be used in datasets for outlier detection or in fraud detection. It has also been used in drug discovery and in cancer tissue detection.

Currently, GANs are somewhat unstable and proper architecture design is essential for convergence due to which the true potential is not always achieved. However, these issues are been tackled by its certain variations. Some of the types of GAN are:

Fig. 5 Outputs obtained from the DCGAN

- Wasserstein GAN

Wasserstein GAN is an alternative to GAN training [9]. Without proper design, instability can cause inadequate training resulting in poor results. Wasserstein GAN tries to eliminate this instability. During training, the loss function is changed to include Wasserstein distance and by which they get rid of several problems such as mode collapse and help to improve the stability of the model. The discriminator, in this case, is trained until convergence, which results in mitigating the need to balance the generator, hence improving the stability.

- Conditional GAN

In traditional GAN, there is no control on the modes of data being generated. Conditional GAN solves these problems by conditioning the model with extra information in order to direct the generation process [10]. Conditional GAN is an extension of GAN framework, in which the generator is provided with some additional information or meta-data along with random noise regarding the expected outputs to be generated by the generator. Some of the keen applications of conditional GAN are in generating images from text annotations given as inputs or in the generation of tags from a given image .

- InfoGAN

InfoGAN is a modification to GAN objective, which makes it learn interpretable and meaningful representations [11]. It does this by encoding meaningful features along with the input noise. Conceptwise it looks quite similar to conditional GANs but the additional information is facilitated within the noise itself. Additionally, it maximizes the mutual information between a small subset of the latent variables and the observation [11]. Applications of InfoGAN can range from tasks that require manipulation of latent codes if InfoGAN can learn the detangled and interpretable representations.

6 Conclusion and Future Scope

Generative adversarial networks truly exemplify the notion of using computers in performing generative tasks. They provide a technique through which generative models are improved as they advance their training, which results in higher quality of results. The case study demonstrated the intricacy of training a GAN; it provided insight regarding the importance of convergence in an adversarial setting. The results of the case study justify the use of GAN for generating images and depict how the proper usage of GAN. Several variations of GAN were looked upon, and their specific applications were illustrated.

In the future, as more research is done in the subject matter GANs will surely be the most appealing state of the art in deep learning. With its growing popularity in the corporate sector, it will further be used in several tasks from music and video generation to creating simulations which could imitate human behavior. The future is bright, and the use cases of GAN are endless.

References

1. Goodfellow, I.J., Pouget-Abadie, J., Warde-Farley, D.: Generative Adversarial Networks, pp. 1–6. Arxiv (2014)
2. Goodfellow, I.: NIPS 2016 Tutorial: Generative Adversarial Networks, pp. 6–13. Arxiv (2016)
3. Welling, M., Kingma, D.P.: Auto-Encoding Variational Bayes, pp. 1–8. Arxiv (2013)
4. Gan, Z., Henao, R., Yuan, X., Li, C., Stevens, A., Carin, L., Pu, Y.: Variational autoencoder for deep learning of images, labels and captions. In: NIPS (2016)
5. Hinton, G., Salakhutdinov, R.: Deep Boltzmann Machines. MLR (2009)
6. Karazeev, A.: Generative Adversarial Networks (GANs): Engine and Applications, 17-8-2017. Available: https://blog.statsbot.co/generative-adversarial-networks-gans-engine-and-applications-f96291965b47 (Online)
7. Luo, P., Wang, X., Tang, X., Liu, Z.: Deep learning face attributes in the wild. In: Proceedings of International Conference on Computer Vision (ICCV) (2015)
8. Metz, L., Chintala, S., Radford, A.: Unsupervised Representation Learning with Deep Convolutional Generative Adversarial Networks, pp. 1–7. Arxiv (2015)
9. Chintala, S., Bottou, L., Arjovsky, M.: Wasserstein GAN, pp. 1–5. Arxiv (2017)

10. Osindero, S., Mirza, M.: Conditional Generative Adversarial Nets, pp. 1–5. Arxiv (2014)
11. Duan, Y., Houthooft, R., Schulman, J., Sutskever, I., Abbeel, P., Chen, X.: InfoGAN: Interpretable Representation Learning by Information Maximizing Generative Adversarial NetsInfoGAN: Interpretable Representation Learning by Information Maximizing Generative Adversarial Nets, pp. 1–12. Arxiv (2016)

Prediction of PhotoVoltaic Power Generation Using Monte Carlo Simulation

Gautam Seth, K. Abhay Prithvi, Arpit Paruthi, Saksham Jain and Umang Soni

Abstract In this paper, Monte Carlo simulation models have been used to forecast the expected amount of energy production from photovoltaic panels. Two MC models have been proposed with traditional Monte Carlo method as their backbone. Model-1 combines the locally weighted scatterplot smoothing (LOESS) with simple Monte Carlo simulation. Model-2 is derived from the first model, and weather forecast data is used as an exogenous input. Also, the effect of weather is considered on traditional MC simulation and compared with Model-2. Using Model-2, the error between generated and predicted data is found to be the least. Results have been generated using Python and are discussed with inference in the manuscript.

Keywords Monte Carlo simulation · Solar energy · Photovoltaic panels Locally weighted scatterplot smoothing

1 Introduction

Owing to the fact that non-renewable energy sources are depleting day by day, there has been a sudden shift toward the alternative energy sources like the solar and wind energy. Events within the past decade have shown that every country is shifting toward

G. Seth · K. A. Prithvi · A. Paruthi · S. Jain (✉) · U. Soni
Netaji Subhas Institute of Technology, Delhi, India
e-mail: jain.saksham01@gmail.com

G. Seth
e-mail: gautamseth21@gmail.com

K. A. Prithvi
e-mail: abhayprithvi97@gmail.com

A. Paruthi
e-mail: paruthiarpit@gmail.com

U. Soni
e-mail: umangsoni.iitd@gmail.com

L. C. Jain et al. (eds.), *Data and Communication Networks*, Advances in Intelligent Systems and Computing 847, https://doi.org/10.1007/978-981-13-2254-9_25

the use of renewable sources of energy to protect the environment. However, there has been a rapid growth in the industrial and domestic consumption of electricity. According to the data released by the government of India, about 418,347 GWh energy is used by the industrial sector, which contributes to 44% of the total electricity consumed whereas about 217,405 GWh energy is consumed by the domestic sector, which contributes to 23% of the total electricity consumed. Since the past decade, this rate has been increasing at 5–8% per annum [1]. India, being one of the largest producers of energy, in its basket has a mix of all the sources available including the renewable sources. The total energy produced in India is about 1,433,392 GWh out of which about 80% is produced in thermal and hydro plants, and only 5.7% of the energy is produced using the renewable sources [2]. Looking at the above-mentioned facts, it can be said that the current energy sector of India is heavily dependent on coal and fossil fuels. This arouses a concern on how to minimize the usage of fossil fuels and save the resources for the future generations [3]. These alternative sources require a huge investment but may lead to very low benefit. Hence, making it important for the nations to collect and analyze the data to forecast how much a plant operating on solar or wind energy will be able to generate. Power generation from renewable sources depends on a number of factors. In the case of solar, factors like irradiation and weather (includes cloud cover along with other conditions like fog, and dust) play an important role in the prediction.

A lot of research has been done in predicting the data using ANN and Fuzzy logic, which can more accurately simulate the system. But the usage of ANN and Fuzzy logic requires higher computing power and takes longer period of time to simulate the system [4, 5]. Monte Carlo simulation, on the other hand, requires lower computational power but the accuracy decreases. Therefore, the two strategies are proposed to improve the accuracy of the traditional Monte Carlo simulation while keeping the computational requirements at a minimum.

The rest of the paper is organized as follows: Methodology in Sect. 2, Results in Sect. 3 and Conclusion in Sect. 4.

2 Methodology

With the advent of powerful computers, the working of a plant/system can be simulated to give its working in real time. Simulation of any system is a powerful tool. It can be used for analysis of new designs and comparing performance of new designs to existing designs. But a simulation model of a complex system can only be an approximation to the actual system [6].

2.1 Monte Carlo Simulation

Monte Carlo simulation is a powerful computerized mathematical technique which is used to model and analyze real-world systems which employ statistical sampling which approximates solutions to quantitative problems. It investigates stochastic permutations of the uncertainty of a system and also quantifies their effect. It uses probability distribution of possible outcomes of a system to randomly select a variable and calculate value of the uncertainty through a number of runs [7] (Fig. 1).

To perform Monte Carlo simulation, the first step is to determine the probability distribution for uncertainty based on history as well as expert opinion. Some of the commonly used probability distributions are:

- Normal or bell distribution
- Lognormal distribution
- Uniform distribution
- Triangular distribution.

2.2 Dataset

The solar irradiance data was obtained from National Solar Radiation Database (NSRDB) website [8]. It consisted of hourly GHI values for our required location coordinates (Delhi, India in the given case). The GHI values were then converted to daily PV generation values using the formula as stated below [9].

$$E = A * r * I_r * P_r$$

Fig. 1 Normal distribution about a non-zero mean

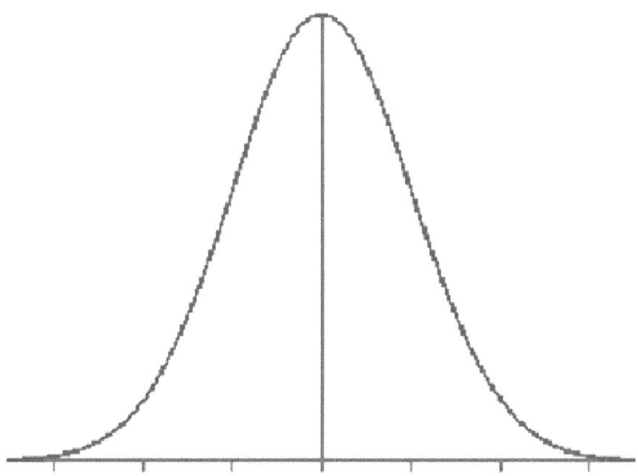

Here, E is the energy generated in kWh, A is the area of PV panels, r is the yield of panel equal to the ratio of electrical power of one PV panel to area of one panel, I_r is irradiation on the panels, and P_r is the performance ratio which ranges between 0.5 and 0.9 depending on the grade of PV panel (Table 1).

We trained the model using this data for the year 2000–2010. Delhi's weather data was also obtained which consisted of hourly weather data. Using weather condition data, approximate cloud cover was calculated. The cloud cover was used as an exogenous input for the solar PV generation and will be discussed in detail in the following sections.

2.3 Problem with MC Simulation and Proposed Model

One of the problems faced while using Monte Carlo simulation was its inability to clearly tell how the uncertainty variable changes. In the case of solar PV generation, it can be easily observed that, for every year, it follows an approximately fixed trend. Power generated is higher during the summer months (May, June), and lower during the winter months (December, January). Therefore, Monte Carlo simulation could give a good overall estimation for the annual PV generation but was not able to give a satisfactory prediction for monthly or daily values (Fig. 2).

To overcome this problem, an approach can be proposed that instead of obtaining the random variable from the probability distribution about a fixed mean (or central tendency), obtain the random variable using the probability distribution about a mean which is a variable of an external input. In this case, it was assumed that the mean is

Table 1 Solar panel specification

Area of each panel	No. of panels	Performance ratio	Yield of a panel
20 m^2	100	0.75	16.16%

Fig. 2 Confidence interval as a function of x for normal distribution about a non-zero mean

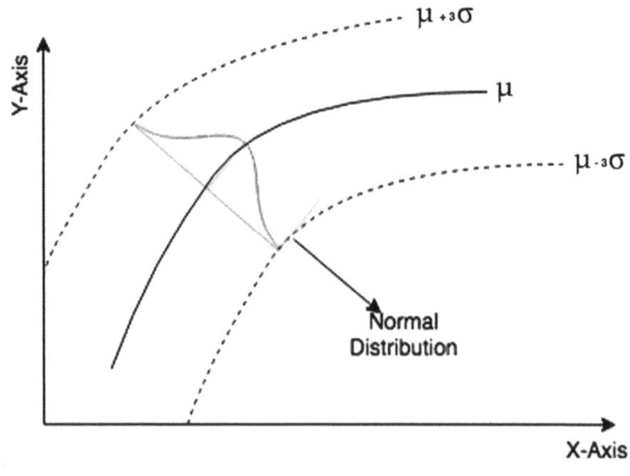

a variable of time (day number, from 1 to 365), keeping in mind the trends observed in the data.

To obtain the relation between the mean PV generation and time, locally weighted scatterplot smoothing (LOESS) [10, 11] was used in the proposed model.

Using the relation between PV generation values and time, mean PV generation value was calculated and Monte Carlo simulations consisting of 10 runs were performed for each day of the year, by selecting the random variable from the probability distribution about the calculated mean. The proposed model could predict both annual PV generation and monthly PV generation with higher degree of accuracy in comparison with the traditional Monte Carlo simulation.

2.4 *Proposed Model Along with Weather Condition as an Exogenous Input Variable*

The solar radiation energy reaching the earth's surface is the sum of the energy directly transmitted from the sun and the radiation energy diffused by the sky. When the sky is clear, directly transmitted radiation affects the total solar radiation while on a cloudy day, scattered radiation plays the major role. During a partly cloudy day, the direct radiation's effect decreases while the scattered radiation's effect increases [12]. Hence, it can be said that the PV generation depends highly on weather conditions (hence they have high correlation). Based on the octants of the sky covered by opaque clouds, the weather conditions can be classified into 9 categories, from 0 to 8, where 0 means 0% cloud coverage and 8 means 100% cloud coverage [13] (Fig. 3; Table 2).

There is a simple relation between the fraction of cloud cover and solar radiation reaching the earth's surface:

$$P\alpha\left(1 - 0.75F^3\right) \text{ watts/m}^2$$

where F = Fraction of cloud cover on a scale of 0.0 (no clouds) to 1.0 (Overcast) and P is the power generated [14]. Using the above expression, we can obtain the

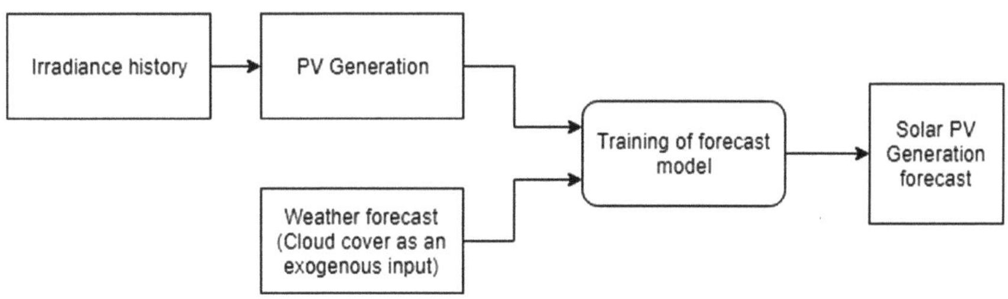

Fig. 3 Flow chart showing weather as an exogenous input

Table 2 Fraction of cloud cover based on weather conditions

Cloud cover	Weather conditions
0/8	Clear/sunny
1/8	Mostly clear
2/8	Partial fog/partly cloudy
3/8	Light drizzle/light fog
4/8	Shallow fog/light rain
5/8	Light freezing rain
6/8	Haze/light thunder storm
7/8	Heavy fog/thunder storm
8/8	Heavy rain

solar PV generation based on cloud cover as well as the value of PV generation obtained from the model explained in the previous section. It must be noted that the value of the mean PV generation in the previous model must be first converted to the maximum possible power, and then be multiplied by the factor $(1–0.75F^3)$, and then the Monte Carlo runs can be performed. This model further increases the accuracy of prediction.

3 Results

As stated above, traditional Monte Carlo simulation generates the probability distribution about a random variable can be seen in Fig. 4. Here the predicted values are nearly constant and do not follow the actual values. This results in monthly predicted also being approximately constant. This leads to a maximum error of about 46% in the month of December. The total annual error is about 1.7% (Fig. 5).

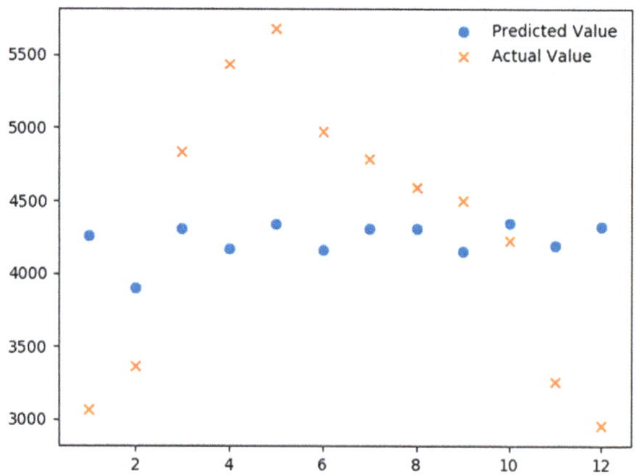

Fig. 4 Daily prediction using trad. MC simulation

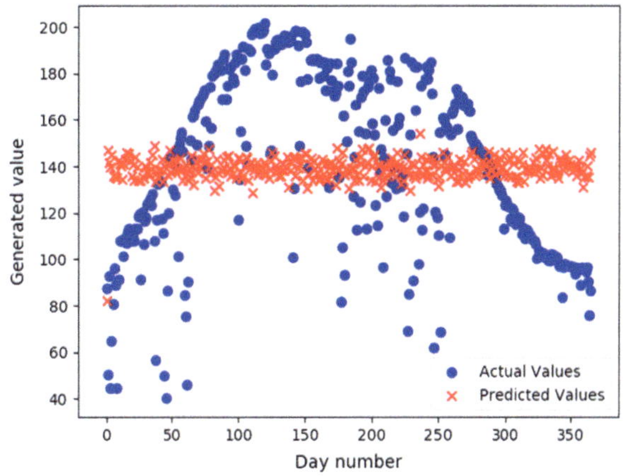

Fig. 5 Monthly prediction using trad. MC simulation

To reduce the error, the random variable is taken about the daily generated values. This reduces the maximum monthly error from 46% to approximately 6%. Further, the daily predicted values now follow the actual values, and hence monthly data is accurately predicted as shown in Figs. 6 and 7.

Table 3 shows the monthly comparison of traditional MC simulation to the proposed Model-1.

Since solar power generation depends on the weather conditions, if the weather conditions are used as an exogenous input, the accuracy can be further increased. The weather conditions are converted into cloud cover code as stated earlier. To compare our proposed model, we find the effects of weather on traditional MC simulation. The daily and monthly predicted values of Traditional + Weather are shown in Figs. 8 and 9. The outputs of Model-2 are shown in Figs. 10 and 11.

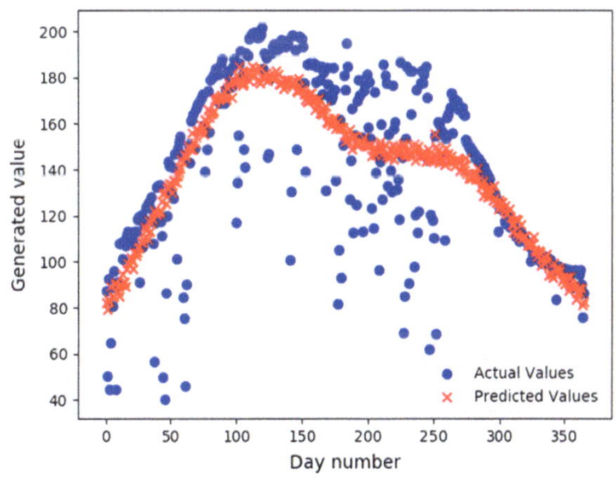

Fig. 6 Daily prediction using Model-1

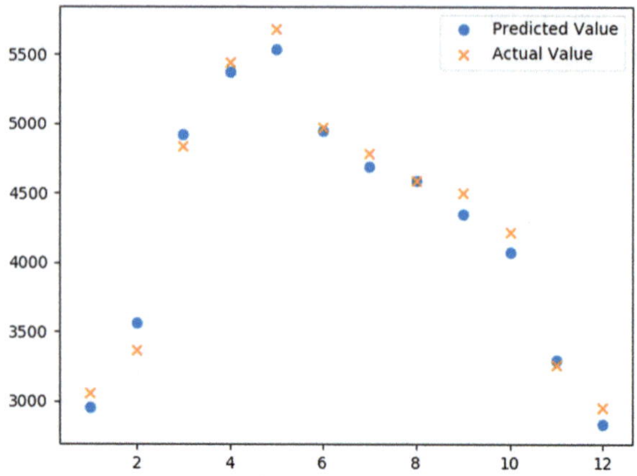

Fig. 7 Monthly prediction using Model-1

Table 3 The monthly comparison of traditional MC simulation to the proposed Model-1

Month	Actual generated (kWh)	MC estimation (kWh)	% error	Model-1 estimation (kWh)	% error
Jan	3060.93	4258.46	39.12	2957.57	3.37
Feb	3365.08	3898.90	15.86	3566.27	5.97
March	4832.60	4305.3	10.91	4919.06	1.78
April	5437.38	4167.1	23.36	5369.64	1.24
May	5678.69	4338.7	23.59	5537.08	2.49
June	4975.20	4157.2	16.44	4943.17	0.64
July	4784.38	4302.7	10.06	4688.60	2.00
Aug	4588.47	4307.9	6.11	4590.30	0.03
Sept	4499.64	4149.5	7.78	4350.85	3.30
Oct	4224.58	4344.2	2.83	4075.98	3.51
Nov	3260.74	4189.8	28.49	3296.75	1.10
Dec	2958.56	4323.8	46.14	2833.58	4.22
Total	51,666.3	50,743.56	1.78	51,128.91	1.04

Table 4 compares the effect of exogenous input of weather on both traditional MC as well as Model-1.

4 Conclusion

The main objective of this study is to compare the two strategies, which are used to obtain the forecast of the solar power generation. This study can help us to simulate a solar power plant model to estimate the amount of solar energy that can be generated

Fig. 8 Daily prediction using Traditional MC + Weather

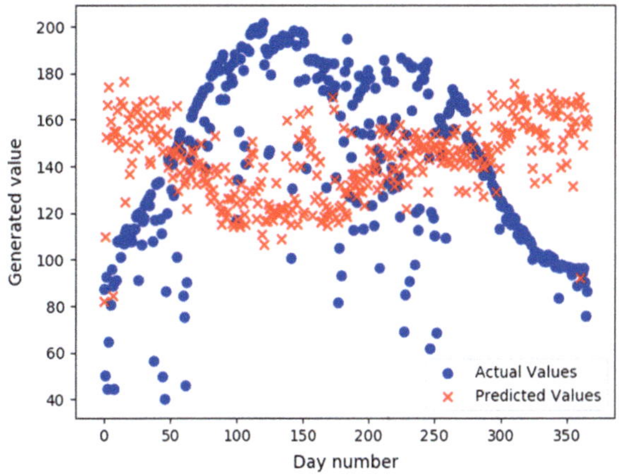

Fig. 9 Monthly prediction using Traditional MC+ Weather

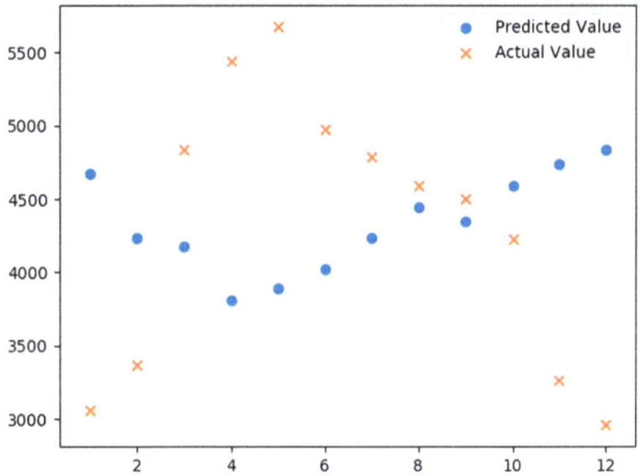

Fig. 10 Daily prediction using Model-2

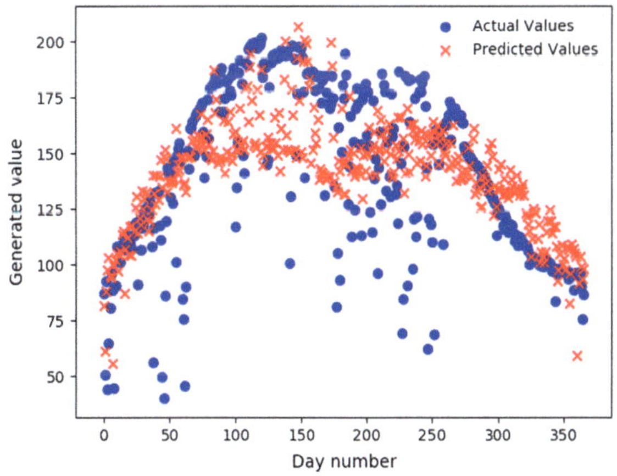

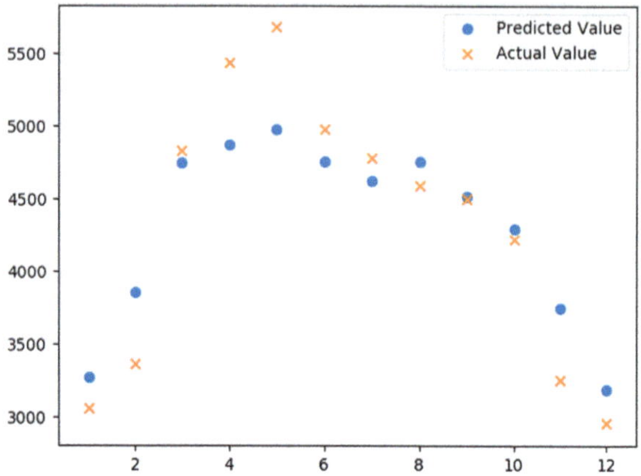

Fig. 11 Monthly prediction using Model-2

Table 4 Comparison of the effect of exogenous input of weather on both traditional MC as well as Model-1

Month	Actual generated (kWh)	Traditional + weather (kWh)	% error	Model-2 estimation (kWh)	% error
Jan	3060.93	4707.56	53.79	3274.77	6.98
Feb	3365.08	4239.34	25.98	3858.40	14.65
March	4832.60	4161.15	13.89	4747.75	1.75
April	5437.38	3812.35	29.88	4874.15	10.35
May	5678.69	3895.26	31.40	4979.72	12.30
June	4975.20	4063.54	18.32	4759.18	4.34
July	4784.38	4265.89	10.83	4625.50	3.32
Aug	4588.47	4519.99	1.49	4752.71	3.57
Sept	4499.64	4351.88	3.28	4522.15	0.50
Oct	4224.58	4537.37	7.40	4297.01	1.71
Nov	3260.74	4699.34	44.11	3748.27	14.95
Dec	2958.56	4861.12	64.30	3188.21	7.76
Total	51,666.3	52,114.86	0.86	51,627.86	0.07

annually, which can help in making decisions based on the demand of electricity, so that the remainder of the required amount that has to be obtained from the electricity grid can be estimated. Previously, a lot of research has been done in predicting the data using ANN and Fuzzy logic. There are some exceptions to Monte Carlo simulation as the use of more complex techniques is limited, therefore only a few have based their study on predicting the solar power generation using the Monte Carlo simulation method.

The results that are obtained from the study show that there is a slight deviation from the actual power generated. The comparison of traditional MC and Model-1 shows that by taking the random variable about a variable point (variable of time), the accuracy of prediction increases exponentially. This is because we are able to replicate the trends that are followed by the actual model throughout the year using a combination of LOESS and MC simulation. Studying the effects of weather as an exogenous input on both traditional MC and Model-1, we see that, even though the accuracy of Traditional MC increases, but still, it was less than Model-1. On comparing Model-1 and Model-2, we see that the annual prediction using Model-2 was much closer to the actual power generated. The annual error using Model-1 was calculated as 1.04%, whereas Model-2 performed with an accuracy of 0.07%. But the monthly prediction was better in the case of Model-1 with maximum error as approx. 6% as compared to the maximum error of 14.65% in the case of Model-2.

As the further result, a better model can be generated with the help of weather information like wind speed and direction, and dew point. Also, ANN and Fuzzy logic seem to be quite promising approach if historical data along with the various weather information are present to generate more effective results.

References

1. ENERGY—Statistical Year Book India. http://mospi.nic.in/statistical-year-book-india/2017/185 (2017)
2. Growth of Electricity Sector in India. http://www.cea.nic.in/reports/others/planning/pdm/growth_2017.pdf
3. Overview of Renewable Energy Resources of India. http://www.rroij.com/open-access/overview-of-renewable-energy-resourcesof-india.php?aid=41988
4. Kim, J.G., Kim, D.H., Yoo, W.S., Lee, J.Y., Kim, Y.B.: Daily prediction of solar power generation based on weather forecast information in Korea. IET Renew. Power Gener. **11**, 1268–1273 (2017)
5. Sivaneasan, B., Yu, C.Y., Goh, K.P.: Solar forecasting using ANN with fuzzy logic pre-processing. Energy Procedia **143**, 727–732 (2017)
6. Tüzüntürk, S., Eren Şenaras, A., Sezen, K.: Forecasting Water Demand by Using Monte Carlo Simulation (2015)
7. Paul, V.K., Basu. C.: A Handbook for Construction Project Planning and Scheduling. Copal Publishing Group (2017)
8. National Solar Radiation Database. https://nsrdb.nrel.gov/
9. Calculations of Solar Energy Output. http://www.academia.edu/9005661/CALCULATIONS_OF_SOLAR_ENERGY_OUTPUT
10. Cleveland, W.S.: Robust locally weighted regression and smoothing scatterplots. J. Am. Stat. Assoc. 829–836 (1979)
11. Cleveland, W.S.: Locally weighted regression: an approach to regression analysis by local fitting. J. Am. Stat. Assoc. 596–610 (1988)
12. Matuszko, D.: Influence of the extent and genera of cloud cover on solar radiation intensity. Int. J. Climatol. **32**, 2403–2414
13. NOAA's National Weather Service—Glossary. https://forecast.weather.gov/glossary.php?word=sky%20condition
14. Cloud Cover Solar Radiation. https://scool.larc.nasa.gov/lesson_plans/CloudCoverSolarRadiation.pdf

Storage and Analysis of Synchrophasor Data for Event Detection in Indian Power System Using Hadoop Ecosystem

Akhilendra Pratap Singh, G. Hemant Kumar, Subhendu Sekhar Paik and Diptendu Sinha Roy

Abstract Synchrophasor devices commonly referred to as phasor measurement units (PMUs) have been rapidly deployed in power grids to get a clearer picture of the events that take place in power grid at very high sampling rates. Thus PMUs enable event predictions to become more accurate. However, high data-sampling rate of PMUs creates huge volumes of data on a GB scale per PMU per day, which in turn makes the storage and analysis of the collected data difficult; giving rise to a Big Data problem use case. In India, National Load Dispatch Centre (NLDC) uses Synchrowave; a centralized software package to collect, store and process, which runs on a single very high-end system for storage and processing. But the centralized architecture of current system gives no room for scalability, neither in computation nor storage, even with costly hardware upgrades. This paper proposes a system model with all the desired features using open-source frameworks and tools like Apache Hadoop and HBase.

Keywords Phasor measurement unit (PMU) · Synchrowave · Hadoop · HBase

1 Introduction

Electrical power system for generation, transmission and distribution is extremely valuable to modern society. It is a difficult task for power sector engineers to keep a sharp vigil on power grid to keep it stable and reliable manually. Automated monitoring and alerting mechanism have been into place for quite sometime with supervisory control and data acquisition (SCADA), but time resolution for SCADA for detection and identification of various events in massively large-scale power system that are interconnected through a wide area measurement system (WAMS). Because of the

A. P. Singh · S. S. Paik · D. Sinha Roy (✉)
National Institute of Technology Meghalaya, Shillong, India
e-mail: diptendu.sr@nitm.ac.in

G. Hemant Kumar
ICTS-TIFR, Bengaluru, India

L. C. Jain et al. (eds.), *Data and Communication Networks*, Advances in Intelligent Systems and Computing 847, https://doi.org/10.1007/978-981-13-2254-9_26

complexity, the nature of power system dynamically changes randomly in real time due to different types of disturbance events such as generation loss, line trips (line-ground, lineline, etc.), faults (transition, persistence, symmetric, Asymmetric, etc.) and blackouts. A wide range of disturbance events in power system like blackouts, line tripping, faults and generation loss are far from the reliability and stability of requirements as they suppose to be. The inaccuracy in analysis of disturbances at different regions of the interconnected smart grid is due to lack of wide-area infrastructures [1].

In many distribution systems primarily due to economic constraints, data acquisition is impeded heavily. However, with the Power Grid Corporation of India Limited's (PGCIL) initiative of synchrophasors that has been eradicated [2]. Along with issues in the current observability of the distribution system, the introductions of newer technologies have resulted in the potential for previously insignificant or non-existing faults and disturbances to occur. The newer grid technology introduced phasor measurement units (PMUs) for continuous monitoring of transmission bus lines and recording the continuous changes in electrical parameters. Those readings are called PMU samples of electrical parameter changes. It helps to analyze the power grid status. Power engineers can detect, analyze and find solutions for electrical power faults, blackouts and grid isolation etc from PMU data, owing to it's high resolution.

PMUs can be thought as descendants of computer relays that employed microprocessors for accounting for power system parameters in earlier days. Computer relays became very popular, and several reliability studies [3, 4], confirmed their robust nature. At the same time, information and communication technology (ICT)-based smart services such as smart energy management for homes [5] and grid computing-based power monitoring systems [6] were also becoming prominent eventually leading to the vision of smart grid. PMUs as smart grid sensors became a very popular topic and several attempts for PMU design [7] were found along with research pertaining to their utility as effective smart grid sensors [8]. Reliability analysis of PMUs has attracted a lot of research attention accounting for different aspects and with various mathematical tools [8–12].

The objective of this paper is to develop an operative event detection model by using the methodology of statistical analyzing on PMU datasets stored. With help of this model, the grid control room engineers present at load despatch centers can monitor and visualize the continuous changes in power grid states (i.e., grid parameters like voltage, current and frequency) which are helpful for early detection of power grid faults/events. Early event detection is much crucial for choosing control operations to overcome the isolated zone, energy loss, blackouts and power outage.

This paper proposes the design of needed system model by implementing distributed, non-relational database—HBase—on a distributed storage system—HDF-S—for structured data with back-end capabilities enabling numbers of features like linear and modular scalability, strictly consistent reads and writes, automatic and configurable shading of tables and automatic fail-over support, thus making the process of data analysis much simpler, robust and reliable. Moreover, this paper brings forth an efficient event detection algorithm using MapReduce framework of Hadoop using PMU datasets.

The rest of paper organization as follows: Sect. 2 gives a brief introduction about the background of the problem and the literature review of different event detection clarification-related works and a brief background of Indian Power Grid PMU projects, grid structure, Synchrowave. Section 3 depicts proposed MapReduce-based data analysis model for power system fault detection with non-relational decentralized storage of PMU data. Section 4 presents implementation, platform setup and details of experimental work done on real PMU datasets collected from NLDC. Section 5 concludes the proposed idea with extensive future work.

2 Background and State of Art

The entire Indian Power Grid had been divided into five regions as North Regional Load Despatch Centre (NRLDC), North East Regional Load Despatch Centre (NERLDC), East Regional Load Despatch Centre (ERLDC), South Regional Load Despatch Centre (SRLDC), West Regional Load Despatch Centre (WRLDC). Each region controls and monitors the substation's power generation unit within that region only. The all five interdependent regions are centrally under control of National Load Dispatch Centre (NLDC) located at New Delhi [13]. NLDC is using Synchrowave Central software package for recoding all 52 PMUs readings. A phasor measurement unit (PMU) or a synchrophasor is a device which measures the electrical waves on an electricity grid, using a common time source for synchronization help of GPS system facility as the block diagram shown in Fig. 1b. Time synchronization allows synchronized real-time measurements of multiple remote measurement points on the grid. In power engineering, these are considered as one of the most important measuring devices in the future of power grid systems. Synchrowave Central software was configured by Synchrowave Central software provider on central server at NLDC central power control or control room. Currently, 52 PMUs were located in Indian grid for electrical dynamic properties monitoring and recording. The entered PMU data samples are in the format shown in Fig. 2. These are collected and transferred to the regional phasor data collector (PDC) located at regional load despatch centres (RLDC) and from there the samples are transferred to the central PDC located at NLDC [13, 14] as shown in Fig. 1a.

Monitoring and analysis of signals of power system parameters like voltage, voltage angle, current, current angle and frequency are typical time series data. A lot of research work was going on in the direction of improving the process of analysis of large time series data sample with a fast reliable model of storage architecture. If it is

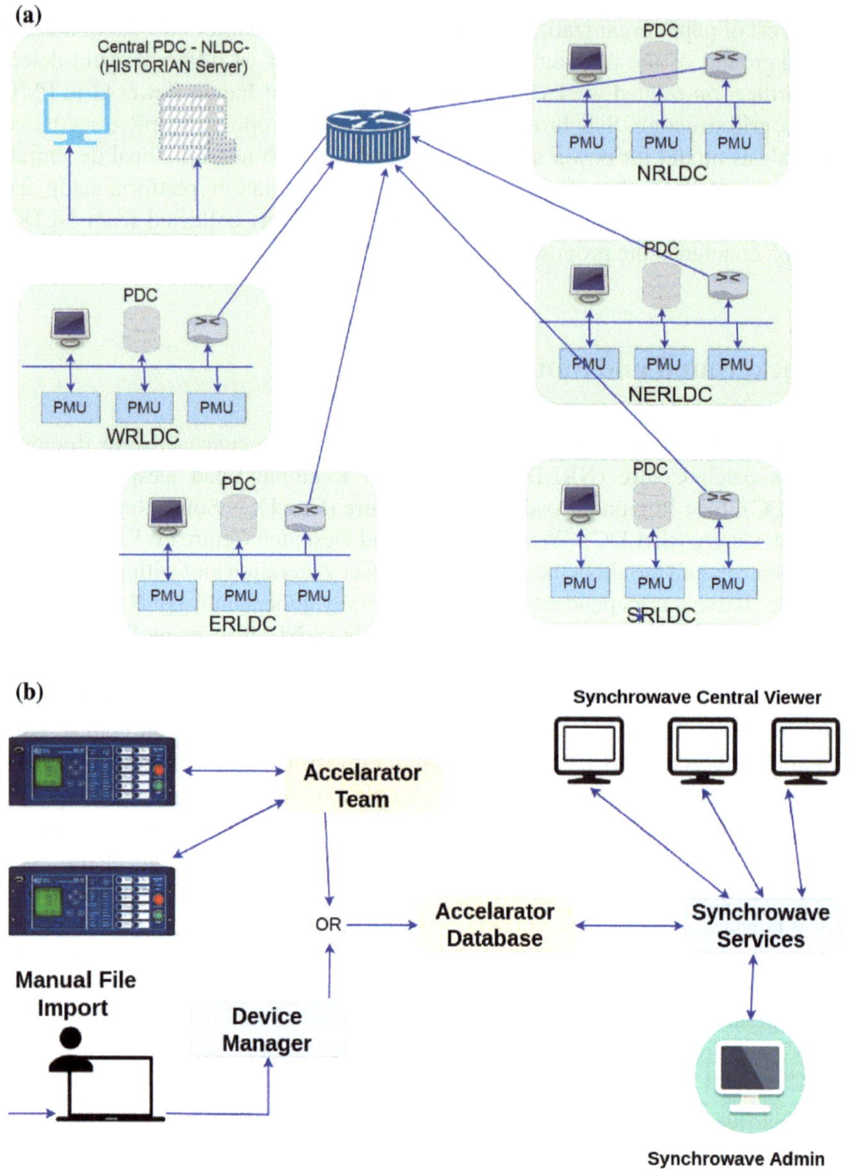

Fig. 1 **a** Indian power system architecture of PMUs; **b** Synchrophasor architecture for Synchrowave Central software

	A	B	C	D
1	Timestamp	AGRA:Voltage A:Magnitude	AGRA:Voltage B:Magnitude	AGRA:Voltage C:Magnitude
2	2015/04/26 07:01:08.040	241458.5781	245718.6406	244248.5625
3	2015/04/26 07:01:08.080	241467.2813	245728.7344	244249.9375
4	2015/04/26 07:01:08.120	241439.2031	245705.0938	244242.3906
5	2015/04/26 07:01:08.160	241457.1563	245714.5	244257.3594
6	2015/04/26 07:01:08.200	241436.6094	245689.625	244243.3125
7	2015/04/26 07:01:08.240	241421.8594	245686.5625	244231.3906
8	2015/04/26 07:01:08.280	241435.7969	245702.7188	244245.7656
9	2015/04/26 07:01:08.320	241422.0156	245683.2656	244219.9375
10	2015/04/26 07:01:08.360	241429.0781	245682.875	244230.5313
11	2015/04/26 07:01:08.400	241443.1406	245712.7031	244248.3594
12	2015/04/26 07:01:08.440	241422.0781	245702.7813	244247.0469
13	2015/04/26 07:01:08.480	241434.8438	245712.4063	244253.9844
14	2015/04/26 07:01:08.520	241424.6094	245690.0781	244248.2969
15	2015/04/26 07:01:08.560	241424.7656	245694.9844	244241.125
16	2015/04/26 07:01:08.600	241426.3594	245695.8438	244239.8906
17	2015/04/26 07:01:08.640	241430.2813	245683.3281	244234.0313
18	2015/04/26 07:01:08.680	241408.9531	245675.4219	244229.5156
19	2015/04/26 07:01:08.720	241408.0781	245686.625	244241.7031
20	2015/04/26 07:01:08.760	241406.1094	245677.4531	244230.1094
21	2015/04/26 07:01:08.800	241434.7813	245696.125	244256.2031
22	2015/04/26 07:01:08.840	241422.0625	245687.4844	244254.0313
23	2015/04/26 07:01:08.880	241409.6875	245680.75	244245.0469
24	2015/04/26 07:01:08.920	241416.5313	245683.7813	244233.5313
25	2015/04/26 07:01:08.960	241410.0625	245674.2969	244235.8438
26	2015/04/26 07:01:09.000	241403.6406	245661.9844	244235.5938
27	2015/04/26 07:01:09.040	241410.4531	245681.5469	244234
28	2015/04/26 07:01:09.080	241377.0625	245672.4063	244206.3281
29	2015/04/26 07:01:09.120	241388.9688	245669.7031	244217.4219
30	2015/04/26 07:01:09.160	241395.3438	245688.7656	244224.0781
31	2015/04/26 07:01:09.200	241418.3438	245687.1406	244252.2656

Fig. 2 PMU raw data sample recorded by Synchrowave

examined thoroughly, then all such research works can be grouped into three classes according to their processing and storage frameworks.

The synchrophasor data is being extensively used for real-time and post-event analysis, moreover, event detection, event analysis and for the purpose of development of analytics and tools to monitor and control the event. A lot of research contribution had been given by many researchers working all across the globe for fault detection and fault classification, to develop smart auto-alarming control system models. To fulfill the needs of the modern age, many types of research and studies are going and some have already developed a smart power grid system. PMU sample data analyzed through any complex mathematical optimization model to extract the features of the data and map them with known disturbance events. The models that can identify disturbance directly from the change in frequency range are designed by Gaussian distribution method [15]. A few approaches to identify events by using some mathematical models are proposed to apply convex optimization method known as ellipsoid model by Dahal et al. [16]. This method is a suitable model for smart grid dynamic behavior identification is well known as minimum volume enclosing ellipsoid (MVEE) [17]. In this model, all the PMU measurement samples are enclosed inside an ellipsoid of minimum volume in n-dimensional space where the number of PMU. The ellipsoid properties like volume, change in volume, center, and semi-major axis are used to classify the power system events and clustering those observed events by applying hierarchical clustering approach, support vector machine (SVM), K nearest neighbor classifier, recursive partitioning and regression tree (RPRT) [14, 18]. Additionally, numerous statistical data analysis models had been developed that employed window-based pattern recognition for tracking power system events [19, 20] and employed experience-based threshold settings for event detection [21].

3 Proposed System Model

Generally, PMUs are located at sub-grid stations working at voltage level of 110 V or above. It is such a sensitive hardware device that it can record all types of minor and major changes observed in power properties caused due to any reason like LL tripping, LG faults and due to fall of tree upon transmission line are observed at load transmission lines and grid stations.

Therefore power properties are dynamic parameters. Any dynamic change of properties result in an alerting generation at control rooms, and this is required to be as soon as possible by the scheme. In this paper, PUM datasets that were collected from NLDC historian storage (as discussed in Sect. 2), and NLDC historian stores a total of 52 PMUs sample records. PMUs are installed over 132, 220 and 400 kV very high voltage transmission bus lines. At each transmission, bus line data was recorded at a sampling rate of 25 samples per second [14–18, 22, 23].

In this proposed designed model, a subset of major power properties like voltage, current, frequency and phasor angle of PMU data sample is considered. All the data are continuously read from the historian to a common sink folder then from a common sink folder to HBase table as Fig. 3a.

In order to handle this architecture given in Fig. 3a, the primary handlers proposed in this paper are Apache Hadoop, Apache HBase and Apache Thrift. Apache Hadoop provides both parallel processing framework, MapReduce (MR) and distributed file system, Hadoop Distributed File System (HDFS) facility to process and store a huge amount of data in decentralized clusters. The clusters can be geographically distributed across different territories. Hadoop cluster supports commodity hardware processing units. Hadoop provides two major frameworks such as [24]. Apache HBase is an open source, distributed non-relational database modeled after Google's Big-table. It allows distributed storage system for structured data with back-end capabilities on top of Hadoop and HDFS. It has numerous numbers of features like linear and modular scalability, strictly consistent reads and writes, automatic and configurable shading of tables, automatic fail-over support between RegionServers and convenient base classes for backing Hadoop MapReduce jobs with Apache HBase tables. The Apache Thrift is a popular software framework, mostly used for scalable cross-language service development, which integrates a fully functional software stack with a core engine to build and provide efficient services and platforms for

(a)

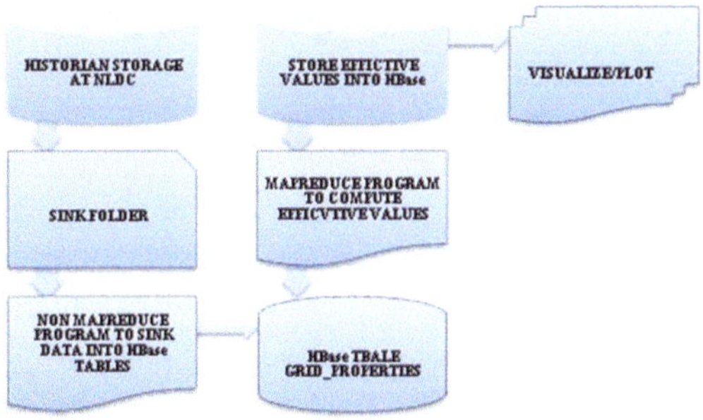

(b)

RowKey (TIMESTAMP)	PMU1					PMU2				
	vol	cur	vAng	cAng	frq	vol	cur	vAng	cAng	frq
2014/04/01 00:00:00.000	78664.7422	65.0123	54.5154	-0.7078	49.892	77764.7489	87.2156	47.3254	- 2.3698	49.891
2014/04/01 00:00:00.040	78710.7864	65.89	53.4567	-6.987	49.893	77710.7077	87.789	45.4577	-3.785	49.875
2014/04/01 00:00:00.080	78720.7897	64.3667	51.345	-9.2869	49.896	77820.8807	86.8596	44.394	-5.564	49.895
***	***	***	***	***	***	***	***	***	***	***
2014/04/01 23:59:59:920	78624.7856	65.2012	48.3908	- 10.4009	49.887	77894.7965	85.3203	43.9870	- 34.568	49.877

Fig. 3 **a** Block diagram of primary power distribution substation: [19]; **b** GRID-PROPERTIES attributes structure

seamlessly various programming languages. Instead of programming a huge lines of code to transport and serialize the objects and invoke remote methods, it helps to get right down to business logics [25].

GRID-PROPERTIES and PMU-PROPERTIES, two logical HBase tables, have been designed to store and manage PMU data in HDFS of Apache Hadoop. The proposed idea of this paper operates in two phases. One sample of has GRID-PROPERTIES table has been shown in Fig. 3b.

Sink data from historian to HBase grid table: This one is designed to read data from OS file system (i.e., .csv file) and put those records of data into HBase table. We designed a normal read-write program by using HBase Client API. This program just scans the sink folder for .csv then reads the .csv file row by row and inserts (i.e., put) into the GRID-PROPERTIES table (Fig. 1a). MapReduce version of statistical trend analysis program for PMU datasets HBase tables.

MapReduce algorithm for characteristic extraction: This one is designed to scan the GRID-PROPERTIES (Fig. 4) frame by frame. The technique of overlapping, moving window approach is used to scan records from Fig. 5. The window size is 12 s long (i.e., 12 * 25 = 300 sample records) and overlapping half window of previously scanned window data (i.e., 6 * 25 = 150 old sample records). Each time the program gets 300 sample records of a PMU and performs MapReduce computation to find out the statistical characteristics of a PMU data trend. It computes mean, maximum,

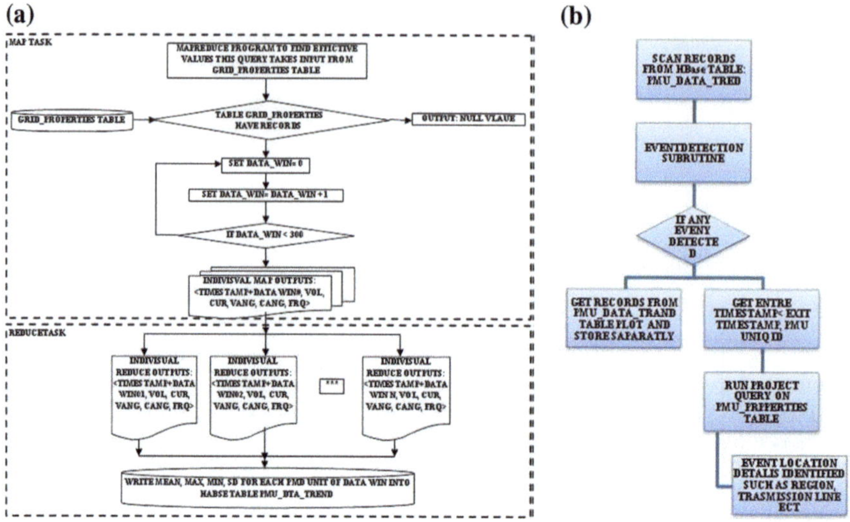

Fig. 4 **a** Stage one process flow for effective values evaluation using MapReduce; **b** Stage two process flow to detect, plot and store events

minimum and standard deviation for each PMU present within the window frame. The results are getting stored into another HBase table PMU-DATA-TREND, and the process workflow can be represented as Fig. 4a, b.

4 Implementation and Experimentation

4.1 TestBed Setup

The proposed idea has been realized in a cluster of six homogeneous nodes with 1 TB SATA Hard Disk, 8 GB DDR3 RAM, Intel Core i3-4790 CPU @ 3.60 GHz 8 processor, Ubuntu 16.04 LTS operating system, Apache Hadoop V2.7.1 Apache HBase V1.4.3. For programming, statistical analysis and result visualization R Studio version 1.1.447 has been configured and Apache thrift V0.11.0. Apache thrift acts as the gateway to HBase server and access HBase table data in R code. A package named rhbase provides necessary method to connect and access HBase table in R. HBase has been configured on HDFS to use its file system [24–26].

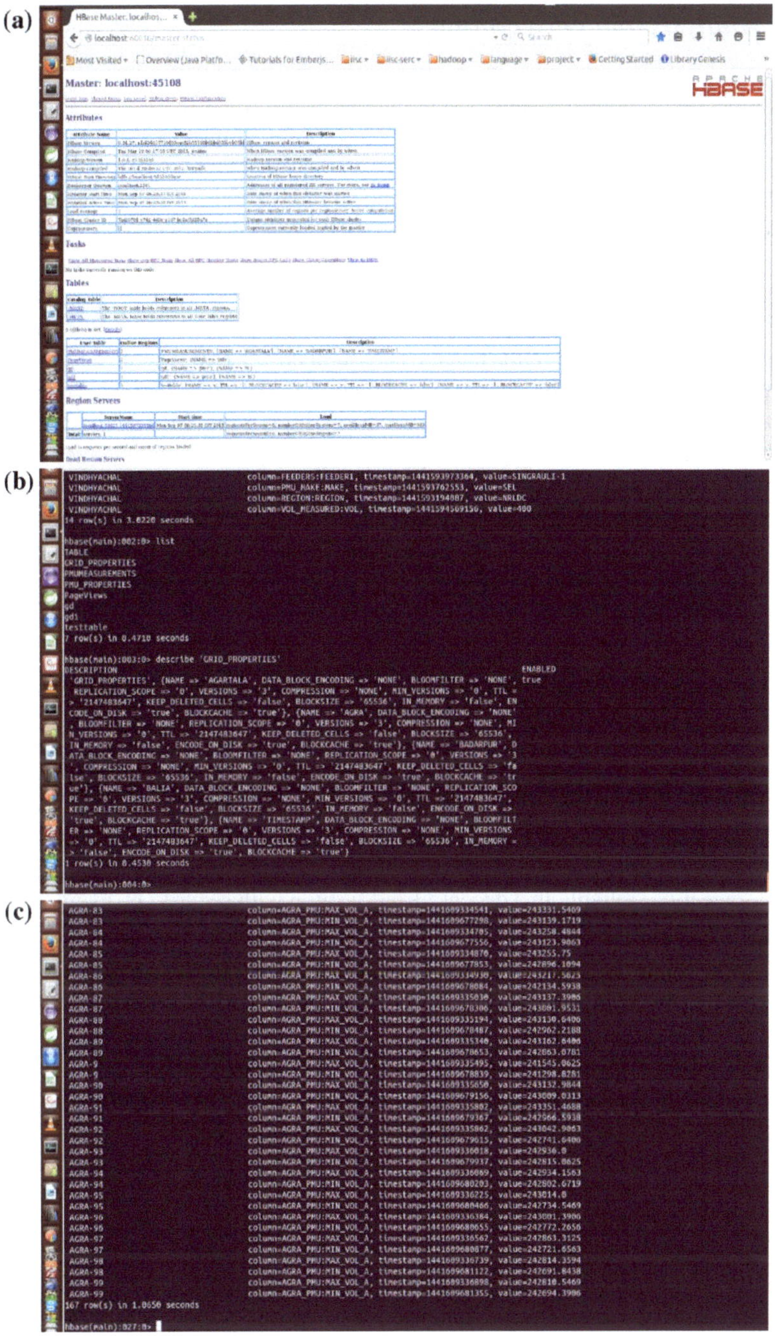

Fig. 5 **a** HBase web UI; **b** List of HBase tables and GRID-PROPERTIES table details; **c** GRID-PROPERTIES table records

4.2 Analysis and Observation

The graph for Max–Min voltage gets data from the records of the table PUM-DATATREND and R Studio IDE was used to plot the graphs as shown in Fig. 5.8. Here, red bots taken for Max voltage and green dots are taken for Min voltage. Window frame counts are taken along x-axis and voltages along y-axis. From the graph, we can observe that at some places difference between Max and Min voltages are likely more. To analyze more on such observations, another graph was plotted by taking frame count along x-axis and voltage difference along y-axis as shown in Fig. 6a. Here one can get clearer and many pick values. Those pick differences may be due to regular electrical faults that are natural in power grid system, and sometime such pick difference may be courage event also.

To analyze in clear and in very sensitive form, here, Max–Min method is used to find the SD of framewise voltage difference and mean. According to the reference of Max–Min method, the more likely and possible event-time window frame's voltage differences always come above the three times of the computed SD. Fault voltage or may be some small electrical property changes can Couse such type of picks as shown in Fig. 6b. Another histogram Fig. 7a was plotted to know the number of window frames which show possibly events. Boxplot also showing some out liars for event Couse window frames as shown in Fig. 7b, in case there is no events are there than Boxplot shows no out liars.

5 Conclusion

The experiments conducted depict the efficacy of the proposed methodology PMU data storage more reliably in a decentralized fashion with security by using low-cost commodity hardware units (i.e., Hadoop clusters and HDFS). This method is cost-efficient than current vertical scalable high-end server systems providing more faster processing. This paper focuses to improve the visualization of computed data point and further analysis and storage and is necessary at load dispatch centers rather than data point and reading of PMUs.

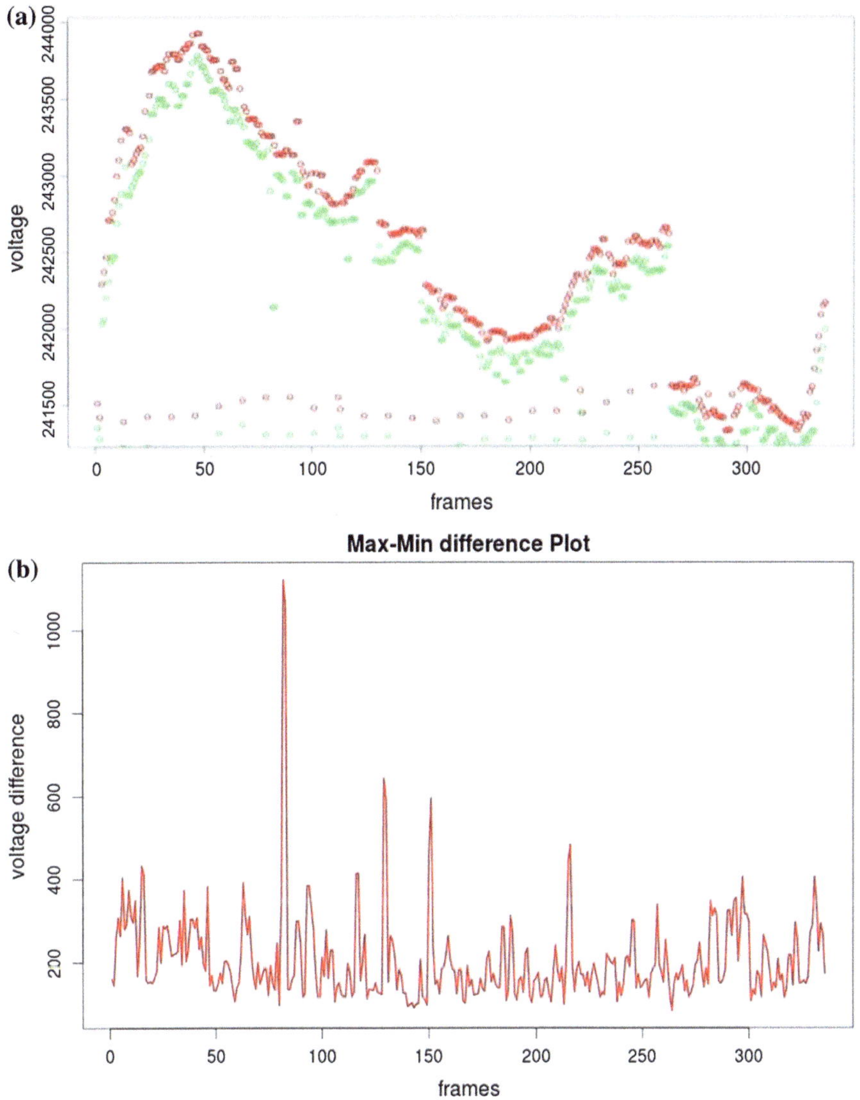

Fig. 6 **a** Max–Min voltage; **b** Max–Min voltage differences

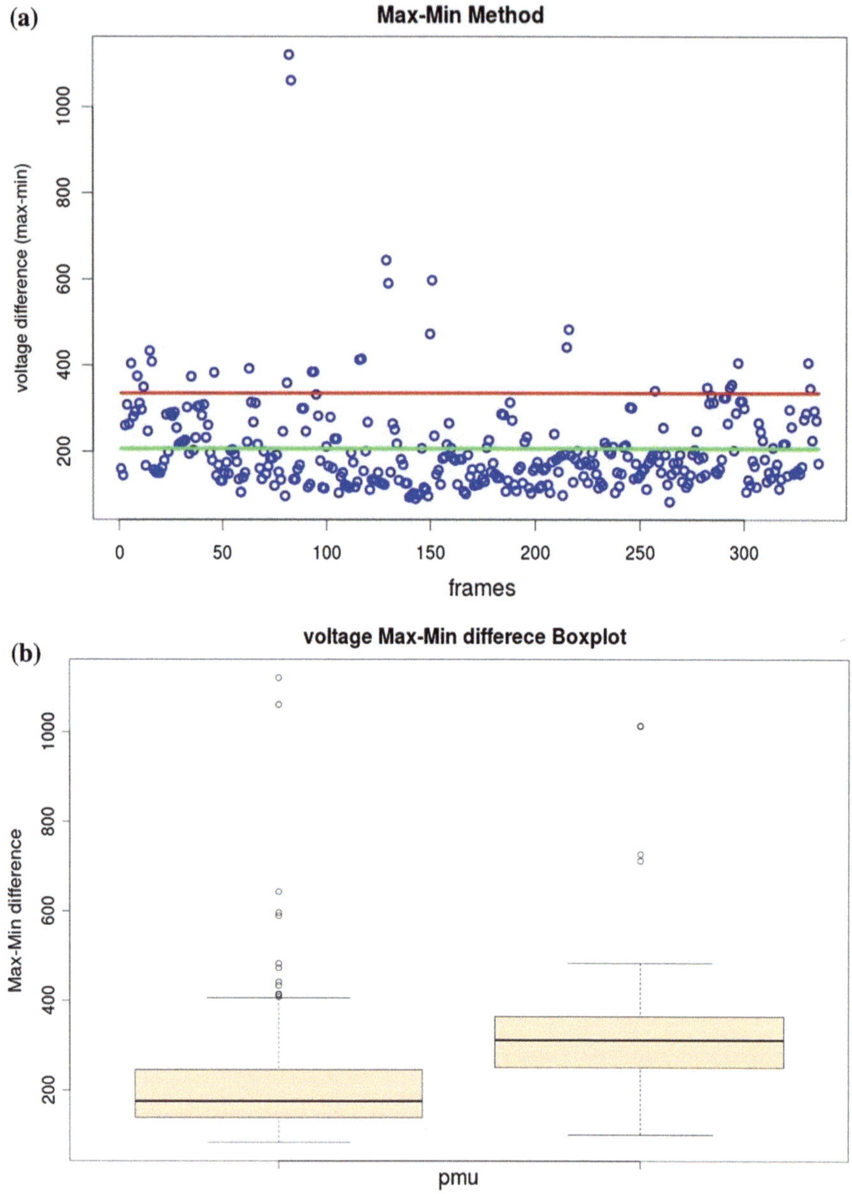

Fig. 7 **a** Max–Min method analysis; **b** Boxplot of voltage difference

Future work will look at the development of MapReduce-based clustering techniques and machine learning techniques for classification of power fault detected as well as the unique events detections. Machine learning is a pattern recognition technique which can help to fetch specific patterns present inside the power signals, every different events or faults develop a unique type of signal pattern [19]. So, development of a self-learning neural network model for signal patent.

Acknowledgements This work has been carried out by financial assistances from SERB-DST (File no. EMR/2017/001508). The authors wish to thank POSOCO, New Delhi for providing necessary synchrophasor data.

References

1. Phadke, A.G.: Synchronized phasor measurements in power systems. IEEE Comput. Appl. Power **6**(2), 10–15 (1993)
2. Aghaei, J., Alizadeh, M.I.: Demand response in smart electricity grids equipped with renewable energy sources: a review. Renew. Sustain. Energy Rev. **18**, 64–72 (2013)
3. Mohanta, D., Roy, D.: Importance and uncertainty analysis in software reliability assessment of computer relay. Proc. Inst. Mech. Eng. Part O J. Risk Reliab. **225**(1), 50–61 (2011)
4. Roy, D.S., Mohanta, D.K., Panda, A.: Software reliability allocation of digital relay for transmission line protection using a combined system hierarchy and fault tree approach. IET Softw. **2**(5), 437–445 (2008)
5. Das, H., Roy, D.: A grid computing service for power system monitoring. Int. J. Comput. Appl. **62**(20) (2013)
6. Polaki, S.K., Reza, M., Roy, D.S.: A genetic algorithm for optimal power scheduling for residential energy management. In: 2015 IEEE 15th International Conference on Environment and Electrical Engineering (EEEIC), IEEE, pp. 2061–2065 (2015)
7. Mondal, S., Murthy, C., Roy, D., Mohanta, D.: Simulation of phasor measurement unit (pmu) using labview. In: 2014 14th International Conference on Environment and Electrical Engineering (EEEIC), IEEE, pp. 164–168 (2014)
8. Mohanta, D.K., Murthy, C., Sinha Roy, D.: A brief review of phasor measurement units as sensors for smart grid. Electr. Power Compon. Syst. **44**(4), 411–425 (2016)
9. Murthy, C., Mishra, A., Ghosh, D., Roy, D.S., Mohanta, D.K.: Reliability analysis of phasor measurement unit using hidden Markov model. IEEE Syst. J. **8**(4), 1293–1301 (2014)
10. Murthy, C., Roy, D.S., Mohanta, D.K.: Reliability evaluation of phasor measurement unit: a system of systems approach. Electr. Power Compon. Syst. **43**(4), 437–448 (2015)
11. Murthy, C., Varma, K.A., Roy, D.S., Mohanta, D.K.: Reliability evaluation of phasor measurement unit using type-2 fuzzy set theory. IEEE Syst. J. **8**(4), 1302–1309 (2014)
12. Roy, D.S., Murthy, C., Mohanta, D.K.: Reliability analysis of phasor measurement unit incorporating hardware and software interaction failures. IET Gener. Transm. Distrib. **9**(2), 164–171 (2014)
13. POSOCO. http://posoco.in/2013-03-12-10-34-42/synchrophasors
14. Dahal, O.P., Brahma, S.M., Cao, H.: Comprehensive clustering of disturbance events recorded by phasor measurement units. IEEE Trans. Power Deliv. **29**(3), 1390–1397 (2014)
15. Liu, L., Chai, J., Qi, H., Liu, Y.: Power grid disturbance analysis using frequency information at the distribution level. In: 2014 IEEE International Conference on Smart Grid Communications (SmartGrid-Comm), IEEE, pp. 523–528 (2014)
16. Dahal, O.P., Brahma, S.M.: Preliminary work to classify the disturbance events recorded by phasor measurement units. In: 2012 IEEE Power and Energy Society General Meeting, IEEE, pp. 1–8 (2012)
17. Todd, M.J., Yıldırım, E.A.: On Khachiyan's algorithm for the computation of minimum-volume enclosing ellipsoids. Discret. Appl. Math. **155**(13), 1731–1744 (2007)
18. Rusitschka, S., Eger, K., Gerdes, C.: Smart grid data cloud: a model for utilizing cloud computing in the smart grid domain. In: 2010 First IEEE International Conference on Smart Grid Communications (SmartGridComm), IEEE, pp. 483–488 (2010)

19. Allen, A.J.: Analysis of Transmission System Events and Behavior Using Customer Level Voltage Synchrophasor Data (2013)
20. Bach, F., Cakmak, H.K., Maass, H., Kuehnapfel, U.: Power grid time series data analysis with pig on a Hadoop cluster compared to multi core systems. In: 2013 21st Euromicro International Conference on Parallel, Distributed and Network-Based Processing (PDP), IEEE, pp. 208–212 (2013)
21. Song, Y., Zhu, Y., Li, L.: Large scale data storage and processing of insulator leakage current using HBase and mapreduce. In: 2014 International Conference on Power System Technology (POW-ERCON), IEEE, pp. 1331–1337 (2014)
22. Allen, A., Singh, M., Muljadi, E., Santoso, S.: PMU Data Event Detection: A User Guide for Power Engineers. National Renewable Energy Laboratory (2014)
23. Foster, I., Zhao, Y., Raicu, I., Lu, S.: Cloud computing and grid computing 360-degree compared. In: Grid Computing Environments Workshop, 2008. GCE'08, IEEE, pp. 1–10 (2008)
24. Borthakur, D., et al.: HDFS architecture guide. Hadoop Apache Proj. **53**, 1–13 (2008)
25. Apache: Apache Thrift. https://thrift.apache.org
26. Apache: Apache HBase. http://hbase.apache.org/

Analysing Trends in Student's Performance Across Maharashtra Through Non-adaptive and Adaptive Online Assessments Based on the Underlying Framework of Classical Test and Item Response Theory

Chandrani Singh and Ajit Pandey

Abstract In today's era with the advancement of information and communication technology, and with the rapid digitization across all arenas, it has become a necessity to identify and map the right resource to a required job profile in minimum time and cost. To identify the right resource, organizations across the world have adopted several online tools, strategies and mechanisms to assess the candidates to check on their potential and suitability. Technologies have even paved ways for creating immersive modes where candidates are required to get acclimatized and play the job role and deduce the fitment for that particular role. The changes brought about by the information and communication technology have been speedy and pervasive and so has the penetration of artificial intelligence and machine learning where human and technology are highly interconnected to achieve the desired outcome. In all segments and sectors, digital assessments (in any form) play a key role in identifying the right asset, where a lot is dependent on the impregnated intelligence and machine learning to help decision-making become easy and accurate. It is said that artificial intelligence could play a role in the growing field of assessment and learning analytics and also can be applied in evaluating the quality of curriculum content so that it can be applied to create unique pathways for individual learning. This in turn will create the desired efficacy in assessing with precision candidate/candidates for the right opportunity. In this regard, the authors in their second paper propose to first identify and extract the performance trends of candidates and item characteristics using non-adaptive and adaptive assessment techniques to create a rationale for measuring their learnability and implementation of their learned skills in specific job roles.

C. Singh · A. Pandey (✉)
Sinhgad Institute of Management, Pune, India
e-mail: awesomeajit@gmail.com

C. Singh
e-mail: singh.chandrani@gmail.com

L. C. Jain et al. (eds.), *Data and Communication Networks*, Advances in Intelligent Systems and Computing 847, https://doi.org/10.1007/978-981-13-2254-9_27

Keywords Performance trends · Adaptive assessment and learning Immersive modes · Resource fitment · Item response theory · Discrimination Pseudoguessing · Latent trait

1 Introduction

With the advancement in the learning and assessment space, the implementation of adaptive form of assessment has come to fore in the higher education. The adaptive form of assessment was developed between 1973 and 1984 by Weiss, Kingsbury. Initially, such form of assessments starts with a large base of questions followed by a selection of individual questions of average complexity for test takers to start off with the test. The difficulty level of the test increases or decreases with each successful/unsuccessful attempt by the test taker, respectively. The adaptive assessment requires the following components to be present: question bank, calibrated to a common measurement scale; an algorithm for item selection, a process to score the candidate's responses; test termination condition and score analytics. Adaptive assessments have been found to take less time than the fixed form tests and provide enough accurate information about the candidates to come to a viable conclusion be it assessing them for employment opportunities or for the course/training/certifications undertaken by them. In the former test, theories such as the classical test theory, there have been certain limitations with regard to the following: single reliability value for the entire test and all participants. The scores of the candidates depend on the items rather than the candidate, and there is always a bias towards the fixed form tests to have questions of average difficulty. On the other hand, when item response theory became the underlying base for adaptive assessments, the latent trait models under this theory helped to look beyond the underlying traits responsible for producing the test performance for each item. The item response theory comprises of majorly three components: item response function, item information function and invariance. Item response function can, in other words, be correlated to item characteristic curve which depicts the item's value in terms of respondent's ability as a function of the probability of recommending an item. For dichotomous items (those scored correct/incorrect), each item has three parameters according to the three-parameter model:

(a) the discrimination parameter that provides the information on how well a question or an item differentiates between the best and poor scorers. While analysing on the item discrimination index is required to segregate the best, average and the poor performers based on the scores received and creating the clusters.
(b) the difficulty parameter, an estimate of the appropriateness of the item to the level of examinees which ranges from −3 to +3, with 0 being an average examinee level.
(c) the pseudoguessing parameter, which is a lower asymptote, with a value of k inverse where k is the number of options.

For an item with discrimination equal to 0.8, difficulty equal to 0.68 and the guess being equal to 0.20, the graph for that question/item would appear as follows (Fig. 1):

This graph has been constructed based on the performance of 20 students across ten items undergoing Knox cube test as a pilot study to understand the probability of answering an item correctly given the latent trait of the candidate. An interpretation of the above item/question states that with increasing ability of the candidates the probability of getting a correct answer for the said question shows a steep rise thus discriminating at 0.8 between weak and strong performers. Also, the probability that the item could be guessed correctly bears a 20-pc chance amongst the low scorers since a guess is inversely proportional to the number of options. The difficulty level is considered to be moderately high, with low performers finding it difficult and high scorers finding it relatively easy with *b* at 0.68.

Other than item response function, the second important component of IRT is the item information function which denotes item's quality i.e., the ability to differentiate among respondents.

Invariance, on the other hand, can be estimated by any items with known IRFs and item characteristics are population independent within a linear transformation.

An item information curve will take a bell shape for the typical reason that the maximum amount of information is achievable at $\theta = \beta i$. This is because the amount of information can be extremely small for ability levels that deviate from βi. Summarizing, the difficulty level of an item, βi when matched to examinee ability, the estimation of the ability is better recorded. Associated with item information function, as shown above, is the test information function which shows the sum over the items on the amount of item information on the ability trait as shown in Fig. 2.

Above is a generalized representation of an item's information and also the test information function assuming that a test contains only one item or many items. The initial rise of the curve and a peak shows that some item/items within the test could provide considerable information with regard to discrimination, difficulty and guess between the low and top scorers, whereas a less steeper zone shows less information disseminated by the items. Considering a more specific case where around 35 kids

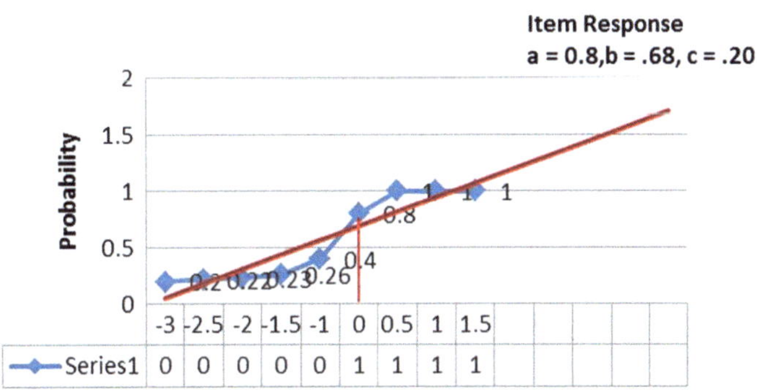

Fig. 1 An item with discrimination 0.8, difficulty 0.68 and guess 0.20

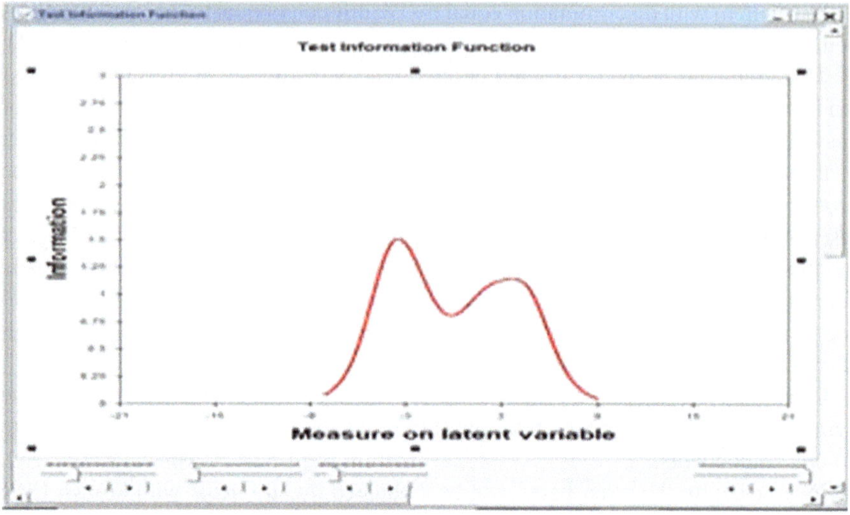

Fig. 2 The test information function

undergo a Knox cube test, the item characteristic curve and the test information function justify the strong correlation to the three-parameter logistic model of the item response theory and the bell shape of the curve. The winstep file specification shows the details of the 35 candidates below who are children ranging from 4 years to 11 years of age and had been subjected to the Knox cube test. The Knox cube test was developed as a non-verbal intelligence test developed by Dr. Howard Andrew Knox, from Ellis Island. It was first published as a paper in 1914 in the Journal of the American Medical Association [1]. In the test 4, black 1″ cubes were placed in a row, each cube separated by 4 in. from its neighbours. The test administrator was instructed to take a smaller cube and tap on the 4 1″ cubes in increasing complicated sequences. The test subject was requested, sometimes only by sign language, to repeat the sequence. The cubes were numbered 1 through 4, with the sequences in order as shown:

(a) 1, 2, 3, 4
(b) 1, 2, 3, 4, 3
(c) 1, 2, 3, 4, 2
(d) 1, 3, 2, 4, 3
(e) 1, 3, 4, 2, 1 and so on.

Dr Knox's inference was that a child of 4 years could imitate the simple sequence of (1–2–3–4) very easily, whereas (1–3–4–2–3–1) could be copied only by a child of 11 years. Taking the above concept into consideration, the 35 kids were subjected to a performance test constituting of 18 items basically to determine their responses to the remembering of correct sequences which would determine their mental ability. Using the test, these kids were assessed on these 18 tapping patterns with the last tapping patterns having seven actions. A correct response was coded as 1 and an incorrect one as a 0 (Fig. 3).

Following the responses by the kids are the components which form the underlying framework of item response theory. The item information function depicts an indication of item quality which tells that based on the level of difficulty of the item the information is disseminated. The level of difficulty is measured based on the responses provided by the candidates. For item 6, we see that the item information function in Fig. 4 provides maximum information at ability level (θ) ranging between -1 and 2 and is about 0.25 in the scale of 0–1. Before and after that the amount of information with regard to the ability is negligible. In other words, the item does not disseminate information on its own attribute such as the level of complexity or the difficulty based on the latent trait of the candidates. An item which does not provide enough information with respect to discrimination or difficulty should be ideally eliminated from the list of questions or item bank. Under item response theory, there exists a single latent trait or ability level of a candidate across all item responses and also the characteristics of one item is not contingent to the others.

This is of considerable importance to both the test constructor and the consumer since it means that the accuracy with which an examinee's ability is estimated depends upon where the examinee's ability is located on the ability scale. Ability and difficulty being synonymous information is received with precision at ability level of a candidate which matches the difficulty level of an item. As like the item information function is the test information function the formula for which is given by the following:

$$1I(\theta) = \sum_{i=1}^{N} \mathrm{Ii}(\theta) - Equation1 \tag{1}$$

where $I(\theta)$ is the summation of test information at an ability level of θ, $\mathrm{Ii}(\theta)$ is the amount of information for item i at ability level θ and N is the number of items in the test. It is extremely useful to have test information function as it tells the

```
This file is EXAM1.TXT - (";" starts a comment) - revised 03-26-2006
&INST                  ; shows this is a control file (optional)
TITLE = 'KNOX CUBE TEST';
NAME1 = 1                      ; First column of person label
NAMELENGTH = 9                 ; Length of person label
ITEM1 = 11                     ; First column of responses in data file
NI = 18                        ; Number of items
CODES = 01                     ; Valid response codes in the data file
CLFILE = *                     ; Labels the observations
0 Wrong                        ; 0 in data is "wrong"
1 Right                        ; 1 in data is "right"
*                              ; "*" is the end of a list
PERSON = KID                   ; Person title: KID means "child"
```

Fig. 3 The winstep text file representing the kid's responses to the items

```
    ITEM = TAP                  ; Item title: TAP means "tapping pattern"
    @GENDER = $C9W1              ; Gender indicator in column 9 of data rec-
ord
    DIF = @GENDER ; Use the Gender for DIF detection PSUBTOTAL =
@GENDER ; Subtotal by gender &END ; Item labels for 18 items follow 1-4
; tapping pattern of first item: cubes 1 then cube 4 are tapped.


    2-3
    1-2-4
    1-3-4
    2-1-4
    3-4-1
    ; last tapping pattern: 7 actions to remember!
    END NAMES ; END NAMES or END LABELS must come at end of list

    Adam     M 111111100000000000     7 Here are the 35 person response
strings
    Anne    F 111111111100000000    10
    Audrey F 111111111100100000    11
    Barbara F 111100100100000000   12
    Bert    M 111110101100000000    10
    Betty   F 111111111000000000     9
    Blaise M 111111111111100000    13
    Brenda F 111111111100000000    10
    Britton F 111011111000000000     8
    Carol   F 111100101000000000     6
    David   M 111111111110100000    12
    Don     M 111111111001000000    10
    Dorothy F 111111111100000000   10
    Elsie   F 111101111100000000     9
    Frank   M 111000000000000000     3 ; worst performance
    Helen   F 111111111100000000    10
    James   M 111111111110000000    11
    Janet   F 111111111100000000    10
    Joe     M 111110011111000000    10
    Kim     F 111111111100100000    11
    Linda   F 111011000000000000    10
    Lisa    F 111111111100000000    10
    Martha F 111111111100000000    10
    Mike    M 111111111111101000    14 best performance
    Pete    M 111111001110010000    10
    Richard M 111111100000000000    7
    Rick    M 111111111010000000    11
    Rod     M 111111111111000000    12
    Ron     M 111111111110000000    11
    Susan   F 111111111000000000    9
    Thomas M 111111111010000000    10
    Tracie F 111111111110100110   ; 14 best performance
```

Fig. 3 (continued)

test's performance of determining the ability of the candidates over a range of ability scores and more the duration of the test better is the precision of measurement. It is to be further stated that much of item information with precision can be derived from

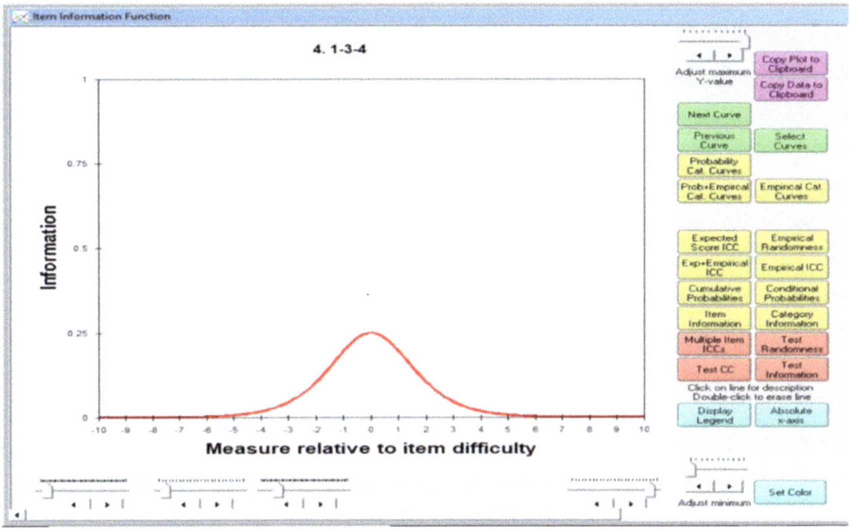

Fig. 4 The item information function

item characteristic curve for the three-parameter model using the following equation where the ability estimate does not get compromised with the guess value:

$$\mathrm{Ii}(\theta) = \mathrm{a2}[\mathrm{Qi}(\theta)/\mathrm{Pi}(\theta)][\mathrm{Pi}(\theta) - \mathrm{c2}]/[1 - \mathrm{c2}] \tag{2}$$

where

$$\mathrm{Pi}(\theta) = c + (1 - c)\,(1/(1 + \mathrm{EXP}(-L)))$$
$$L = a(\theta - b) \quad \mathrm{Qi} = 1.0 - \mathrm{Pi}(\theta).$$

The computation of the item information at ability level θ for $a = 0.8$, $b = 0.68$ and $c = 0.2$ for 6th item from the set of 18 items is given as follows (Table 1):

From the above result extracted from the group of 18 items, the item information provided by the 6th item is moderate (Fig. 5).

Table 1 The computation of the item information at ability level θ

θ	L	EXP($-L$)	1/1+ EXP(−L)	$P_6(-3)$	$Q_6(-3)$	$Q6/P6$
−3	−2.94	18.91	0.05	0.040	0.96	24
$((P_6(3) - c)^2)$	$I_6(-3)$					
0.0256	***0.61***					

The bold results are moderate in nature as compared to others

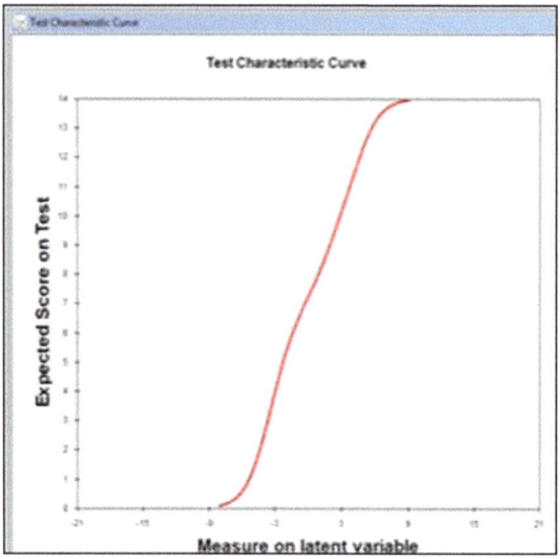

Fig. 5 Test characteristic curve

Also, once information of the item is extracted the test characteristic curve gives the information on how well the test performs in estimating the ability of candidates over a given range of scores.

The test characteristic curve generated on the above data of 35 kids shows the sum of the item characteristic curve of the respective items/questions attempted by the kids and predicts the scores of the examinee at the given ability level given by the equation:

$$\mathrm{TCC}(\theta) = \sum \mathrm{Pi}(\theta) \tag{3}$$

If we see the above curve, it shows that there is a significant difference shown by a steep rise in the scores earned on total number of questions between the high and the low-performing kids. The immediate inference is that as the ability trait increases a little so does the probability of answering the question rises abruptly but then the test information should show with precision the probability of correct attempts across all score ranges.

The underlying benefit of assessing items/questions across the parameters of difficulty, discrimination and guess provides a discrete judgement of a candidate who is being assessed. The added edge of item response theory over classical test theory is that while classical test theory relies on the performance of the test as a whole, IRT chooses to assess the performance of each item based on the candidate's attempt to answer it. In other words, we can say that it is a two-way handshake between item and candidate and as a candidate is assessed so also an item is assessed. In CTT an item's difficulty, discrimination is based on the test takers and person's ability parameter is dependent on the test. But with IRT instead of a single test score model there are item score models and a person's ability does not vary with the test he or she takes, neither an item's difficulty nor discrimination changes with the attempts made by the candidate. Hence to have a fair judgement of prospective job aspirants, the adaptive assessment technique which has a strong correlation with the item response theory was implemented by the authors in this paper. The authors also extracted zonewise performance trends of candidates using the classical test framework to have a broader picture of the level of the candidates.

The Development Environment
For the implementation of non-adaptive and adaptive assessment, an application and an interface were designed and developed and an environment was set up for the assessment of the candidates (Table 2).

The development environment constitutes of a spring and hibernate framework at the application layer, with the user interface developed in JavaScript and angular JS and MySQL at the database layer. The entire application and the database are hosted on the cloud across the servers. The assessment application is interlinked with the job application platform wherein a candidate can apply for jobs and get assessed. The platform on which the candidates apply for the job is shown in Fig. 7. This platform is integrated with analytics dashboard for recruiters and candidates as the stakeholders.

2 The Methodology

The data for this study are the responses received of 2016 and 2017 fresher candidates both pursuing graduate or post-graduate professional course and aspiring to get employed in some organization. The assessment comprising of items are of multiple choice categories, and this assessment of a particular organization had been conducted across the various zones of Maharashtra for the same category of candidates. The population of this study constitutes a random sampling of 450 student data that was selected.

2.1 The Instrument

The researchers collected the data for this study from the candidate base who participated in the assessment conducted by an organization across Maharashtra. The candidate base used for experimentation was around 450 in number from each zone of Maharashtra. The total numbers of questions considered were 85, and the duration of the test was around 90 minutes. The assessment constitutes of objective questions with four options and the guessing parameter can be computed as the inverse of the numbers. If the numbers of options are four, then the guessing parameter can be

Table 2 Snapshot of the database

S. No.	online_test_question_master_id	Question	Total no. of candidate attempt the question	No. of candidates gives the right answer	No. of candidates gives the wrong answer
1	2490	Out of 7 consonants and 4 vowels, how many words of 3 consonants and 2 vowels can be formed?	368	93	275
2	2491	In how many ways can the letters of the word 'LEADER' be arranged?	371	139	232
3	2492	In a group of 6 boys and 4 girls, four children are to be selected. In how many different ways can they be selected such th	364	82	282
4	2493	In a shower, 5 cm of rain falls. The volume of water that falls on 1.5 ha of ground is:	358	156	202
5	2494	A hall is 15 m long and 12 m broad. If the sum of the areas of the floor and the ceiling is equal to the sum of the areas of f	363	122	241
6	2495	66 cubic centimetres of silver is drawn into a wire 1 mm in diameter. The length of the wire in metres will be	357	59	298
7	2496	A hollow iron pipe is 21 cm long and its external diameter is 8 cm. If the thickness of the pipe is 1 cm and iron weighs 8 g/	347	145	202
8	2497	A boat having a length 3 m and breadth 2 m is floating on a lake. The boat sinks by 1 cm when a man gets on it. The mass	358	189	169
9	2498	A boat can travel with a speed of 13 km/h in still water. If the speed of the stream is 4 km/h, find the time taken by the	361	208	153
10	2499	A man's speed with the current is 15 km/h and the speed of the current is 2.5 km/h. The man's speed against the current	367	111	256

(continued)

Table 2 (continued)

S. No.	online_test_question_master_id	Question	Total no. of candidate attempt the question	No. of candidates gives the right answer	No. of candidates gives the wrong answer
11	2500	If one-third of one-fourth of a number is 15, then three-tenth of that number is	375	153	222
12	2501	I saw a __________ of cows in the field	376	104	272
13	2502	Success in this examination depends _____________ hard work alone	380	351	29
14	2503	choose the word which is the exact OPPOSITE of ENORMOUS	372	195	177
15	2504	choose the word which is the exact OPPOSITE of EXODUS	364	136	228
16	2505	Tanya is older than EricCliff. is older than TanyaEric. is older than CliffIf. the first two statements are true, the third statement is not correct	378	318	60
17	2506	All the trees in the park are flowering treesSome. of the trees in the park are dogwoodsAll. dogwoods in the park are floweri	375	224	151
18	2507	Mara runs faster than GailLily. runs faster than MaraGail. runs faster than LilyIf. the first two statements are true, the third statement is not correct	375	316	59
19	2508	All the tulips in Zoe's garden are whiteAll. the pansies in Zoe's garden are yellowAll. the flowers in Zoe's garden are either white or yellow	380	106	274
20	2509	Form the sentence: 1. left 2. the 3. house 4. he 5. suddenly	374	344	30
21	2510	Form the sentence: 1. of 2. we 3. heard 4. him 5. had	375	274	101

computed as 0.25 or 0.20. In the scheme of assessment, the content to be covered for assessments was already notified to the candidates.

To examine the difficulty, discrimination and guessing parameter using the examinees' response, of test items, the analyses were performed on only the items that fitted the 3PL model. The result concludes that around 63 items can fit the three PL model of the item response theory using chi-square goodness of fit and a sample ICC for item 12 is shown in Figs. 6 and 7.

Items		3PL	
	Chi-square	p	df
12	19.9	0.0182	9.0
13	67.4	0.0000	9.0
14	6.7	0.6718	9.0
15	16.4	0.0596	9.0
16	42.9	0.0000	9.0
17	35.0	0.0000	8.0
18	19.1	0.0242	9.0
19	28.7	0.0007	9.0
20	20.6	0.0146	9.0
21	6.6	0.6812	9.0
22	39.2	0.0000	8.0
23	12.8	0.1736	9.0
24	12.7	0.1747	9.0
25	14.1	0.1196	9.0
26	5.8	0.7557	9.0
27	15.6	0.0749	9.0
28	27.7	0.0011	9.0
29	6.0	0.7404	9.0
30	8.3	0.4999	9.0
31	10.3	0.3234	9.0
32	9.0	0.4404	9.0
33	4.6	0.8708	9.0
34	12.8	0.1723	9.0
35	14.3	0.1126	9.0
36	40.8	0.0000	9.0
37	24.8	0.0032	9.0
38	14.1	0.1204	9.0
39	8.4	0.4980	9.0
40	15.1	0.0893	9.0

Fig. 6 Item fit snapshot using chi-square test

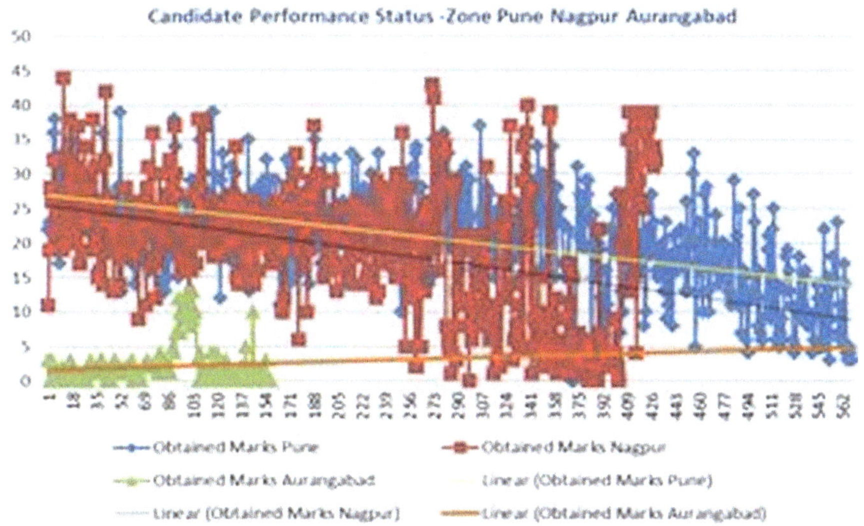

Fig. 7 Candidate's performance scenario—Zonewise

After the candidates' application on the platform, the assessment environment for these candidates enables them to take up both the forms of tests as shown below.

2.1.1 Analytics with CTT and IRT Framework (Item, Candidate, Performance Trends)

The analytics is based on CTT and IRT framework for the non-adaptive and adaptive assessments for the candidates across the zones. The classical test theory is based on the fact that CTT assumes that the true score plus the error gives the observed score.

$$X = T + E \tag{4}$$

where T = the true score and E = the error component but the CTT has its own limitations in terms of analysis of the examinees' proficiency and quality of test items. The IRT, on the other hand,takes as explained above the three parameters into consideration:

a as in item discrimination
b as in item difficulty
c as in pseudo guessing

The IRT model which has been taken into consideration in this paper is the 3. Parameter logistic models which take the following form:

$$\mathrm{Pi}(\theta) = \mathrm{ci} + (1 - \mathrm{ci}) * 1/(1 + \mathrm{e} - \mathrm{Daj}(\theta - \mathrm{bi}) - \tag{5}$$

where c, a and b have already been explained before and D being an arbitrary constant with a value of 1.7 and θ being the ability level of the examinee.

These logistic equations when displayed in the form of graphs produced plots that are called item characteristic curves as discussed in the introductory section where the ability of the examinee is denoted by theta (θ) on the x-axis, while the probability of an examinee correctly answering the question is denoted by P (θ) on the y-axis which is denoted by an s-shaped curve.

The environment till now has facilitated the assessment of around 20,000 students mostly from the various districts of Maharashtra of which 15,000 candidates have been assessed on tests whose underlying framework is based on classical test theory, whereas the rest 5,000 have been subjected to adaptive assessments. Prior to putting forth the analysis of item/candidates, a broad scenario of the zonewise performance of the candidates is shown below. An estimate varying between 450 to 500 students from each zone of Maharashtra had appeared for the non-adaptive assessment, and the following observations were noted:

The overall performance as clearly known from the graph shows candidates from Pune zone performing better than Nagpur and Aurangabad.

Using the concept of item response theory and also to test the item's fitness to the entire assessment, the discrimination and the difficulty index was also calculated. For this reason, the entire student segment was segregated into three segments. Around 30 pc each were considered poor, average and good performers. For each question answered correctly, the discrimination index was computed as shown for the zones below.

To calculate the discrimination and difficulty index, the following steps were observed. Candidates with highest overall scores were allowed to occupy the top positions in the excel file followed by candidates with average and lower scores. For the difficulty index, the computation of each item/question was done as follows:

Difficulty Index ($Q1$) = Total no of students who answered from the top segment of students + total no of students who answered from the bottom segment/total no of students who gave the exam. For $Q1$, it was $b = 0.155$.

The discrimination index was obtained by subtracting the number of candidates in the lower segment who got the item/question correct from the number of students in the upper segment who got the item correct. Then, divided by the number of students in each group. In this case, it would be 150 students in each group. For Q1, it was discrimination index ($Q1$) was 0.06. In the same way, the discrimination and difficulty index of the other questions has been computed for the students who attended the assessments. If we observe the discrimination and difficulty index of the 50th question, we see that difficulty being high the question does not do very well discriminate the top and the poor performers. Reason being that a fair number of poor performers could answer the question correctly depicting they are prepared for difficult questions. It was also seen that questions those were less difficult had higher discrimination power but were less difficult where the question as in question number 28 had the discrimination and the difficulty in the same range. Around 40 pc of the questions of the assessment conducted have high-to-moderate discrimination

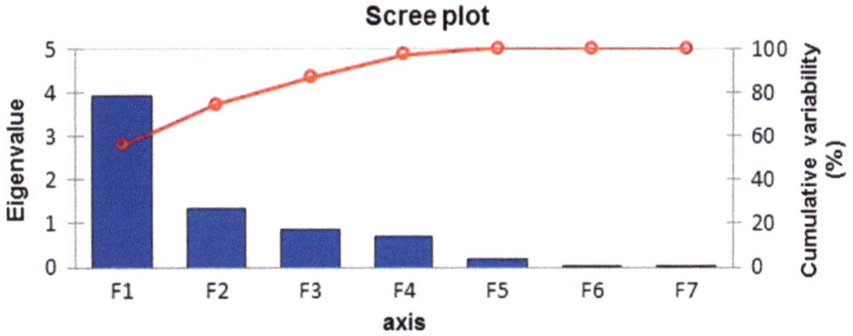

Fig. 8 Scree plot showing contribution of factors

Fig. 9 Item characteristic curve for item 12

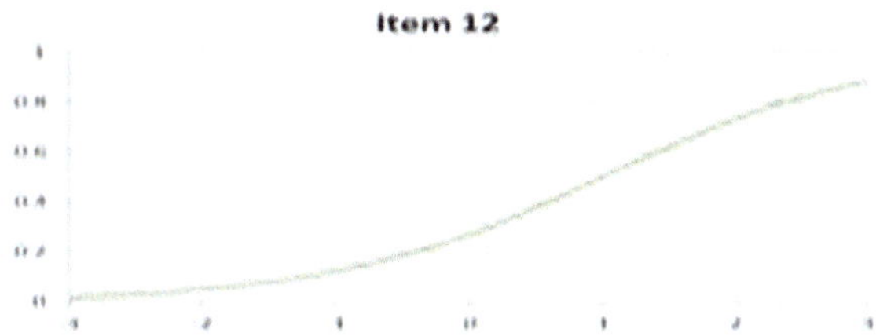

powers, whereas the difficulty index of 60 pc of the questions ranges from high to moderate.

An interesting revelation is that while performing a factor analysis on the data set the authors could identify the contribution of the various factors involved through the scree plot as shown below (Figs. 8 and 9; Table 3):

The performance trend-line of Aurangabad candidates depicts a real need for creating unique pathways of learning.

The percentile score trends for the candidates, using classical test framework, from across the various districts of Maharashtra as shown in Fig. 10 depict that highest count of candidates from Pune falls in the category of 0–10‰ below, whereas the count of candidates from Nagpur is the least when we consider the percentile range of (0–10) and are moderate on their overall performance. The data has been collated across the three zones taking into consideration strength of 450 students each and considering the last five performances in the assessment conducted by more than one organization. Students were chosen from a base of around 20,000 students of which around 4500 students had attended more than two assessments and around 1500 more than four tests. Using excel random sampling software, the sampling was conducted (Fig. 11).

Extracting the moving average, for the year 2016, from a base of candidates as elaborated in the preceding section, who had participated in the assessment across the zones of Maharashtra, the forecast depicts that the average performance scores would increase across the candidate base for the first three segments and then would take a downward trend for the last segment of candidate's assuming that performance

Table 3 Discrimination and difficulty index of items/questions

S. No	online_test_question_master_id	Total	No. of candidates gives the right answer	Total no. of candidate attempt the question	No. of candidates gives the right answer	Discrimination (a)	Difficulty (b)
1	2495	122	23	116	14	0.067834935	0.155462185
2	2498	120	87	121	52	0.295247934	0.576763485
3	2497	121	77	119	59	0.140565317	0.566666667
4	2537	123	33	122	19	0.112554978	0.212244898
5	2525	124	79	117	38	0.312310449	0.485477178
6	2494	123	48	117	28	0.150927663	0.316666667
7	2496	121	57	113	39	0.125941637	0.41025641
8	2499	124	51	121	23	0.221207678	0.302040816
9	2506	124	83	126	67	0.137608807	0.6
10	2508	125	52	128	23	0.2363125	0.296442688
11	2569	118	37	107	22	0.107951845	0.262222222
12	2573	120	95	108	65	0.189814815	0.701754386
13	2514	125	95	124	52	0.340645161	0.590361446
14	2513	123	37	125	26	0.092813008	0.254032258
15	2515	122	74	122	57	0.139344262	0.536885246
16	2503	124	81	125	41	0.325225806	0.489959839
17	2504	125	52	118	40	0.077016949	0.378600823
18	2517	125	119	126	102	0.14247619	0.880478088
19	2539	124	44	121	37	0.049053586	0.330612245
20	2516	124	104	126	79	0.21172555	0.732

(continued)

Table 3 (continued)

S. No	online_test_question_master_id	Total	No. of candidates gives the right answer	Total no. of candidate attempt the question	No. of candidates gives the right answer	Discrimination (a)	Difficulty (b)
21	2558	121	54	120	33	0.171280992	0.360995851
22	2521	123	95	116	61	0.246495655	0.652719665
23	2520	120	39	115	24	0.116304348	0.268085106
24	2509	125	124	126	111	0.111047619	0.93625498
25	2510	125	107	124	76	0.243096774	0.734939759
26	2527	122	54	125	30	0.202622951	0.340080972
27	2523	124	57	128	32	0.209677419	0.353174603
28	2501	125	48	126	22	0.209396825	0.278884462
29	2522	124	68	124	52	0.129032258	0.483870968
30	2534	123	72	120	37	0.27703252	0.448559671
31	2500	124	70	125	30	0.324516129	0.401606426
32	2530	125	73	126	37	0.290349206	0.438247012
33	2556	124	51	125	37	0.115290323	0.353413655
34	2492	123	41	120	22	0.15	0.259259259
35	2528	122	45	119	28	0.133558341	0.302904564
36	2493	120	54	123	54	0.01097561	0.444444444
37	2491	124	70	122	33	0.294024326	0.418699187
38	2512	125	109	128	77	0.2704375	0.735177866
39	2507	125	115	126	98	0.142222222	0.848605578
40	2545	124	100	126	71	0.242959549	0.684
41	2529	123	59	122	42	0.135412502	0.412244898

(continued)

Table 3 (continued)

S. No	online_test_question_master_id	Total	No. of candidates gives the right answer	Total no. of candidate attempt the question	No. of candidates gives the right answer	Discrimination (a)	Difficulty (b)
42	2519	125	82	129	54	0.237395349	0.535433071
43	2490	125	37	122	26	0.082885246	0.255060729
44	2518	125	54	125	36	0.144	0.36
45	2542	124	63	123	36	0.215381589	0.400809717
46	2541	124	115	126	79	0.300435228	0.776
47	2533	124	77	129	46	0.264378595	0.486166008
48	2532	125	87	125	48	0.312	0.54
49	2531	124	86	122	68	0.136171338	0.62601626
50	2502	125	119	129	114	0.06827907	0.917322835
51	2505	125	120	129	89	0.270077519	0.822834646
52	2540	123	55	124	49	0.051993181	0.421052632
53	2526	122	30	122	11	0.155737705	0.168032787
54	2548	125	76	124	51	0.196709677	0.510040161
55	2524	123	12	120	9	0.022560976	0.086419753
56	2555	124	61	125	46	0.123935484	0.429718876
57	2535	123	42	121	30	0.09352953	0.295081967
58	2554	125	107	122	77	0.224852459	0.744939271
59	2561	120	32	123	21	0.095934959	0.218106996
60	2560	123	15	124	9	0.049370574	0.097165992
61	2552	125	35	123	18	0.133658537	0.213709677
62	2511	125	107	128	72	0.2935	0.707509881

(continued)

Table 3 (continued)

S. No	online_test_question_master_id	Total	No. of candidates gives the right answer	Total no. of candidate attempt the question	No. of candidates gives the right answer	Discrimination (a)	Difficulty (b)
63	2574	120	19	106	5	0.111163522	0.10619469
64	2546	125	106	124	74	0.251225806	0.722891566
65	2547	124	16	122	14	0.01427816	0.12195122
66	2551	125	56	126	61	−0.036126984	0.466135458
67	2550	124	58	120	42	0.117741935	0.409836066
68	2567	119	40	105	30	0.050420168	0.3125
69	2559	123	64	124	53	0.092905848	0.473684211
70	2566	119	63	107	22	0.323804288	0.376106195
71	2562	125	38	127	17	0.170141732	0.218253968
72	2570	121	21	106	15	0.032044285	0.158590308
73	2568	119	28	107	22	0.029686641	0.221238938
74	2557	123	49	125	18	0.254373984	0.27016129
75	2572	120	48	103	34	0.069902913	0.367713004
76	2571	118	52	105	23	0.221630347	0.33632287
77	2565	116	35	108	24	0.079501916	0.263392857
78	2538	124	95	121	59	0.278525726	0.628571429
79	2536	123	35	124	29	0.050681878	0.259109312
80	2564	125	72	124	53	0.148580645	0.502008032
81	2553	124	62	125	39	0.188	0.40562249
82	2549	125	91	126	55	0.291492063	0.581673307
83	2563	123	63	125	45	0.152195122	0.435483871
84	2544	124	75	124	66	0.072580645	0.568548387
85	2543	122	57	125	51	0.059213115	0.437246964

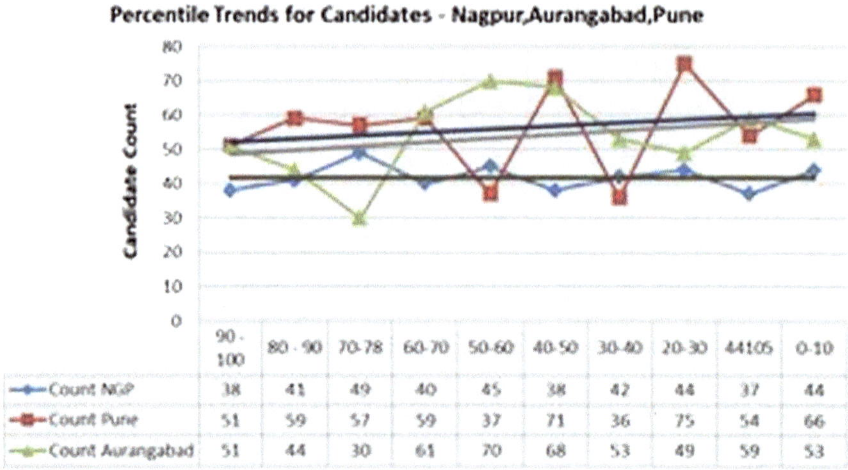

Fig. 10 Zonewise percentile trends

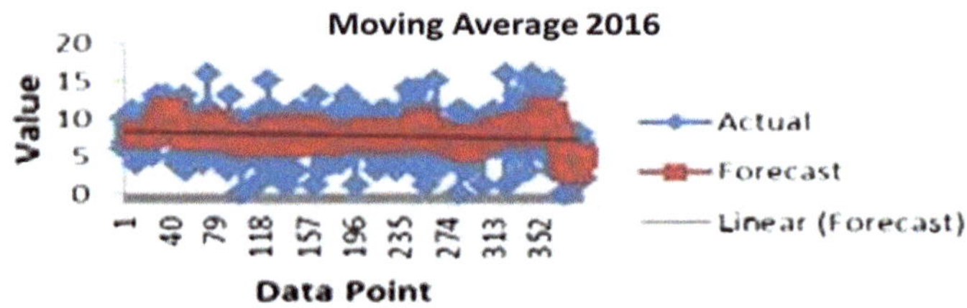

Fig. 11 Performance trend of candidates—moving average

across three student segments had been made for a group of 150 students each across the zones of Maharashtra for based on the previous five assessments.

3 Conclusion and Future Work

The findings from this study indicate that for assessing the candidates it is highly recommended that the items should be fairly good in their characteristics and also should possess the necessary value to make the overall test performance well. If the quality or the standard of the test items are improved so will increase the precision with which the candidate can be measured or assessed by educators or recruiters. It is crucial that the candidate's response to test item is proportional to the overall development of the test and measures the facets of learning with the greatest precision and accuracy and good quality test items are indicators for those measures. The growth in computer adaptive testing, in particular, has supported the growing interest in the use of IRT and IRT has now become a necessity in assessing learning skills and learnability and in developing improved measures for assessing change over time. Educational/employment tests are the main source of information about candidate's achievement; hence, the analysis of test data is crucial in determining the test quality and test information. Hence, the prime task for any educator or recruiter is to choose quality items for better precision. In the research conducted by the authors, the adaptive assessment framework will enable recruiters to align items more to job profiles and candidate's ability traits so that selection and retention ratio can increase thus giving lesser turnaround time and more employment satisfaction.

Reference

1. Adedoyin, O.O.: University of Botswana, Botswana T Mokobi Department of Educational Planning and Research Services, Botswana, USING IRT PSYCHOMETRIC ANALYSIS IN EXAMINING THE quality of junior certificate mathematics multiple choice examination test items, journal homepage: http://www.aessweb.com/journal-detail.php?id=5007

Author Index

L. C. Jain et al. (eds.), *Data and Communication Networks*, Advances in Intelligent Systems and Computing 847, https://doi.org/10.1007/978-981-13-2254-9

MIX
Papier aus verantwortungsvollen Quellen
Paper from responsible sources
FSC® C105338

If you have any concerns about our products,
you can contact us on
ProductSafety@springernature.com

In case Publisher is established outside the EU,
the EU authorized representative is:
Springer Nature Customer Service Center GmbH
Europaplatz 3, 69115 Heidelberg, Germany

Printed by Libri Plureos GmbH
in Hamburg, Germany